PHYSICS LABORATORY EXPERIMENTS

Eighth Edition

"What is the meaning of it all, Mr. Holmes?"
"Ah, I have no data. I cannot tell," he said

Arthur Conan Doyle (1859–1930), *The Adventures of Sherlock Holmes,* 1892

Contents

[Key: GL (Guided Learning), TI (Traditional Instruction), and CI (Computer Instruction), GL is associated only with TI experiments. See Preface.]

Preface

Physics Laboratory Experiments was written for students of introductory physics—in fact, it was originally written at the request of students. The main purpose of laboratory experiments is to augment and supplement the learning and understanding of basic physical principles, while introducing laboratory procedures, techniques, and equipment.

The eighth edition of *Physics Laboratory Experiments* has 38 experiments, with 15 additional customized experiments. All 53 experiments are available for customization at TextChoice.com. (See Experiments Available for Customized Publishing.) This provides an ample number of experiments to choose from for a two-semester or three-quarter physics course. Those features that proved effective in previous editions have been retained, along with **Guided Learning (GL)**. Basically, GL is an effort to supplement the "cookbook" style. For better learning and understanding, an Experimental Planning section gives a brief introduction and guides the students through the basics of an experiment by a series of related questions that they answer.

The GL Experimental Planning is limited to selected Traditional Instruction (TI) experiments, about which students should have some knowledge. These are labeled **GL** in the table of contents.

Pre-Lab Demos

An added feature to the eighth edition of *Physics Laboratory Experiments* is a set of pre-lab demonstrations. The purpose of these demos is to "break the ice," so to speak. Students often come to the laboratory lacking enthusiasm. If something attention-getting (a demo) is presented initially that is related to the experiment or associated class material, interest is aroused. This being done, the instructor can lead into the current lab experiment. Abbreviated demo descriptions are given at the beginning of the experiments. A complete list of demonstrations (descriptions and explanations) is available online, along with the Instructor's Resource Manual.

Traditional Instruction (TI) and Computerized Instruction (CI)

The use of computerized instruction and equipment has become increasingly popular in introductory physics laboratories. To accommodate this, 10 experiments have both TI and CI sections, the latter of which describes an experiment using computerized equipment.* The TI and CI components generally treat the same principles but from different perspectives. These experiments give the instructor the option of doing the TI experiment, the CI experiment, or *both*.

It is suggested that in some instances students do the hands-on TI experiment first, so as to gain a basic knowledge of what is being measured. It is here that the physical parameters of the experiment are clearly associated with principles and results. Once students have this type acquaintance with experimental concepts, they can better perform the CI experiment (or view it as a demonstration if limited CI equipment is available).

The student can then better understand the computer procedure and analysis of electronic recorded data. This is particularly important in graphical analysis, where graphs are immediately plotted on monitor screens without a firm understanding of the parameters involved.

Experiments Available for Customized Publishing

These provide a handy, customizable option—a way for instructors to build their own lab manual that fits the need of their specific courses. All 38 experiments available in the printed manual, and an additional 15 experiments that include four TI–CI experiments, are available through TextChoice.

*Four more TI/CI experiments are available in the customized listing in the Table of Contents.

Cengage Learning's digital library, TextChoice, enables you to build your custom version of *Physics Laboratory Experiments* from scratch. You may pick and choose the content you want to be included in your lab manual and even add your own original materials creating a unique, all in one learning solution. Visit www.textchoice.com to start building your book today.

A list of the additional experiments can be seen in the Table of Contents.

Organization of the Eighth Edition
Both the TI and CI experiments are generally organized into the following sections:

- (In some instances, **TI Experimental Planning for Guided Learning**)
- Advance Study Assignment
- Introduction and Objectives
- Equipment Needed
- Theory
- Experimental Procedure
- Laboratory Report
- Post-lab Questions

Features include:

Laboratory Safety. Safety is continually stressed and highlighted in the manual. This critical issue is expanded upon in the Introduction to the manual.

Advance Study Assignments. Students often come to the laboratory unprepared, even though they should have read the experiment before the lab period to familiarize themselves with it. To address this problem, an Advance Study Assignment precedes each experiment. The assignment consists of a set of questions drawn from the Theory and Experimental Procedures sections of the experiment. To answer the questions, students must read the experiment before the lab period; consequently, they will be better prepared. It is recommended that the Advance Study Assignment be collected at the beginning of the laboratory period.

Example Calculations. In the Theory section of some experiments, sample calculations that involve the equations and mathematics used in the experiment have been included where appropriate. These demonstrate to the student how experimental data are applied.

Illustrations. Over 200 photographs and diagrams illustrate experimental procedures, equipment, and computer programs. To allow for variation in laboratory equipment, different types of equipment that can be used are often illustrated.

Laboratory Reports. Because a standardized format for Laboratory Reports greatly facilitates grading by the instructor, a Laboratory Report is provided for both TI and CI experiments. These reports provide a place for recording data, calculations, experimental results, and analyses. Only the Laboratory Report and post-lab Questions that follow it need to be submitted for grading. The Laboratory Report tables are organized for easy data recording and analysis. Students are reminded to include the units of measurement.

Maximum Application of Available Equipment. Laboratory equipment at many institutions is limited, and often only standard equipment, purchased from scientific suppliers, is available. The TI experimental procedures in this manual are described for different types of common laboratory apparatus, thus maximizing the application of the manual.

Instructor's Resource Manual
The Instructor's Resource Manual is a special feature and resource for the instructor. It is available online on the instructor Web site prepared to accompany the eighth edition of *Physics Laboratory Experiments.* To view a sampling of instructor materials, go to www.cengage.com/Physics, and click on the link for Algebra and Trigonometry Based Lab Manuals. For the eighth edition of *Physics Laboratory Experiments*, clicking the About This Product link will allow you to view online resources including the Instructor's Resource Manual. You may contact your Cengage representative if you need new access to this password-protected material.

Professor Fred B. Otto, previously of the Maine Maritime Academy, who has over 20 years of teaching and laboratory experience, has revised this manual. He retained the general format of the previous edition. For each experiment, there are (1) Comments and Hints, (2) Answers to post-Experiment Questions, and (3) Post-lab Quiz Questions [completion and multiple-choice (with answers), and essay]. The Instructor's Resource Manual also includes laboratory safety references, lists of scientific equipment suppliers and physics software suppliers, and graph paper copy masters.

Of course, the publication of this manual would not have been possible without a great deal of help. The in-depth review of TI experiments, along with independent testing of new experiments and demonstrations, by Professor Paige Ouzts of Lander University, resulted in many helpful suggestions and recommendations. Professor Hernandez and I would like to thank the people at PASCO—in particular, Paul A. Stokstad, and Jon and Ann Hanks—for their support. We are grateful to Charlie Hartford, Senior Product Manager; Alyssa White, Content Developer; Brandi Kirksey, Associate Content Developer; and Christopher Robinson, Product Assistant.

We both hope that you will find the eighth edition of *Physics Laboratory Experiments* helpful, educational, and enjoyable. And we urge anyone—student or instructor—to pass on to us any suggestions that you might have for improvement.

Jerry D. Wilson
Emeritus Professor of Physics
Lander University
Greenwood, South Carolina
jwilson@greenwood.net

Cecilia A. Hernández-Hall
Professor of Physics
American River College
Sacramento, California
hernanc@arc.losrios.edu

Introduction

WHY WE MAKE EXPERIMENTAL MEASUREMENTS

When you can measure what you are speaking about and express it in numbers, you know something about it; but when you cannot measure it, when you cannot express it in numbers, your knowledge is of a meager and unsatisfactory kind.

LORD KELVIN
(1824–1907)

As Lord Kelvin so aptly expressed, we measure things to know something about them—so that we can describe objects and understand phenomena. Experimental measurement is the cornerstone of the *scientific method,* which holds that no theory or model of nature is tenable unless the results it predicts are in accord with experiment. More on this in Experiment 1: The Scientific Method and Thought.

The main purpose of an introductory physics laboratory is to provide "hands-on" experiences of various physical principles. In so doing, one becomes familiar with laboratory equipment, procedures, and the scientific method.

In general, the theory of a physical principle will be presented in an experiment, and the predicted results will be tested by experimental measurements. Of course, these well-known principles have been tested many times before, and there are accepted values for certain physical quantities. Basically, you will be comparing your experimentally measured values to accepted theoretical or measured values. Even so, you will experience the excitement of the scientific method. Imagine that you are the first person to perform an experiment to test a scientific theory.

GENERAL LABORATORY PROCEDURES

Safety

The most important thing in the laboratory is your safety and that of others. Experiments are designed to be done safely, but proper caution should always be exercised.

A potential danger comes from a lack of knowledge of the equipment and procedures. Upon entering the physics lab at the beginning of the lab period, you will probably find the equipment for an experiment on the laboratory table. Restrain your curiosity, and do not play with the equipment. You may hurt yourself and/or the equipment. A good general rule is:

> *Do not touch or turn on laboratory equipment until it has been explained and permission has been given by the instructor.*

Also, certain items used in various experiments can be particularly dangerous, for example, hot objects, electricity, mercury lamps, and radioactive sources. In some instances, such as with hot objects and electricity, basic common sense and knowledge are required.

However, in other instances, such as with mercury lamps and radioactive sources, you may not be aware of the possible dangers. Mercury lamps may emit ultraviolet radiation that can be harmful to your eyes. Consequently, some sources need to be properly shielded. Some radioactive sources are solids and are encapsulated to prevent contact. Others are in liquid form and are transferred during an experiment, so there is a danger of spillage. Proper handling is therefore important.

In general, necessary precautions will be given in the experiment descriptions. *Note them well.* When you see the arrow symbol in the margin as illustrated here, you should take extra care to follow the procedure carefully and adhere to the precautions described in the text. As pointed out earlier, experiments are designed to be done safely. Yet a common kitchen match can be dangerous if used improperly. Another good rule for the laboratory is:

> *If you have any questions about the safety of a procedure, ask your instructor before doing it.*

The physics lab is a place to learn and practice safety.

Equipment Care

The equipment provided for the laboratory experiment is often expensive and, in some instances, quite delicate. If used improperly, certain pieces of apparatus can be damaged. The general rules given above concerning personal safety also apply to equipment care.

Even after familiarizing oneself with the equipment, it is often advisable or required to have an experimental setup checked and approved by the instructor before putting it into operation. This is particularly true for electrical experiments. Applying power to improperly wired circuits can cause serious damage to meters and other pieces of apparatus.

If a piece of equipment is broken or does not function properly, it should be reported to the laboratory instructor. Also, after you complete an experiment, the experimental setup should be disassembled and left neatly as found, unless you are otherwise instructed.

If you accidentally break some equipment or the equipment stops working properly during an experiment, *report it to your instructor.* Otherwise, the next time the equipment is used, a great deal of time may be wasted trying to get good results.

Laboratory Reports

A Laboratory Report form is provided for each experiment in which experimental data are recorded. This should be done *neatly.* Calculations of experimental results should be included. Remember, the neatness, organization, and explanations of your measurements and calculations in the Laboratory Report represent the quality of your work.

PHYSICS LABORATORY EXPERIMENTS

Eighth Edition

Name ______________________ Section ______________ Date ______________

Lab Partner(s) ______________________________

EXPERIMENT 1

The Scientific Method and Thought

TI *Advance Study Assignment*

Read the experiment and answer the following questions.

1. What are the elements of the scientific method?

2. How is a hypothesis tested?

3. When does a theory become a law?

4. What happens when a scientific law is broken?

(continued)

5. We use our senses to make observations. Comment on their limitations and reliability.

6. Is scientific thinking only applicable to sciences such as physics or chemistry? Explain.

PRE-LAB DEMONSTRATION | Skewed Balloon (and it doesn't burst)

The purpose of this demonstration is to get you thinking and to apply the scientific method (or scientific thinking). A wooden (bamboo) skewer is pushed completely through an inflated balloon without it collapsing (● Fig. D1). How is this possible?

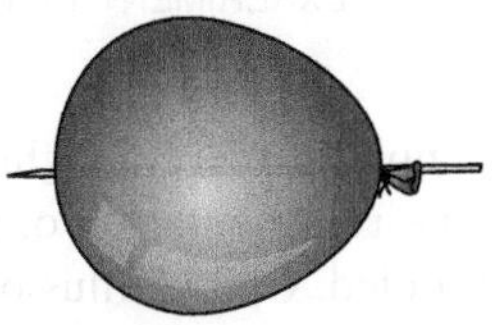

Figure D1 Skewed Balloon

EXPERIMENT 1

The Scientific Method and Thought

"It is a capital mistake to theorize before one has data. Insensibly one begins to twist the facts to suit the theories, instead of the theories to suit the facts."

ARTHUR CONAN DOYLE (1859–1930), *THE ADVENTURES OF SHERLOCK HOLMES*

INTRODUCTION AND OBJECTIVES

The physics laboratory is a learning place—a place to observe and analyze. To investigate the physical world, scientists use what is known as the **scientific method**, which states: *No hypothesis, theory, or law of nature is valid unless its predictions are in agreement with experimental results (quantitative measurements).*

In this experiment, you will examine several situations using the scientific method so as to become familiar with its aspects.

After performing the experiment and analyzing the results, you should be able to do the following:

1. Know how to formulate a hypothesis.
2. Explain how a hypothesis is tested.
3. Distinguish between a theory and a law.
4. Describe how instruments can aid in measurements and perceptions.

THEORY

A general breakdown of the elements of the scientific method is as follows:

- *Observation.* Scientists are curious folks. This curiosity leads them to ask questions about the observed occurrences in the world around them.
- *Hypothesis.* Scientists give a possible explanation of some observation(s), in other words, a tentative answer or an educated guess.
- *Experiments.* The hypothesis is tested under controlled conditions to see if the results confirm the hypothetical assumptions and can be duplicated.
- *Theory.* If a hypothesis passes enough experimental tests and generates new predictions that also prove correct, it takes on the status of a theory—a well-tested explanation of observed physical phenomena.
- *Law.* If a theory has withstood the test of many valid experiments with great regularity, that theory may be accepted by scientists as a law. This is a concise statement in words or mathematical equations that describes a fundamental relationship of nature. Scientific laws are somewhat analogous to legal laws, which may be repealed or modified if inconsistencies are later discovered. Unlike legal laws, however, scientific laws are not meant to regulate but to describe.

Galileo (1564–1642) is said to be the father of the scientific method. He was one of the first to use experimental results to test hypotheses. As you will probably study in class, legend has it that one of his experiments was to drop cannon balls of different weights from the Leaning Tower of Pisa. This was to investigate his hypothesis that the speed of a falling object did not depend on weight or mass. The cannon balls of different weights dropped simultaneously from the same height hit the ground together, giving experimental credence to Galileo's hypothesis.

Of course, you will not be using the scientific method to make new discoveries in the laboratory. For the most part, you will be investigating established theories and laws so as to become familiar with experimental procedures and analysis. However, in this experiment, we'll have some fun in focusing on the first three elements of the scientific method: observation, hypothesis, and experiment (testing).

And then there is perception. Our senses have limitations. For example, the unaided eye cannot see the majority of stars and galaxies. We cannot immediately distinguish the visible stars of our galaxy from the planets of our solar system, all of which appear as points of light. (The planets, however, move relative to the stars. The word *planet* comes from the Greek word meaning "wanderer.")

Not only do our senses have limitations, but they can also be deceived and may provide false information. For example, perceived sight information may not always be a true representation of the facts because the brain can be fooled. Optical illusions are examples of this; to demonstrate, some of the classic illusions will be examined. An optical illusion may be defined as a visual perception of a real object in such a way as to misinterpret its actual nature. Some people may be quite convinced that what they see in such drawings actually exists as they perceive it.

However, in experiments, one must observe carefully and objectively and guard against making such errors. In general, such deceptions may be eliminated by using instruments.

Finally, a word about scientific thought or scientific thinking. Using the scientific method is in essence practicing scientific thinking—observing, reasoning, reaching logical conclusions, and so forth. However, this is not restricted just to science. Methods and principles of scientific thinking are used in everyday life and in the study of other disciplines—history, philosophy, literature, economics, and so on. This is sometimes called *critical thinking* and is used to obtain reliable answers to questions and solutions to problems by critically analyzing them.

Have you ever seen a magician do a magic trick and tried to figure out and explain what actually happened (or didn't happen)? If so, you were trying to do so by applying scientific or critical thinking, which is based on logic and experience.

EQUIPMENT NEEDED

- Crisp, unfolded dollar bill (larger denominations work too).
- Meterstick and ruler (with centimeter scale).
- Four paper muffin cups. (Paper coffee filters, as used in coffee makers, may also be used.)
- Optical illusions (in experiment ● Fig. 1.2).

INVESTIGATIONS

A. Dollar Drop

Observation. Hold your hand with the thumb and forefinger about 1.3 cm (0.5 in.) apart. The thumb and forefinger should be horizontal and parallel. The idea is to form a slot through which a dollar bill will be dropped. Have a lab partner hold the dollar bill vertically with the short side just even with the top of your fingers (see ● Fig. 1.1).

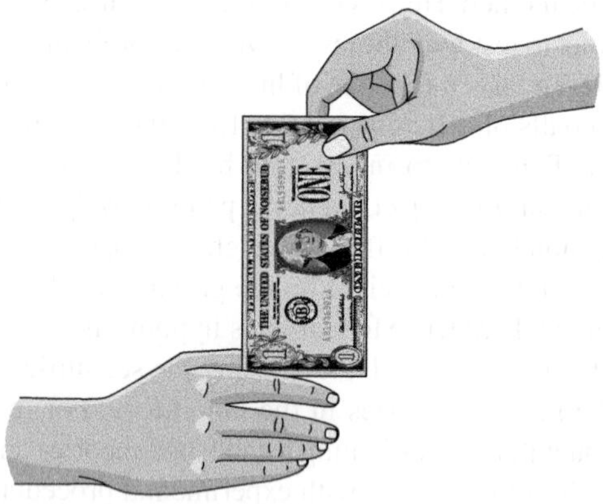

Figure 1.1 Have the bill held just right. See text for description.

Then, without telling you when it will be released, have the partner drop the bill; you try to catch it. (Dropping the bill yourself isn't cricket as you would be inadvertently aware of when it was released.) You may want to repeat this process several times, and also with your lab partner doing the catching and you the dropping. (If you are like most of us, you will not be able to catch the bill. This is sometimes used as a party joke—if you catch the bill, you can have it.)

Hypothesis. Why couldn't you catch the bill falling under the influence of gravity? Closing the fingers must be too slow; that is, it involves a person's reaction time—the time it takes the muscles to react to the brain's signal and complete the task.

Experiment. So how does one measure the reaction time to prove this? There is no stop watch available, and even so, you have to use reaction times to start and stop the watch. Quantities are often measured by indirect means. This can be done by using a falling ruler. Repeat the dollar bill procedure with the ruler, and note the centimeter mark at the top of your fingers on catching it. (If you start with the zero end of the ruler downward, this is the length the ruler fell through your fingers during the reaction time.) Repeat three times with each lab partner and record the results in the Laboratory Report, and take the average of the set of measurements.

Analysis. The reaction time may be computed from the equation

$$d = \tfrac{1}{2}gt^2$$

This is the distance (d) an object falls when released from rest; t is the time it takes to fall that distance. Solve for t and, using the average ruler drop distance, compute an average reaction time. (The acceleration due to gravity $g = 9.8\ \text{m/s}^2$, and to be consistent, d must be in meters.) Record and answer the questions in the lab report.

B. Muffin Cup Drop

In section A, the falling ruler was assumed to be in free fall (only under the influence of gravity), and there was no consideration of air resistance. This is a good approximation over relatively short distances. However, in falling, air resistance occurs when an object collides with air molecules. Therefore, air resistance depends on an object's size and shape (which determines the area exposed to collisions). The more exposed surface area, the more collisions there will be with air molecules.

The air resistance builds up as a falling object gains speed. Then, when the downward weight force is balanced by the upward resistance force (a frictional force), the object falls at a constant rate that is called the *terminal velocity*.

Observation. Suppose Galileo had simultaneously dropped two cannon balls of identical size but of different masses (weights) from a high-altitude hot air balloon. In this case, it is observed that the heavier ball strikes the ground first.

Hypothesis. The balls have identical shapes, so the exposed areas to the air molecules are the same. The heavier ball must speed up ahead of the light ball to get the ground first, but they both experience the same acceleration due to gravity. The masses of the balls are different, so this must have something to do with it. What is the answer? Sometimes experiments are needed to formulate a hypothesis that will help clarify an observation.

Experiment. Let's try to recreate the observation in the lab. Hot air balloons and cannon balls are out, but we have paper muffin cups that fit nicely into the laboratory budget. The cups may be nested (put inside each other), which gives approximately the same size and shape in falling when one, two, or more nested cups are dropped. However, the masses are different. Does this have an effect?

(a) Drop one cup from shoulder height (bottom side down) and notice that it simulates terminal velocity (falls at a constant rate).

(b) Simultaneously drop a single cup and two nested cups from the same shoulder height. Does the number of cups (mass) make a difference?

(c) Simultaneously drop the single cup from a 1 m height and the two nested cups from a greater height until a height is found for which both hit the floor at the same time. Note and measure the height at which the nested cups were released to accomplish this and record it in the Laboratory Report.

(d) Repeat the procedure for three nested cups.

Analysis. When a falling object reaches terminal velocity, the downward gravitational force on the object ($F = mg$) is balanced by the upward force of air resistance that is proportional to the square of the speed, $f = kv^2$. Thus, a falling object can reach terminal velocity relatively quickly. This being the case, the distance of fall may be approximated by $d = vt$.*

Note that with $f = kv^2$, and since f and $F = mg$ are equal in magnitude at terminal velocity, we have $v = \sqrt{f/k} = \sqrt{F/k} = \sqrt{mg/k}$. The ratio of the distances of fall of two objects of different masses is then

$$\frac{d_2}{d_1} = \frac{v_2 t}{v_1 t} = \frac{v_2}{v_1} = \left(\frac{F_2}{F_1}\right) = \left(\frac{m_2}{m_1}\right)^{\frac{1}{2}} \tag{1.1}$$

(1) Using the mass ratio of two cups to one, $m_2 = 2m_1$, compute the distance ratio (Eq. 1.1). Using $d_1 =$ 1 m (as in the experiment), compute d_2, the height at which two nested cups should be released to hit the floor simultaneously with the single cup. Record this in the Laboratory Report and compare with the experimental height.

(2) Repeat the calculation for three (3) nested cups. Record in the Laboratory Report and answer the questions.

C. Perception: Can You Be Fooled?

Look at the optical illusions in Fig. 1.2 and answer the associated questions. You may use instruments if you like. (There is no need to record; the instructor will take your word for it.) See if you can explain why there is an illusion in each case.

*$d = vt$ is not strictly true because there is some initial acceleration. However, scientists often make assumptions in analyses to simplify and make understanding easier. If the assumption is too far afield, the results of the analysis will not agree with experimental data (which violates the scientific method).

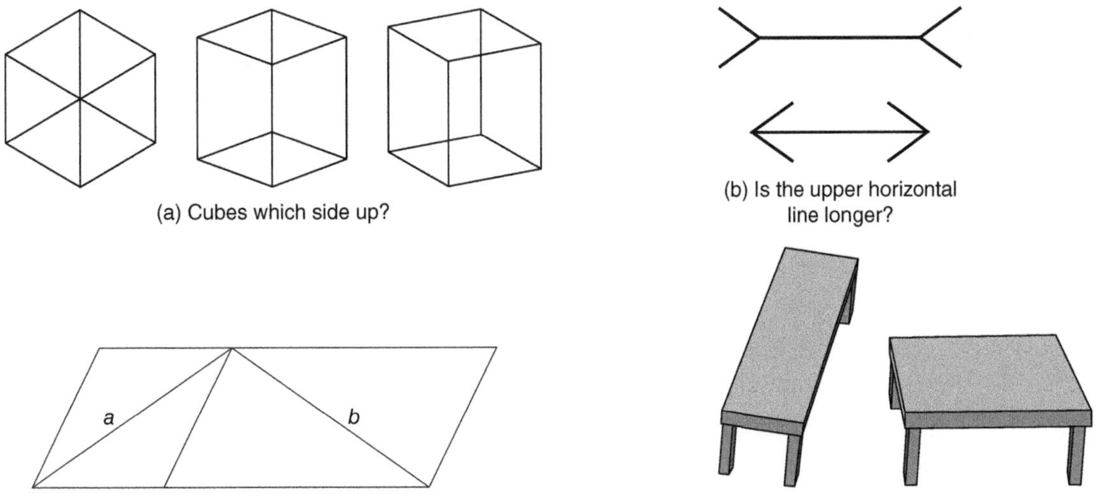

(a) Cubes which side up?

(b) Is the upper horizontal line longer?

(c) Is the diagonal line *b* longer than the diagonal line *a*?

(d) Which table top is larger?

Figure 1.2 We can be fooled by what we see. Answer the questions under the drawings.

(e) Are the diagonal lines parallel?

(f) Are the horizontal lines parallel or do they slope?

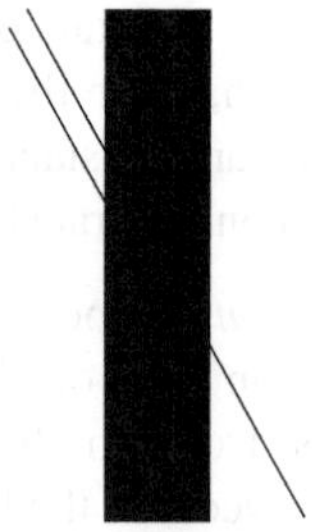

(g) With which of the upper lines does the line on the right connect?

(h) Going down?

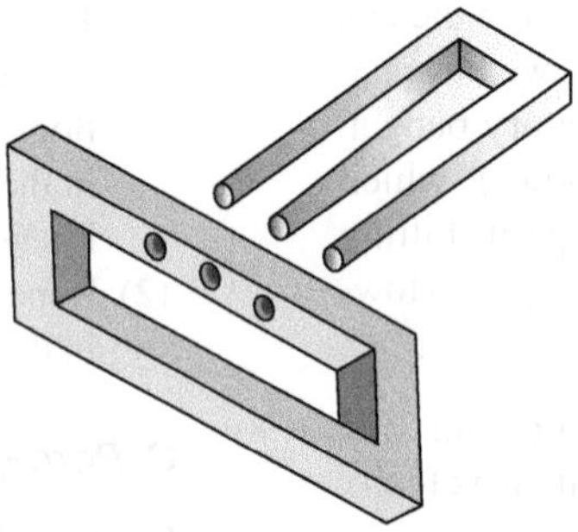

(i) Is something dimensionally wrong here?

Figure 1.2 We can be fooled by what we see. Answer the questions under the drawings. (*Continued*)

Name ______________________ Section ______________ Date ______________

Lab Partner(s) ______________________

EXPERIMENT 1

The Scientific Method and Thought

TI Laboratory Report

A. Dollar Drop

DATA TABLE 1

Purpose: To determine reaction time.

Distance of fall ( )		
Average		

Calculations
(show work)

Reaction
time: ______________ ______________ ______________

A. Data Table

(a) Measure the length of the dollar bill and compare with the ruler drop distance. What does this tell you?

(b) What would your reaction time have to be to catch the bill?

(continued)

B. Muffin Cup Drop

DATA TABLE 2

Purpose: To determine terminal velocity distances.

Experimental

height d_2 ( )	2 cups	3 cups

Calculations
(show work)

Computed d_2 ______________ ______________
2 cups 3 cups

Calculations:

How do the heights compare?

____

____

Name ______________________ Section ______________ Date ______________

Lab Partner(s) ____

EXPERIMENT 1 **The Scientific Method and Thought** *Laboratory Report*

TI QUESTIONS

1. If you dropped the dollar bill (or ruler) yourself, how might this affect the reaction time?

2. Suppose three nested muffin cups and five nested cups were used in the experiment. From what height should the five cups be dropped to hit the floor together with the three cups dropped from a height 1 m? (Show calculations.)

3. Why is using instruments so important in taking experimental data, rather than using our senses directly?

Name ______________________ Section ______________ Date ______________

Lab Partner(s) ____

EXPERIMENT 2

Experimental Uncertainty (Error) and Data Analysis

TI Advance Study Assignment

Read the experiment and answer the following questions.

1. Do experimental measurements give the true value of a physical quantity? Explain.

2. Distinguish between random (statistical) error and systematic error. Give an example of each.

3. What is the difference between measurement accuracy and precision? Explain the general dependence of these properties on the various types of errors.

(continued)

4. What determines how many figures are significant in reported measurement values? What would be the effect of reporting more or fewer figures or digits than are significant?

5. In expressing experimental error or uncertainty, when should (a) experimental error and (b) percent difference be used?

6. How could the function $y = 3t^2 + 4$ be plotted on a Cartesian graph to produce a straight line? What would be the numerical values of the slope and intercept of the line?

PRE-LAB DEMONSTRATION | Problem in Data Analysis (there's a dollar missing)

A student wants to buy a new cell phone for $97 and borrows $50 from each of two friends. He buys the phone and gets $3 back. He gives each friend $1 and keeps the other $1.

Now he owes his two friends $49 apiece. But $49 plus $49 is $98; and when you add the $1 he kept, you get $99. Where's the missing dollar? (Sounds like the data need to be analyzed.)

EXPERIMENT 2

Experimental Uncertainty (Error) and Data Analysis

INTRODUCTION AND OBJECTIVES

Laboratory investigations involve taking measurements of physical quantities, and the process of taking any measurement always involves some experimental uncertainty or error.* Suppose you and another person independently took several measurements of the length of an object. It is highly unlikely that you both would come up with exactly the same results. Or you may be experimentally verifying the value of a known quantity and want to express uncertainty, perhaps on a graph. Therefore, questions such as the following arise:

- Whose data are better, or how does one express the degree of uncertainty or error in experimental measurements?
- How do you compare an experimental result with an accepted value?
- How does one graphically analyze and report experimental data?

In this experiment, types of experimental uncertainties will be examined, along with some methods of error and data analysis that may be used in subsequent experiments.

After performing the experiment and analyzing the data, you should be able to do the following:

1. Categorize the types of experimental uncertainty (error), and explain how they may be reduced.
2. Distinguish between measurement accuracy and precision, and understand how they may be improved experimentally.
3. Define the term *least count* and explain the meaning and importance of significant figures (or digits) in reporting measurement values.
4. Express experimental results and uncertainty in appropriate numerical values so that someone reading your report will have an estimate of the reliability of the data.
5. Represent measurement data in graphical form so as to illustrate experimental data and uncertainty visually.

*Although *experimental uncertainty* is more descriptive, the term *error* is commonly used synonymously.

EQUIPMENT NEEDED

- Rod or other linear object less than 1 m in length
- Four-meter-long measuring sticks with calibrations of meter, decimeter, centimeter, and millimeter, respectively†
- Pencil and ruler
- Hand calculator
- 3 sheets of Cartesian graph paper
- French curve (optional)

†A four-sided meterstick with calibrations on each side is commercially available from PASCO Scientific.

THEORY

A. Types of Experimental Uncertainty

Experimental uncertainty (error) generally can be classified as being of two types: (1) random or statistical error and (2) systematic error. Let's take a closer look at each type of experimental uncertainty.

Random or Statistical Error

Random errors result from unknown and unpredictable variations that arise in all experimental measurement situations. There is no way to determine the magnitude or sign (+, too large; –, too small) of the error in any individual measurement. Conditions in which random errors can result include:

1. Unpredictable fluctuations in temperature or line voltage.
2. Mechanical vibrations of an experimental setup.
3. Unbiased estimates of measurement readings by the observer.

Repeated measurements with random errors give slightly different values each time. *The effect of random errors may be reduced and minimized by improving and refining experimental techniques.*

Systematic Errors

Systematic errors are associated with particular measurement instruments or techniques, such as an improperly calibrated instrument or bias on the part of the observer. The term *systematic* implies that the same magnitude and sign of experimental uncertainty are obtained when the measurement is repeated several times. The magnitude and sign of the uncertainty can be determined if the error is identified. Conditions from which systematic errors can result include

1. An improperly "zeroed" instrument, for example, an ammeter as shown in ● Fig. 2.1.
2. A faulty instrument, such as a thermometer that reads 101 °C when immersed in boiling water at standard atmospheric pressure. This thermometer is faulty because the reading should be 100 °C.
3. Personal error, such as using a wrong constant in calculation or always taking a high or low reading of a scale division. Reading a value from a measurement scale generally involves aligning a mark on the scale. The alignment—and hence the value of the reading—can depend on the position of the eye (parallax). Examples of such personal systematic error are shown in ● Fig. 2.2.

Avoiding systematic errors depends on the skill of the observer to recognize the sources of such errors and to prevent or correct them.

B. Accuracy and Precision

Accuracy and *precision* are commonly used synonymously, but in experimental measurements there is an important distinction. The **accuracy** of a measurement signifies how close it comes to the true (or accepted) value—that is, how nearly correct it is.

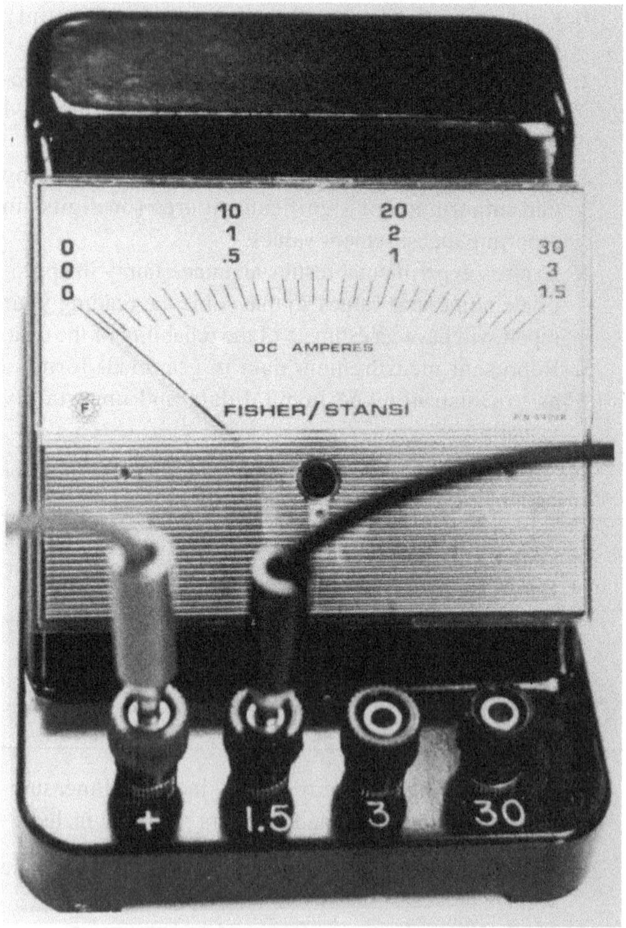

Figure 2.1 Systematic error. An improperly zeroed instrument gives rise to systematic error. The ammeter, which measures electric current, has positive nonzero reading with no current through it. This would systematically give an incorrect reading larger that the true value. (After correcting the error by zeroing the meter, which scale would you read when using the ammeter?) (© Cengage Learning)

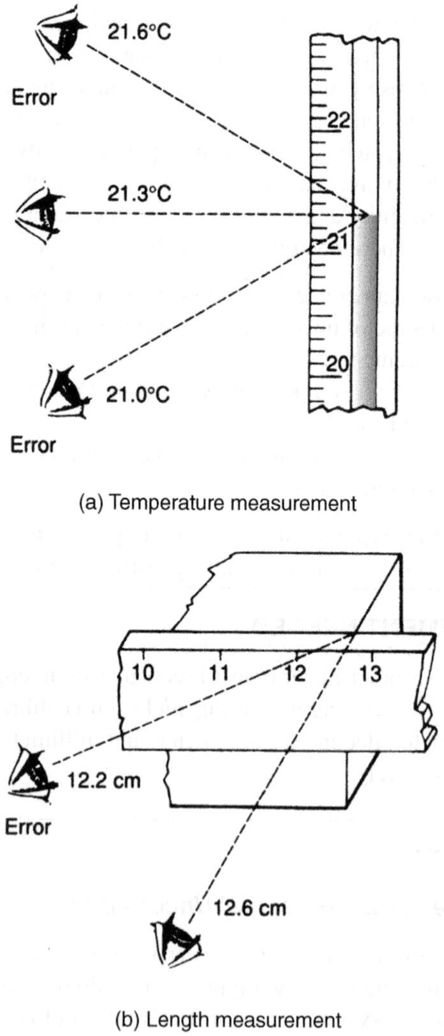

Figure 2.2 Personal error. Examples of personal error due to parallax in reading (a) a thermometer and (b) a meterstick. (Parallax is the apparent displacement of a point or mark due to a change in position of the observer.) Readings may systematically be made either too high or too low.

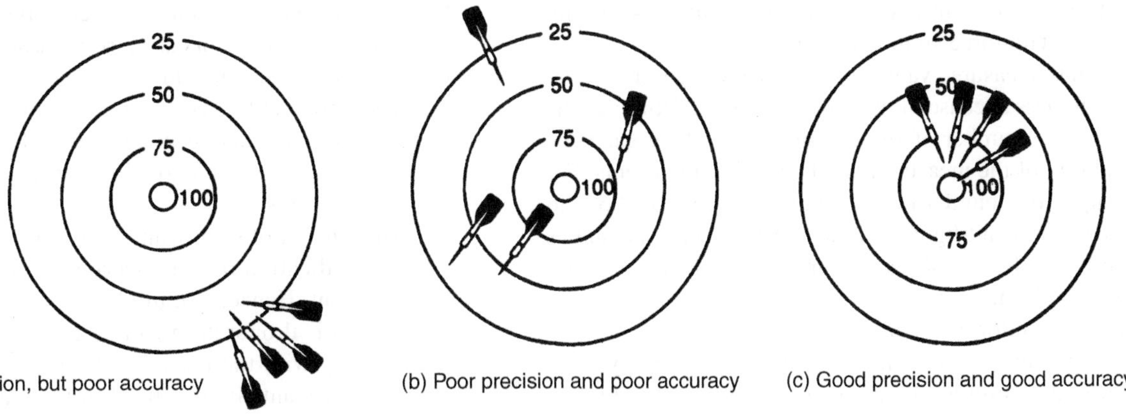

Figure 2.3 Accuracy and precision. The true value in this analogy is the bull's eye. The degree of scattering is an indication of precision—the closer together a dart grouping, the greater the precision. A group (or symmetric grouping with an average) close to the true value represents accuracy.

Example 2.1 Two independent measurement results using the diameter *d* and circumference *c* of a circle in the determination of the value of π are 3.140 and 3.143. (Recall that $\pi = c/d$.) The second result is more accurate (more nearly correct) than the first because the true value of π, to four figures, is 3.142.

Precision refers to the agreement among repeated measurements—that is, the "spread" of the measurements or how close they are together. The more precise a group of measurements, the closer together they are. However, a large degree of precision does not necessarily imply accuracy, as illustrated in ● Fig. 2.3.

Example 2.2 Two independent experiments give two sets of data with the expressed results and uncertainties of 2.5 ± 0.1 cm and 2.5 ± 0.2 cm, respectively.

The first result is more precise than the second because the spread in the first set of measurements is between 2.4 cm and 2.6 cm, whereas the spread in the second set of measurements is between 2.3 cm and 2.7 cm. That is, the measurements of the first experiment are less uncertain than those of the second.

Note: Obtaining *greater accuracy* for an experimental value depends in general on *minimizing systematic errors*. Obtaining *greater precision* for an experimental value depends on *minimizing random errors*.

C. Least Count and Significant Figures

In general, there are *exact* numbers and *measured* numbers (or quantities). Factors such as the 100 used in calculating percentage and the 2 in $2\pi r$ are exact numbers. Measured numbers, as the name implies, are those obtained from measurement instruments and generally involve some error or uncertainty.

In reporting experimentally measured values, it is important to read instruments correctly. The degree of uncertainty of a number read from a measurement instrument depends on the quality of the instrument and the fineness of its measuring scale. When reading the value from a calibrated scale, only a certain number of figures or digits can properly be obtained or read. That is, only a certain number of figures are *significant*.

This depends on the **least count** of the instrument scale, which is the smallest subdivision on the measurement scale. This is the unit of the smallest reading that can be made without estimating. For example, the least count of a meterstick is usually the millimeter (mm). We commonly say "the meterstick is calibrated in centimeters (numbered major divisions) with a millimeter least count." (See ● Fig. 2.4.)

The **significant figures** (sometimes called **significant digits**) of a measured value include all the numbers that can be read directly from the instrument scale, *plus* one doubtful or estimated number—the fractional part of the least count smallest division. For example, the length of the rod in Fig. 2.4 (as measured from the zero end) is 2.64 cm. The rod's length is known to be between 2.6 cm and 2.7 cm. The estimated fraction is taken to be 4/10 of

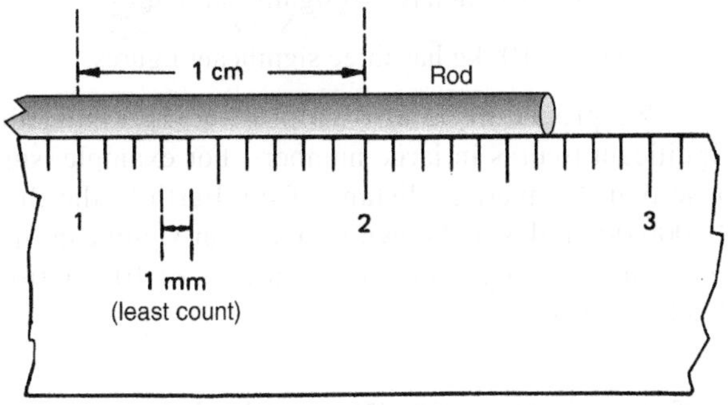

Figure 2.4 Least count. Metersticks are commonly calibrated in centimeters (cm), the numbered major divisions, with a least count, or smallest subdivision, of millimeters (mm).

the least count (mm), so the doubtful figure is 4, giving 2.64 cm with three significant figures.

Thus, measured values contain inherent uncertainty or doubtfulness because of the estimated figure. However, the greater the number of significant figures, the greater the reliability of the measurement the number represents. For example, the length of an object may be read as 3.65 cm (three significant figures) on one instrument scale and as 3.5605 cm (five significant figures) on another. The latter reading is from an instrument with a finer scale (why?) and gives more information and reliability.

Zeros and the decimal point must be properly dealt with in determining the number of significant figures in a result. For example, how many significant figures does 0.0543 m have? What about 209.4 m and 2705.0 m? In such cases, the following rules are generally used to determine significance:

1. Zeros at the beginning of a number are not significant. They merely locate the decimal point. For example,

   0.0543 m has three significant figures (5, 4, and 3).

2. Zeros within a number are significant. For example,

   209.4 m has four significant figures (2, 0, 9, and 4).

3. Zeros at the end of a number after the decimal point are significant. For example,

   2705.0 has five significant figures (2, 7, 0, 5, and 0).

Some confusion may arise with whole numbers that have one or more zeros at the end without a decimal point. Consider, for example, 300 kg, where the zeros (called trailing zeros) may or may not be significant. In such cases, it is not clear which zeros serve only to locate the decimal point and which are actually part of the measurement (and hence significant). That is, if the first zero from the left ($3\underline{0}0$ kg) is the estimated digit in the measurement, then only two digits are reliably known, and there are only two significant figures.

Similarly, if the last zero is the estimated digit ($30\underline{0}$ kg), then there are three significant figures. This ambiguity is be removed by using *scientific (powers of 10) notation*:

$$3.0 \times 10^2 \text{ kg has two significant figures.}$$

$$3.00 \times 10^2 \text{ kg has three significant figures.}$$

This procedure is also helpful in expressing the significant figures in large numbers. For example, suppose that the average distance from Earth to the Sun, 93,000,000 miles, is known to only four significant figures. This is easily expressed in powers of 10 notation: 9.300×10^7 mi.

D. Computations with Measured Values

Calculations are often performed with measured values or numbers, and any error or uncertainty is "propagated" by the mathematical operations—for example, multiplication or division. That is, errors are carried through to the results by the mathematical operations.

The error can be better expressed by statistical methods; however, a widely used procedure for *estimating* the uncertainty of a mathematical result involves the use of significant figures.

The number of significant figures in a measured value gives an indication of the uncertainty or reliability of a measurement. Hence, you might expect that the result of a mathematical operation can be no more reliable than the quantity with the least reliability, or smallest number of significant figures, used in the calculation. That is, *reliability cannot be gained through a mathematical operation.*

It is important to report the results of mathematical operations with the proper number of significant figures. This is accomplished by using rules for (1) multiplication and division and (2) addition and subtraction. To obtain the proper number of significant figures, one rounds off the results. The general rules used for mathematical operations and rounding follow.

SIGNIFICANT FIGURES IN CALCULATIONS

1. When multiplying and dividing quantities, leave as many significant figures in the answer as there are in the quantity with the least number of significant figures.
2. When adding or subtracting quantities, leave the same number of decimal places (rounded) in the answer as there are in the quantity with the least number of decimal places.

RULES FOR ROUNDING*

1. If the first digit to be dropped is less than 5, leave the preceding digit as is.
2. If the first digit to be dropped is 5 or greater, increase the preceding digit by one.

Notice that in this method, five digits (0, 1, 2, 3, and 4) are rounded down and five digits (5, 6, 7, 8, and 9) are rounded up.

These rules come into play frequently when doing mathematical operations with a hand calculator that may give a string of digits. ● Fig. 2.5 shows the result of the division of 374 by 29. The result must be rounded off to two significant figures—that is, to 13. (Why?)

Example 2.3 Applying the rules.

Multiplication:

$$\underset{(2\ sf)}{2.5 \text{ m}} \times \underset{(4\ sf)}{1.308 \text{ m}} = 3.27 \text{ m}^2 = \underset{(2\ sf)}{3.3 \text{ m}^2}$$

*It should be noted that these rounding rules give an approximation of accuracy, as opposed to the results provided by more advanced statistical methods.

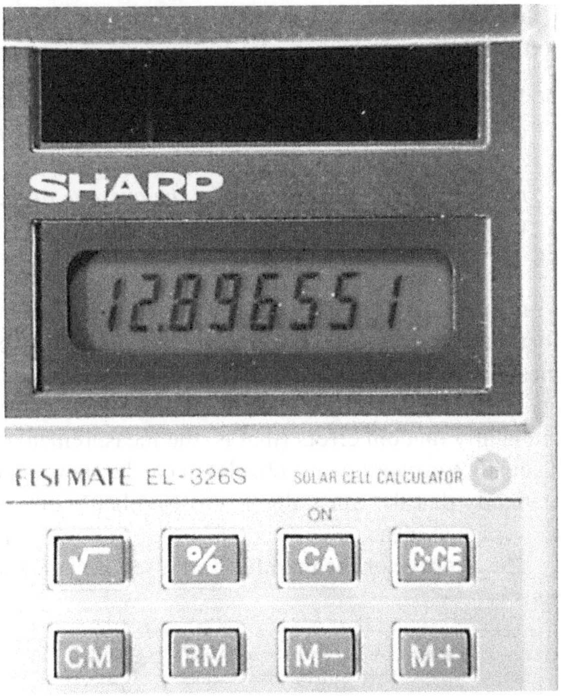

Figure 2.5 Insignificant figures. The calculator shows the result of the division operation 374/29. Because there are only two significant figures in the 29, a reported result should have no more than two significant figures, and the calculator display value should be rounded off to 13. (© Jerry D. Wilson)

Division:

(4 sf)

$$\frac{882.0 \text{ s}}{0.245} = 3600 \text{ s} = 3.60 \times 10^3 \text{ s}$$

(3 sf) *(represented to three significant figures; why?)*

Addition:

$$\begin{array}{r} 46.4 \\ 1.37 \\ 0.505 \\ \hline 48.275 \rightarrow 48.3 \end{array}$$

(rounding off)

(46.4 has the least number of decimal places)

Subtraction:

$$\begin{array}{r} 163 \\ -4.5 \\ \hline 158.5 \rightarrow 159 \end{array}$$

(rounding off)

(163 has the least number of decimal places, none)

E. Expressing Experimental Error and Uncertainty

Percent Error

The object of some experiments is to determine the value of a well-known physical quantity—for example, the value of π.

The **accepted or "true" value** of such a quantity found in textbooks and physics handbooks is the most accurate value (usually rounded off to a certain number of significant figures) obtained through sophisticated experiments or mathematical methods.

The **absolute difference** between the experimental value E and the accepted value A, written $|E - A|$, is the *positive* difference in the values, for example, $|2 - 4| = |-2| = 2$ and $|4 - 2| = 2$. Simply subtract the smaller value from the larger, and take the result as positive. For a set of measurements, E is taken as the average value of the experimental measurements.

When an accepted value of a physical quantity is known, the **percent error** is calculated for a comparison of an experimental value with the accepted value. Percent error is given by the ratio of the absolute difference between the experimental value E and the accepted value A over the accepted value, expressed as a percent.*

$$\text{Percent error} = \frac{\text{absolute difference}}{\text{accepted value}} \times 100\%$$

or

$$\boxed{\text{Percent error} = \frac{|E - A|}{A} \times 100\%} \quad \textbf{(2.1)}$$

Example 2.4 A cylindrical object is measured to have a diameter d of 5.25 cm and a circumference c of 16.38 cm. What are the experimental value of π and the percent error of the experimental value if the accepted value of π to two decimal places is 3.14?

Solution with $d = 5.25$ cm and $c = 16.38$ cm,

$$c = \pi d \quad \text{or} \quad \pi = \frac{c}{d} = \frac{16.38 \text{ cm}}{5.25 \text{ cm}} = 3.12$$

Then $E = 3.12$ and $A = 3.14$, so

$$\text{Percent error} = \frac{|E - A|}{A} \times 100\%$$

$$= \frac{|3.12 - 3.14|}{3.14} \times 100\%$$

$$= \frac{0.020}{3.14} \times 100\% = 0.64\%$$

*It should be noted that percent error only gives a measure of experimental error or uncertainty when the accepted or standard value is highly accurate. If an accepted value itself has a large degree of uncertainty, then the percent error does not give a good measure of experimental uncertainty.

Note: To avoid rounding errors, the preferred order of operations is addition and subtraction before multiplication and division.*

If the uncertainty in experimentally measured values as expressed by the percent error is large, you should check for possible sources of error. If found, additional measurements should then be made to reduce the uncertainty. Your instructor may wish to set a maximum percent error for experimental results.

Percent Difference

It is sometimes instructive to compare the results of two measurements when there is no known or accepted value. The comparison is expressed as a **percent difference**, which is the ratio of the absolute difference between the experimental values E_2 and E_1 and the average or mean value of the two results, expressed as a percent.

$$\text{Percent difference} = \frac{\text{absolute difference}}{\text{average}} \times 100\%$$

or

$$\text{Percent difference} = \frac{|E_2 - E_1|}{(E_2 + E_1)/2} \times 100\% \qquad (2.2)$$

Dividing by the average or mean value of the experimental values is logical, because there is no way of deciding which of the two results is better.

Example 2.5 What is the percent difference between two measured values of 4.60 cm and 5.00 cm?

Solution With $E_1 = 4.60$ cm and $E_2 = 5.00$ cm,

$$\text{Percent difference} = \frac{|E_2 - E_1|}{(E_2 + E_1)/2} \times 100\%$$

$$\text{Percent difference} = \frac{|5.00 - 4.60|}{(5.00 + 4.60)/2} \times 100\%$$

$$= \frac{0.40}{4.80} \times 100\% = 8.0\%$$

As in the case of percent error, when the percent difference is large, it is advisable to check the experiment for errors and possibly make more measurements.

*Although percent error is generally defined using the absolute difference $|E - A|$, some instructors prefer to use $(E - A)$, which results in positive (+) or negative (−) percent errors, for example, −0.64% in Example 2.4. In the case of a series of measurements and computed percent errors, this gives an indication of systematic error.

In many instances there will be more than two measurement values.

When there are three or more measurements, the percent difference is found by dividing the absolute value of the difference of the extreme values (that is, the values with greatest difference) by the average or mean value of all the measurements.

Average (Mean) Value

Most experimental measurements are repeated several times, and it is very unlikely that identical results will be obtained for all trials. For a set of measurements with predominantly random errors (that is, the measurements are all equally trustworthy or probable), it can be shown mathematically that the true value is most probably given by the average or mean value.

The **average** or **mean value** $\bar{x}$ of a set of N measurements is

$$\bar{x} = \frac{x_1 + x_2 + x_3 + \cdots + x_N}{N} = \frac{1}{N}\sum_{i=1}^{N} x_i \qquad (2.3)$$

where the summation sign Σ is a shorthand notation indicating the sum of N measurements from x_1 to x_N. ($\bar{x}$ is commonly referred to simply as the *mean.*)

Example 2.6 What is the average or mean value of the set of numbers 5.42, 6.18, 5.70, 6.01, and 6.32?

$$\bar{x} = \frac{1}{N}\sum_{i=1}^{N} x_i$$

$$= \frac{5.42 + 6.18 + 5.70 + 6.01 + 6.32}{5}$$

$$= 5.93$$

There are other, more advanced methods to express the dispersion or precision of sets of measurements. Two of these are given in the appendices. Appendix C: "Absolute Deviation from the Mean and Mean Absolute Deviation," and Appendix D: "Standard Deviation and Method of Least Squares."

F. Graphical Representation of Data

It is often convenient to represent experimental data in graphical form, not only for reporting but also to obtain information.

Graphing Procedures

Quantities are commonly plotted using rectangular Cartesian axes (X and Y). The horizontal axis (X) is called the *abscissa,* and the vertical axis (Y), the *ordinate.* The location of a point on the graph is defined by its coordinates x and y, written (x, y), referenced to the origin O, the intersection of the X and Y axes.

It is customary to say that y is plotted versus x, which indicates that in the mathematical relationship, y is a function of x; for example, $y = mx + b$, that is, the value of y depends on that of x.

When plotting data, choose axis scales that are easy to plot and read. The graph in ● Fig. 2.6A, a plot of force (F) versus displacement (x), shows an example of scales that are too small. This "bunches up" the data, making the graph too small, and the major horizontal scale values make it difficult to read intermediate values. Also, the dots or data points should not be connected. Choose scales so that most of the graph paper is used. The graph in ● Fig. 2.6B shows data plotted with more appropriate scales.*

Also note in Fig. 2.6A that scale units on the axes are not given. For example, you don't know the units of force or whether the units of displacement are feet, meters, kilometers, or whatever. *Scale units should always be included,* as in Fig. 2.6B. It is also acceptable, and saves time, to use standard unit abbreviations, such as N for newton and m for meter. This will be done on subsequent graphs.

With the data points plotted, draw a smooth line described by the data points. *Smooth* means that the line does not have to pass exactly through each point but connects the general areas of significance of the data points (*not* connecting the data points as in Fig. 2.6A). The

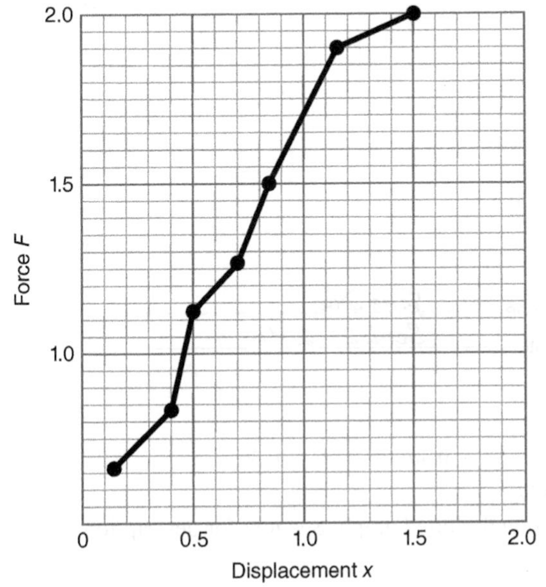

Figure 2.6A Poor graphing. An example of an improperly labeled and plotted graph. See text for description.

*As a general rule, it is convenient to choose the unit of the first major scale division to the right or above the origin or zero point as 1, 2, or 5 (or multiples or submultiples thereof, for example, 10 or 0.1) so that the minor (intermediate) scale divisions can be easily interpolated and read.

Table 2.1 Data for Figure 2.7

Mass (kg)	Period (s)	±	$\bar{d}$
0.025	1.9	±	0.40
0.050	2.7	±	0.30
0.10	3.8	±	0.25
0.15	4.6	±	0.28
0.20	5.4	±	0.18
0.25	6.0	±	0.15

graph in Fig. 2.6B with an approximately equal number of points on each side of the line gives a "line of best fit."†

In cases where several determinations of each experimental quantity are made, the average value is plotted and the mean deviation or the standard deviation may be plotted as *error bars*. For example, the data for the period of a mass oscillating on a spring given in Table 2.1 are plotted in ● Fig. 2.7, period (T) versus mass (m). (The $\bar{d}$ is the mean deviation, shown here for an illustration of error bars. See Appendix C.)* A smooth line is drawn so as to pass within the error bars. (Your instructor may want to explain the use of a French curve at this point.)

Graphs should have the following elements (see Fig. 2.7):

1. Each axis labeled with the quantity plotted.
2. The units of the quantities plotted.
3. The title of the graph on the graph paper (commonly listed as the y-coordinate versus the x-coordinate).
4. Your name and the date.

Linear (Straight-Line) Graphs

Two quantities (x and y) are often linearly related; that is, there is an algebraic relationship of the form $y = mx + b$, where m and b are constants. When the values of such quantities are plotted, the graph is a straight line, as shown in ● Fig. 2.8.

The m in the algebraic relationship is called the **slope** of the line and is equal to the ratio of the intervals $\Delta y/\Delta x$. Any set of intervals may be used to determine the slope of a straight-line graph; for example, in Fig. 2.8,

$$m = \frac{\Delta y_1}{\Delta x_1} = \frac{15 \text{ cm}}{2.0 \text{ s}} = 7.5 \text{ cm/s}$$

$$m = \frac{\Delta y_2}{\Delta x_2} = \frac{45 \text{ cm}}{6.0 \text{ s}} = 7.5 \text{ cm/s}$$

Points should be chosen relatively far apart on the line. For best results, points corresponding to data points should not be chosen, even if they appear to lie on the line.

†The straight line of "best fit" for a set of data points on a graph can be determined by a statistical procedure called *linear regression*, using what is known as the *method of least squares*. This method determines the best-fitting straight line by means of differential calculus, which is beyond the scope of this manual. The resulting equations are given in Appendix D, along with the procedure for determining the slope and intercept of a best-fitting straight line.
*The mean deviation and standard deviation are discussed in Appendix C and D, respectively. They give an indication of the dispersion of a set of measured values. These methods are optional at your instructor's discretion.

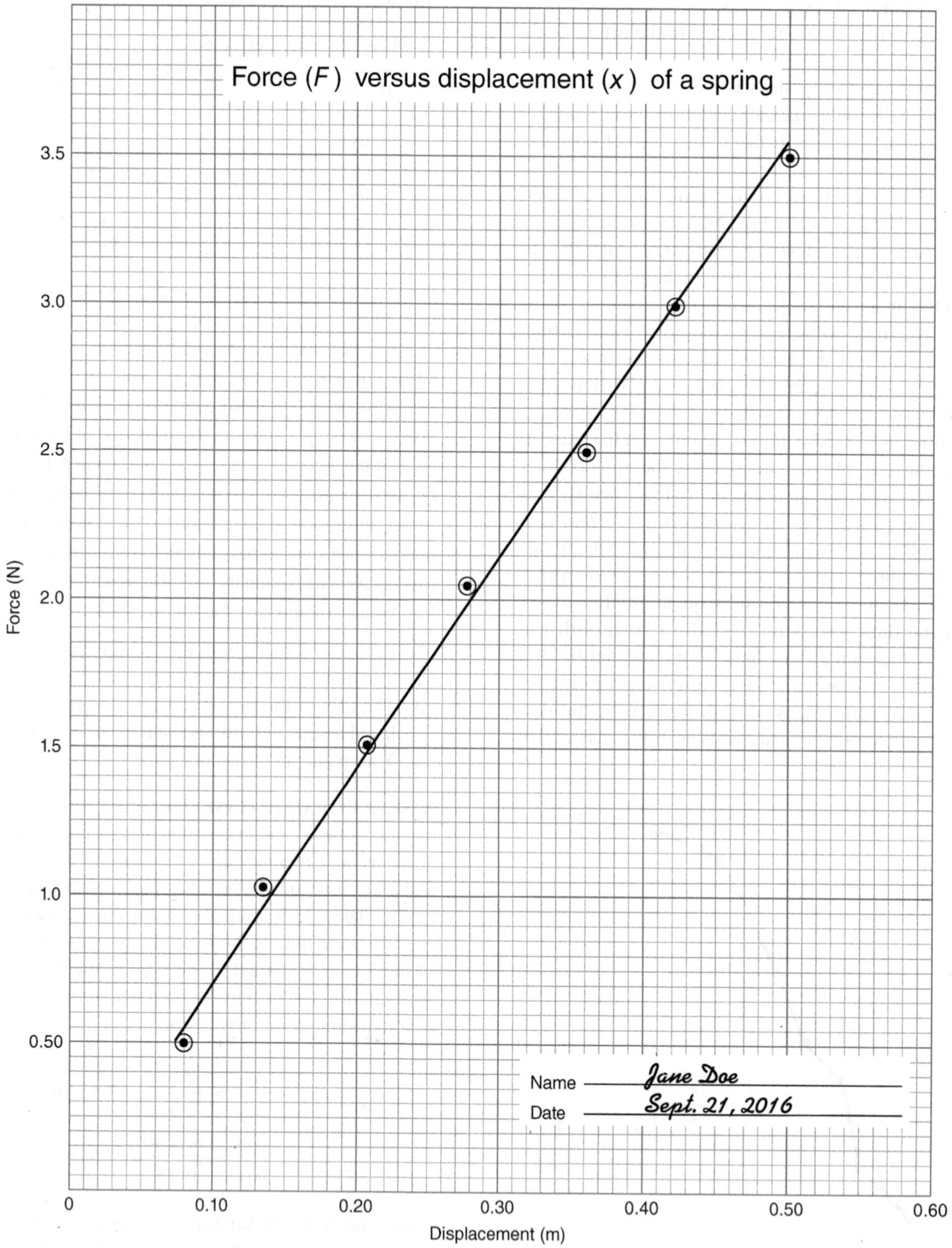

Figure 2.6B Proper graphing. An example of a properly labeled and plotted graph. See text for description.

The b in the algebraic relationship is called the **y-intercept** and is equal to the value of the y-coordinate where the graph line intercepts the Y-axis. In Fig. 2.8, $b = 3.0$ cm. Notice from the relationship that $y = mx + b$, so that when $x = 0$, then $y = b$. If the intercept is at the origin (0, 0), then $b = 0$.

The equation of the line in the graph in Fig. 2.8 is $d = 7.5t + 3.0$. The general equation for uniform motion has the form $d = vt + d_o$. Hence, the initial displacement $d_o = 3.0$ cm and the speed $v = 7.5$ cm/s.

Some forms of nonlinear functions that are common in physics can be represented as straight lines on a

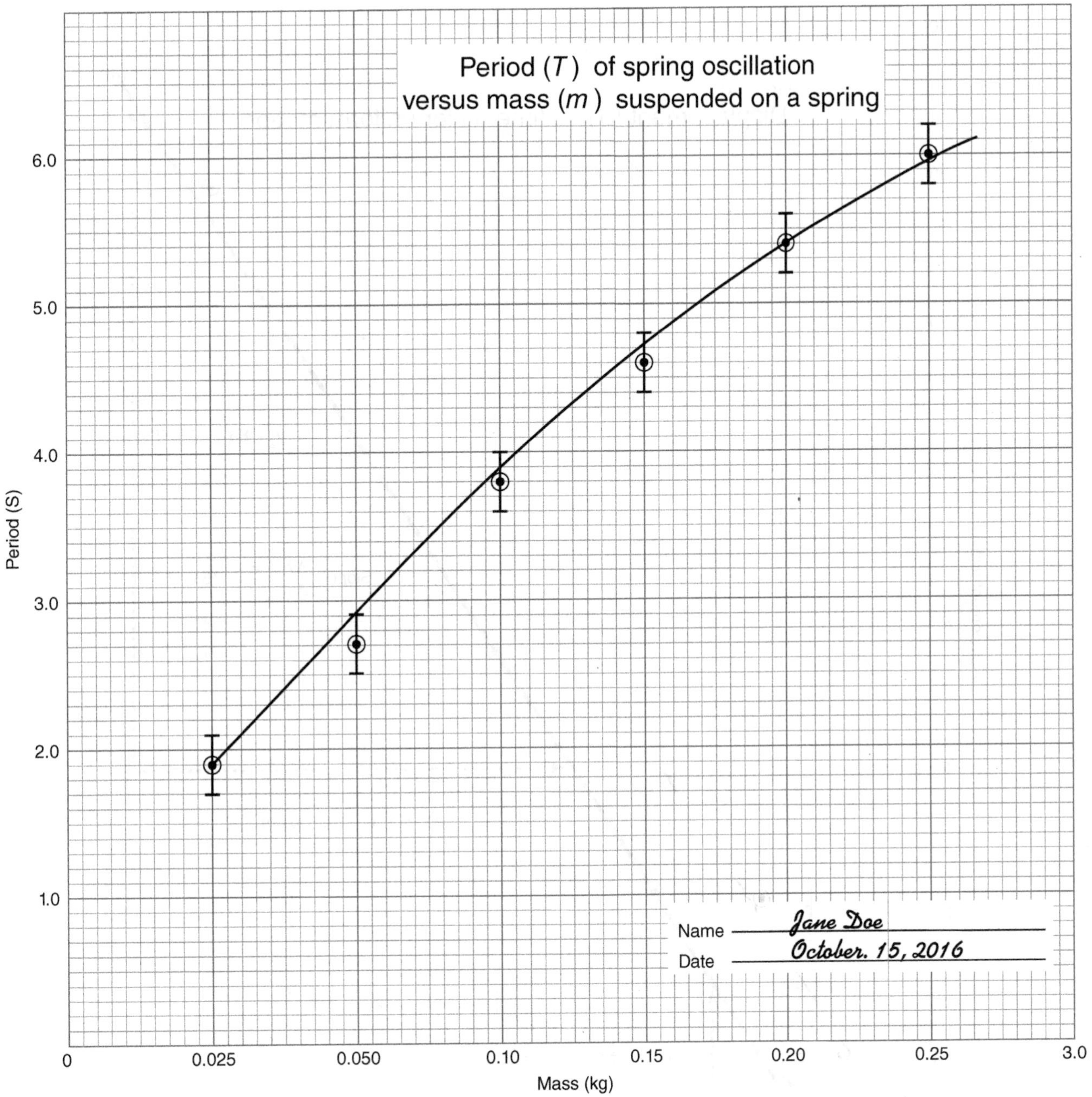

Figure 2.7 Error bars. An example of graphically presented data with error bars. An error bar indicates the precision of a measurement. In this case, the error bars represent mean deviations.

Cartesian graph. This is done by plotting nonlinear values. For example, if

$$y = ax^2 + b$$

is plotted on a regular y-versus-x graph, a parabola would be obtained. But if $x^2 = x'$ were used, the equation becomes

$$y = ax' = b$$

which has the form of a linear graph, or straight line.

This means plotting y versus x' would give a straight line. Since $x' = x^2$, the squared values of x must be plotted. That is, square all the values of x in the data table, and plot these numbers with the corresponding y values.

Other functions can be "straightened out" by this procedure, including an exponential function:

$$y = Ae^{ax}$$

In this case, taking the natural logarithm of both sides:

$$\ln y = \ln A + \ln e^{ax}$$

or

$$\ln y = ax + \ln A$$

(where $\ln e^x = x$)

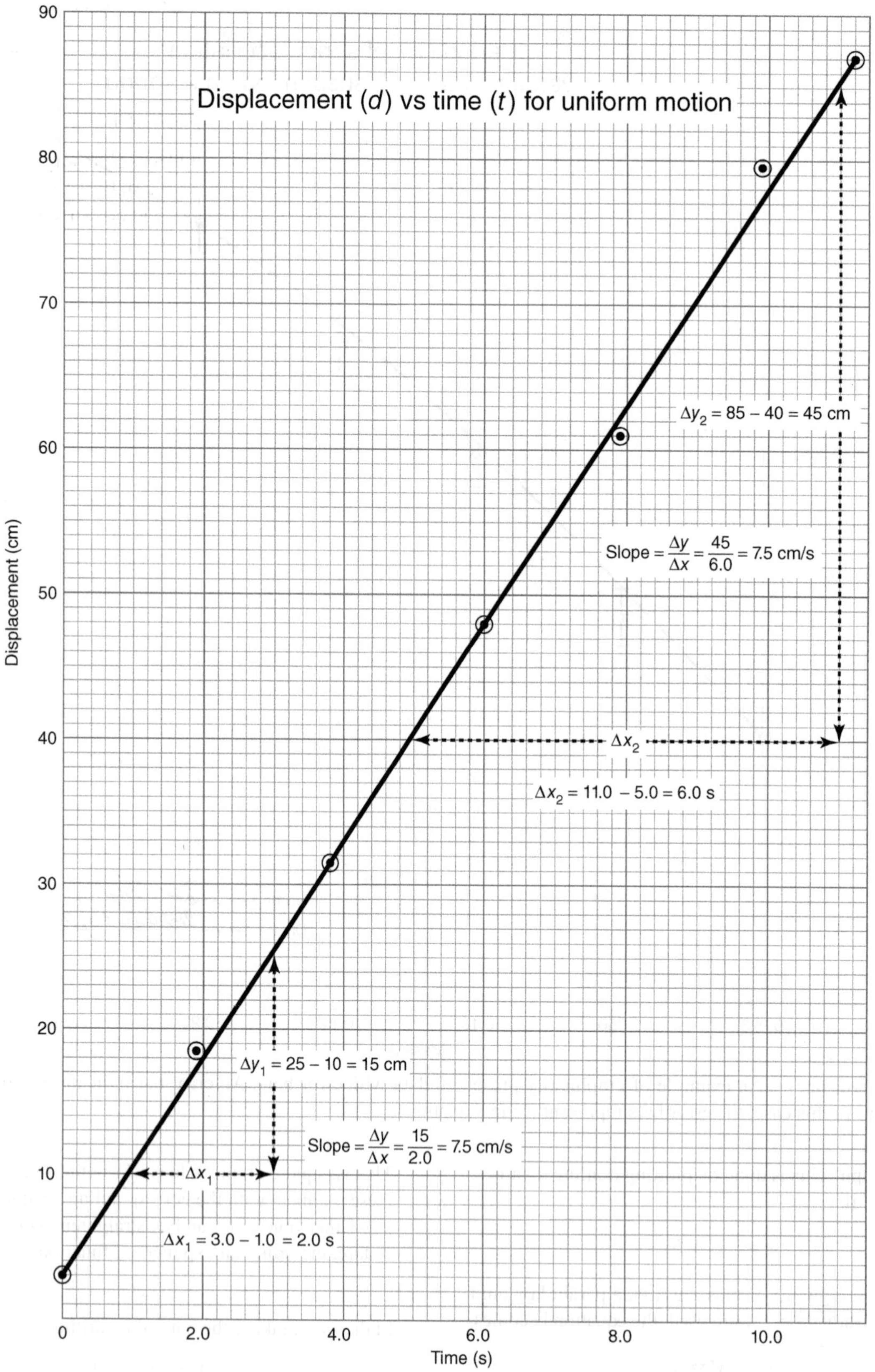

Figure 2.8 Linear (straight-line) slope. Examples of intervals for determining the slope of a straight line. The slope is the ratio of $\Delta y/\Delta x$ (or $\Delta d/\Delta t$). Any set of intervals may be used, but the endpoints of an interval should be relatively far apart, as for $\Delta y_2/\Delta x_2$.

Plotting the values of the natural (base e) logarithm versus x gives a straight line with slope a and an intercept ln A.

Similarly, for

$$y = ax^n$$

using the common (base 10) logarithm,

$$\log y = \log a + \log x^n$$

and

$$\log y = n \log x + \log a$$

(where $\log x^n = n \log x$).

Plotting the values of log y versus log x gives a straight line with slope n and intercept log a. (See Appendix E.)

EXPERIMENTAL PROCEDURE

Complete the exercises in the Laboratory Report, showing calculations and attaching graphs as required. (*Note*: In this experiment and throughout, attach an additional sheet for calculations if necessary.)

2a page 26

normal — 1440

scientific — 1.00×10^{-4}

both 3 sig figs

percent error

Name ______________________ Section ______________ Date ______________

Lab Partner(s) ______________________________________

EXPERIMENT 3

Measurement Instruments (Mass, Volume, and Density)

TI Advance Study Assignment

Read the experiment and answer the following questions.

1. What is the least count of a measurement instrument, and how is it related to the number of significant figures of a measurement reading?

2. Does a laboratory balance measure weight or mass? Explain.

3. What is the function of the vernier scale on the vernier caliper? Does it extend accuracy or precision? Explain.

4. Distinguish between positive and negative zero errors and how corrections are made for such errors. For what kind of error does a zero correction correct?

(continued)

5. What is the purpose of the ratchet mechanism on a micrometer caliper?

6. Explain how readings from 0.00 through 1.00 mm are obtained from the micrometer thimble scale when it is calibrated only from 0.00 through 0.50 mm.

7. If the density of one object is greater than that of another, what does this indicate? Do the sizes of the objects affect their densities? Explain.

8. Explain how the volume of a heavy, irregularly shaped object may be determined experimentally. Are there any limitations?

PRE-LAB DEMONSTRATION | Mixed Liquids Become More Dense

Equal volumes of methyl alcohol (density = 0.790 g/cm^3) and water (density = 1.000 g/cm^3) are mixed together. The resultant density of the mixture can be easily calculated theoretically using the known volumes and densities ($\rho = m/V$). Do this using volumes of 500 mL. (Remember mL and cm^3 are the same volumes.)

Yet when this is actually done in the lab and experimental values used to calculate the mixture density, it is found to be greater than theoretically predicted. How can that be? (Make sure the experimental measurements are done accurately.)

EXPERIMENT 3

Measurement Instruments (Mass, Volume, and Density)

INTRODUCTION AND OBJECTIVES

Common laboratory measurements involve the determination of the fundamental properties of mass and length. Most people are familiar with the use of scales and rulers or metersticks. However, for more accurate and precise measurements, laboratory balances and vernier calipers or micrometer calipers are often used, particularly in measurements involving small objects.

In this initial experiment on measurement, you will learn how to use these instruments and what advantages they offer. Density, the ratio of mass to volume, will also be considered, and the densities of several materials will be determined experimentally.

After performing this experiment and analyzing the data, you should be able to do the following:

1. Use the vernier caliper and read the vernier scale.
2. Use the micrometer caliper and read its scale.
3. Distinguish between mass and density, and know how to determine experimentally the density of an object or substance.

EQUIPMENT NEEDED

- Laboratory balance
- Vernier caliper
- Micrometer caliper (metric)
- Meterstick
- Graduated cylinder
- Cylindrical metal rod (for example, aluminum, brass, or copper)
- Sphere (metal or glass, for example, a ball bearing or marble)
- Short piece of solid copper wire
- Rectangular piece of metal sheet (for example, aluminum)
- Irregularly shaped metal object

THEORY

A. Laboratory Balances

Some common types of laboratory balances are shown in ● Fig. 3.1. Mechanical balances or "scales" are used to balance the weight of an unknown mass m against that of a known mass m_1 (that is, $mg = m_1g$ or $m = m_1$). The mass of the unknown is then read directly in mass units, usually grams. The weight w of an object is its mass m times a constant g, the acceleration due to gravity; $g = 9.80 \text{ m/s}^2 = 980 \text{ cm/s}^2$ (that is, $w = mg$ or $m = w/g$). Some scales, such as bathroom scales, are commonly calibrated in weight (force) units, such as pounds, rather than in mass units (kilograms).

A set of known masses is used to balance an unknown mass on a platform balance (Fig. 3.1a). On a beam balance, the riders on the beams are used to balance the unknown mass on the platform (Fig. 3.1b). The common laboratory beam balance is calibrated in grams. In this case, the least count is 0.1 g and a reading can be estimated to 0.01 g.* (See Experiment 2 for a review of least count.)

Before making a mass determination, a balance should be checked without a mass to make sure the scale is zeroed (reads zero). Adjustments can be made by various means on different scales.

Balances with digital readouts are common (Fig. 3.1c). These have the advantages of accuracy and ease of operation. However, electronic balances are much more delicate (Fig. 3.1d). The mass value is displayed automatically, and the accuracy or number of significant figures depends on the particular balance. Some electronic

*The official abbreviation of the gram unit is g (roman). The standard symbol for acceleration due to gravity is g (italic), where weight is given by mg, which is not to be confused with mg for milligram. Look closely so as to avoid confusion with these symbols.

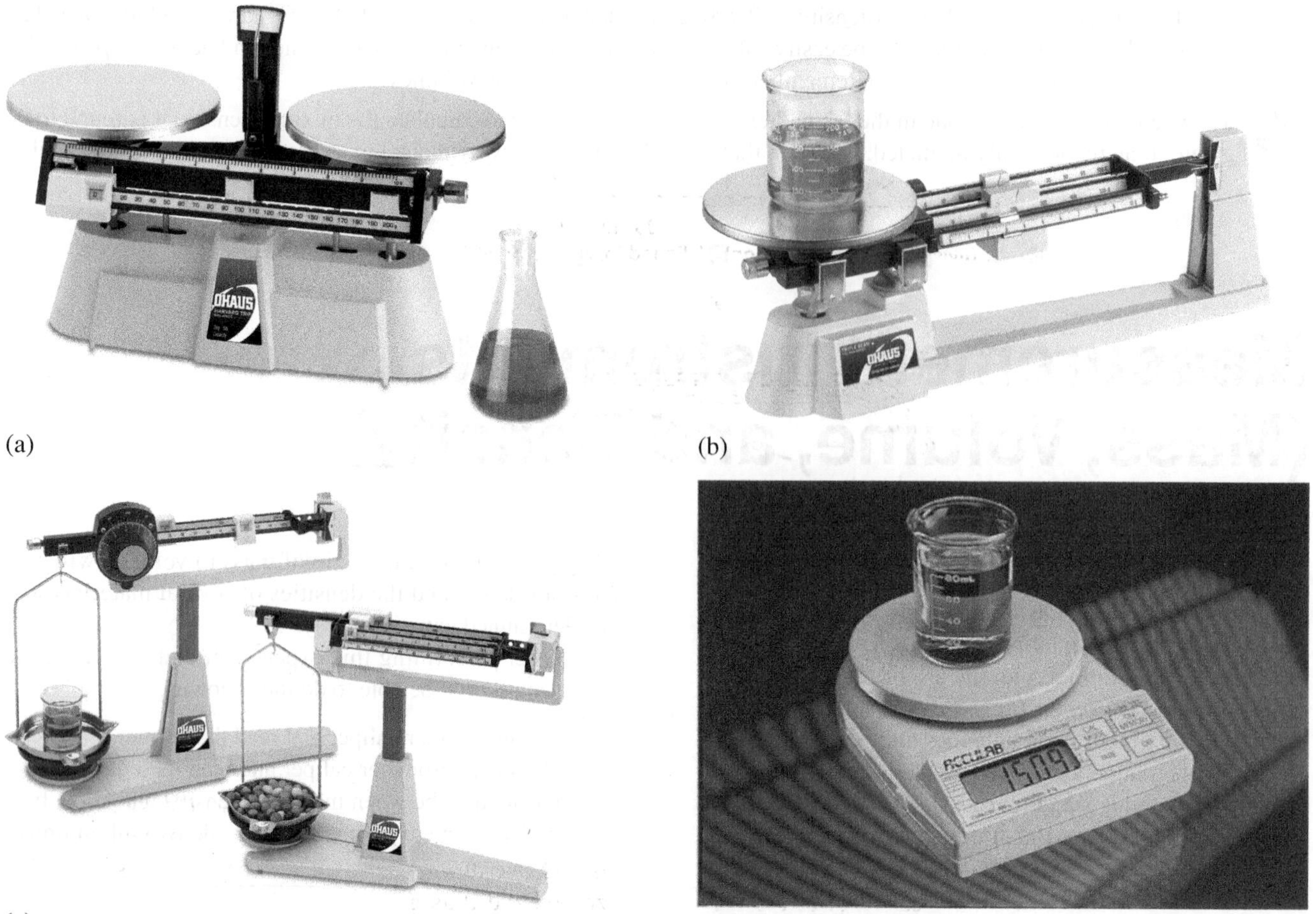

(a) (b) (c) (d)

Figure 3.1 Laboratory balances. (a) A double-beam, double-platform Harvard trip balance, which is also called an *equal-arm balance.* (b) A single-platform, triple-beam balance. (c) High-form beam balances. The balance on the left has a dial mechanism that replaces the lower-mass beams. (d) A digital electronic balance. (Courtesy of Sargent-Welch.)

balances have autocalibration and other have a keypad for calibration by the user.

Most electronic balances are zeroed by pressing a "tare" button. This has the advantage that one can place an empty dish on the balance before pressing the "tare" button, and then, when the material is added to the dish, the balance displays the mass of the contents alone.

Because of the wide variety of electronic balances available, if you are using one in this experiment you should first familiarize yourself with its operation. Your instructor may brief you, or an operation manual should be available.

B. The Vernier Caliper

In 1631, a French instrument maker, Pierre Vernier, devised a way to improve the precision of length measurements. The **vernier caliper**, commonly called a **vernier**, consists of a rule with a main engraved scale and a movable jaw with an engraved vernier scale (● Fig. 3.2). The span of the lower jaw is used to measure length and is particularly convenient for measuring the diameter of a cylindrical object. The span of the upper jaw is used to measure distances between two surfaces, such as the inside diameter of a hollow cylindrical object.

The main scale is calibrated in centimeters with a millimeter least count, and the movable vernier scale has 10 divisions that cover 9 divisions on the main scale. When making a measurement with a meterstick, it is necessary to estimate, or "eyeball," the fractional part of the smallest scale division (tenth of a millimeter). The function of the vernier scale is to assist in the accurate reading of the fractional part of the scale division, thus increasing the precision.

The leftmost mark on the vernier scale is the zero mark (lower scale for metric reading and upper scale for inches) (see ● Fig. 3.3a). The zero mark is often unlabeled. A measurement is made by closing the jaws on the

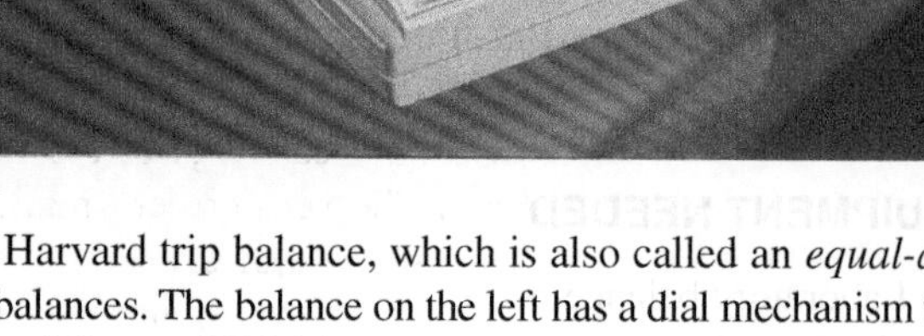

Figure 3.2 A vernier caliper. A good instrument for measuring rectangular dimensions and circular diameters. This caliper has scales for both metric and British measurements. See text for description. (Courtesy of Sargent-Welch.)

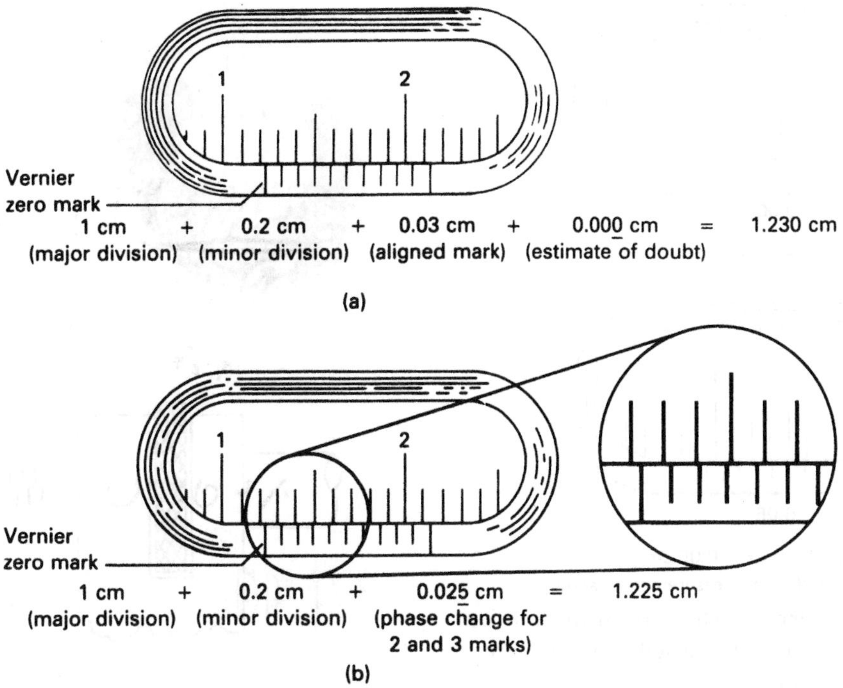

Figure 3.3 The vernier scale. An example of reading the vernier scale on a caliper. See text for description.

object to be measured and reading where the zero mark on the vernier scale falls on the main scale (see ● Fig. 3.3b). Some calipers, as the one in Fig. 3.2, have vernier scales for both metric and British units.

In Fig. 3.3, the first two significant figures are read directly from the main scale. The vernier zero mark is past the 2-mm line after the 1-cm major division mark, so there is a reading of 1.2 cm for both (a) and (b). The next significant figure is the fractional part of the smallest subdivision on the main scale. This is obtained by referring to the vernier scale markings below the main scale.

If a vernier mark coincides with a mark on the main scale, then the vernier mark number is the fractional part of the main-scale division (see Fig. 3.3a). In the figure, this is the third mark to the right of the vernier zero, so the third significant figure is 3 (0.03 cm). Finally, since the 0.03-cm reading is known exactly, a zero is added as the doubtful figure for a reading of 1.230 cm or 12.30 mm. Note how the vernier scale gives more significant figures or extends the precision.

However, a mark on the vernier scale may not always line up exactly with one on the main scale (Fig. 3.3b). In this case, there is more uncertainty in the 0.001-cm or 0.01-mm figure, and we say there is a change of "phase" between two successive vernier markings.

Notice how in Fig. 3.3b the second vernier mark after the zero is to the right of the closest main-scale mark, and the third vernier mark is to the left of the next main-scale mark. Hence, the marks change "phase" between the 2 and 3 marks, which means the reading is between 1.22 cm and 1.23 cm. Most vernier scales are not fine enough for us to make an estimate of the doubtful figure, so a suggested method is to take the middle of the range. Thus a 5 would be put in the thousandth-of-a-centimeter digit for a reading of 1.225 cm.

Zeroing

Before making a measurement, one should check the zero of the vernier caliper with the jaws completely closed. It is possible that through misuse the caliper is no longer zeroed and thus gives erroneous readings (systematic error). If this is the case, a zero correction should be made for each reading.

In zeroing, if the vernier zero lies to the right of the main-scale zero, measurements will be too large and the error is taken to be *positive.* In this case, the zero correction is made by subtracting the zero reading from the measurement reading. For example, the "zero" reading in ● Fig. 3.4 is +0.05 cm, and this amount must be subtracted from each measurement reading for more accurate results.

Similarly, if the error is *negative*, or the vernier zero lies to the left of the main-scale zero, measurements will be too small, and the zero correction must be added to the measurement readings.

Summarizing these corrections in equation form,

$$\text{Corrected reading} = \text{actual reading} - \text{zero reading}$$

For example, for a *positive* error of $+0.05$ cm as in Fig. 3.4,

$$\text{Corrected reading} = \text{actual reading} - 0.05\text{ cm}$$

If there is a *negative* correction of -0.05 cm, then

$$\begin{aligned}\text{Corrected reading} &= \text{actual reading} - (-0.05)\text{ cm}\\ &= \text{actual reading} + 0.05\text{ cm}\end{aligned}$$

C. The Micrometer Caliper

The **micrometer caliper** (● Fig. 3.5a), commonly called a **mike**, provides for accurate measurements of small lengths. A mike is particularly convenient in measuring the

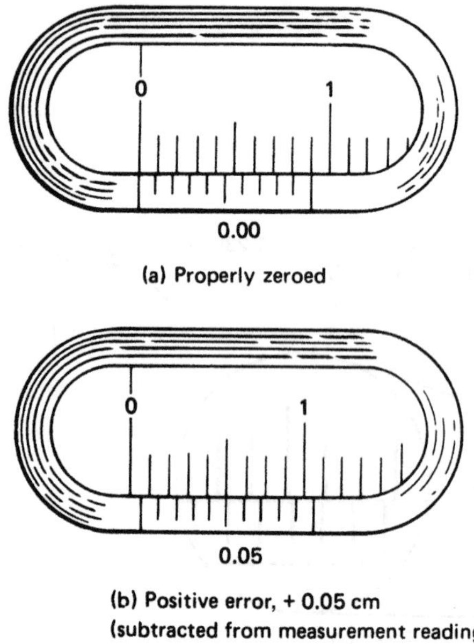

Figure 3.4 Zeroing and error. The zero of the vernier caliper is checked with the jaws closed. (a) Zero error. (b) Positive error, +0.05 cm.

diameters of thin wires and the thicknesses of thin sheets. It consists of a movable spindle (jaw) that is advanced toward another, parallel-faced jaw (called an anvil) by rotating the thimble. The thimble rotates over an engraved sleeve (or "barrel") mounted on a solid frame.

Most micrometers are equipped with a ratchet (ratchet handle is to the far right in the figure) that allows slippage of the screw mechanism when a small and constant force is exerted on the jaw. This permits the jaw to be tightened on an object with the same amount of force each time. Care should be taken not to force the screw (particularly if the micrometer does not have a ratchet mechanism), so as not to damage the measured object and/or the micrometer.

The axial main scale on the sleeve is calibrated in millimeters, and the thimble scale is calibrated in 0.01 mm (hundredths of a millimeter). The movement mechanism of the micrometer is a carefully machined screw with a pitch of 0.5 mm. The pitch of a screw, or the distance between screw threads, is the lateral linear distance the screw moves when turned through one rotation (Fig. 3.5b).

The axial line on the sleeve main scale serves as a reading line. Since the pitch of the screw is 0.5 mm and there are 50 divisions on the thimble, when the thimble is turned through one of its divisions, the thimble moves (and the jaws open or close) $\frac{1}{50}$ of 0.5 mm, or 0.01 mm ($\frac{1}{50} \times 0.5 \text{ mm} = 0.01 \text{ mm}$).

One complete rotation of the thimble (50 divisions) moves it through 0.5 mm, and a second rotation moves it through another 0.5 mm, for a total of 1.0 mm, or one scale division along the main scale. That is, the first rotation moves the thimble from 0.00 through 0.50 mm, and the second rotation moves the thimble from 0.50 mm through 1.00 mm.

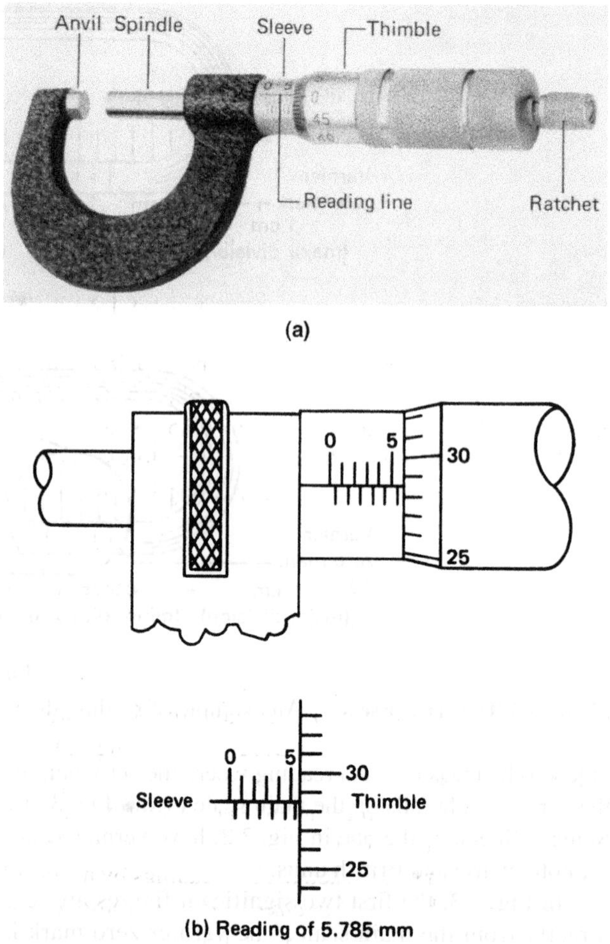

Figure 3.5 A micrometer caliper and an example of a micrometer reading. (a) This particular mike has the 1.0-mm and 0.5-mm scale divisions below the reading line. (b) In this diagram, as on some mikes, the 1.0-mm divisions are above the reading line and the 0.5-mm divisions are below it. The thimble in the diagram is in the second rotation of millimeter movement, as indicated by its being past the 0.5-mm mark. The reading is 5.500 + 0.285 mm, or 5.785 mm, where the last 5 is the estimated figure. (Photo courtesy of Sargent-Welch.)

It is sometimes instructive to think of the 1-mm main-scale divisions as analogous to dollar ($) divisions and of the thimble scale divisions as cents ($0.01). The first rotation of the thimble corresponds to going from $0.00 to $0.50 (50 cents), and the second rotation corresponds to going from $0.50 to $1.00, so that two complete rotations go through 100 cents, or $1.00, of the main scale.

Some micrometers have a scale that indicates the 0.5-mm marks of the main-scale divisions and hence tells which rotation the thimble is in (see Fig. 3.5). Cheaper mikes do not have this extra graduation, and the main scale must be closely examined to determine which rotation the thimble is in.

If a mike does not have the 0.5-mm scale, you must determine whether the thimble is in its first rotation, in which case the thimble reading is between 0.00 and 0.50 mm (corresponding to the actual engraved numbers on the thimble), or in the second rotation, in which case the reading is between 0.50 mm and 1.00 mm (the actual thimble scale reading plus 0.50). This can be done by judging whether the edge of the thimble is in the first or the second half of the main-scale division. Notice that the zero mark on the thimble is used to indicate both 0.00 mm (beginning of the first rotation) and 0.50 mm (beginning of the second rotation).

Measurements are taken by noting the position of the edge of the thimble on the main scale and the position of the reading line on the thimble scale. For example, for the drawing in Fig. 3.5, the mike has a reading of 5.785 mm. On the main scale is a reading of 5.000 mm plus one 0.500-mm division (scale below reading line), giving 5.500 mm.

That is, in the figure, the thimble is in the second rotation of a main-scale division. The reading on the thimble scale is 0.285 mm, where the 5 is the estimated or doubtful figure. That is, the reading line is estimated to be midway between the 28 and the 29 marks. (Some mikes have vernier scales on the sleeves to help the user read this last significant figure and further extend the precision.)

As with all instruments, a zero check should be made and a zero correction applied to each reading if necessary, as described in Section B. A zero reading is made by rotating the screw until the jaw is closed or the spindle comes into contact with the anvil. The contacting surfaces of the spindle and anvil should be clean and free of dust. (Micrometers can be adjusted to zero readings by means of a spanner wrench. *Do not attempt to do this* without your instructor's permission or supervision.)

D. Density

The **density** (ρ) of a substance is defined as the mass m per unit volume V (that is, $\rho = m/V$). Thus, the densities of substances or materials provide comparative measures of the amounts of matter in a particular (unit) space. Note that there are two variables in density—mass and volume. Hence, densities can be affected by the masses of atoms and/or by their compactness (volume).

As can be seen from the defining equation ($\rho = m/V$), the SI units of density are kilogram per cubic meter (kg/m^3). However, measurements are commonly made in the smaller metric units of grams per cubic centimeter (g/cm^3), which can easily be converted to standard units.*

Density may be determined experimentally by measuring the mass and volume of a sample of a substance and calculating the ratio m/V. The volume of regularly shaped objects may be calculated from length measurements. For example,

Rectangle

$V = l \times w \times h$	(length × width × height)

Cylinder

$V = Al = (\pi r^2)l$	(circular cross-sectional area $A = \pi r^2$, where r is the radius and l is the length of the cylinder)

Sphere

$V = \frac{4}{3}\pi r^3$	(where r is the radius of the sphere)

*In the British fps (foot–pound–second) system, density is expressed in terms of weight rather than mass. For example, the weight density of water is 62.4 lb/ft^3.

Figure 3.6 Density, mass, and volume. The marble and the Styrofoam ball have equal masses but different densities ($\rho = m/V$). Because the volume of the ball is greater than that of the marble, its density is less. (© Cengage Learning.)

To illustrate how density provides a measure of compactness of matter, consider the marble and Styrofoam ball in ● Fig. 3.6. Both have the same mass (5.0 g), but the marble has greater density. (Why?) With measured radii of $r_m = 0.75$ cm and $r_b = 6.0$ cm for the marble and ball, respectively, the calculated densities are

$$\rho_m = \frac{m_m}{V_m} = \frac{m_m}{\frac{4}{3}\pi r_m^3} = \frac{5.0 \text{ g}}{\frac{4}{3}\pi(0.75 \text{ cm})^3} = 2.8 \text{ g/cm}^3$$

$$\rho_b = \frac{m_b}{V_b} = \frac{m_b}{\frac{4}{3}\pi r_b^3} = \frac{5.0 \text{ g}}{\frac{4}{3}\pi(6.0 \text{ cm})^3} = 0.0055 \text{ g/cm}^3$$

(Notice that the calculated results have only two significant figures. Why?) In standard SI units, these results are 2.8×10^3 kg/m^3 and 5.5 kg/m^3, respectively.

But how does one find the volume of an irregularly shaped solid object? This may be done by immersing it in water (or some other liquid) in a graduated container. Since the object will displace a volume of water equal to its own volume, the difference in the container readings before and after immersion is the volume of the object. Cylinders commonly have scale divisions of milliliters (mL) and 1 mL = 1 cm^3.* [cm^3 (cubic centimeter) is sometimes written on glassware as cc.]

The physical property of density can be used to identify substances in some cases. If a substance is not pure or is not homogeneous (that is, its mass is not evenly distributed), an average density is obtained, which is generally different from that of a pure or homogeneous substance.

EXPERIMENTAL PROCEDURE

A. Least Count of an Instrument Scale

1. List the least count and the estimated fraction of the least count for each of the measuring instruments in Data Table 1 of the laboratory report. For example, for a meterstick, these would be 1 mm and 0.1 mm, respectively. (Review Experiment 2C if necessary.)

B. Thickness Measurements

2. Using the micrometer caliper, take a zero reading and record it in Data Table 2. Then take several measurements of a single page of this manual, incorporating the zero correction if necessary, to determine the average thickness per page. Record the data and result in Data Table 2.

3. With the micrometer, take thickness measurements of a group of several pages together [for example, 10 pages (sheets of paper)], and record the data in Data Table 2. Calculate the average thickness per page.

4. With the vernier caliper, take several measurements of the total thickness of the manual (*excluding* covers).† Record the data in Data Table 2, and compute the average overall thickness of the manual. (Did you remember to take a zero reading and record in Data Table 2?)

5. Using the values of the average thickness per page determined in Procedures 2 and 3 and the overall average thickness of the manual from Procedure 4, compute the number of pages (sheets of paper) in your manual. For example, if the average thickness per page is 0.150 mm and the average overall thickness is 35.5 mm (3.55 cm), the calculated number of papers is

$$\frac{35.5 \text{ mm}}{0.150 \text{ mm/page}} = 236.6666 = 237 \text{ pages}$$

6. Determine the actual number of pages (sheets of paper) in the manual. (Remember to subtract any pages handed in from previous experiments, the Advance Study Assignment for this experiment, and any others that might be missing.) Compute the percent error for each of the two experimentally determined values.

C. Density Determinations

7. The densities of the materials of the various objects are to be determined from mass and volume (length) measurements. Taking the mass and length measurements will give you experience in using the laboratory balance and the vernier and micrometer calipers.

8. Using the appropriate measuring instrument(s), take several measurements to determine the average dimensions of the regularly shaped objects so that their volumes can be calculated. Record the data in Data Table 3. Remember to make a zero correction for each reading if necessary.

9. Calculate the volume of each of the objects, and record in Data Table 4.

10. Determine the volume of the irregularly shaped metal object by the method described in Theory section D. Record the volume in Data Table 4.*

11. Using a laboratory balance, determine the mass of each object, and record the results in Data Table 4.

12. Calculate the density of the material of each object, and find the percent error of each experimental result. (Accepted density values are given in Appendix A, Table A1.)

*Milliliter is abbreviated both ml and mL. The mL abbreviation is generally preferred in order to avoid confusion of a lowercase l ("ell") with the number 1.

†Be sure the pages are compacted as much as possible before you take the measurements.

*The water will have a curved, bowl-shaped upper surface called a *meniscus*. Make readings at the bottom of the curved surface.

Name ______________________ Section ______________ Date ______________

Lab Partner(s) ______________________________

TI QUESTIONS

1. Explain the probable source of error(s) in the experimental determination of the number of manual pages.

2. In the first four density determinations in Data Table 4, what major factors might account for the experimental errors that were obtained?

3. In determining the volume of the irregularly shaped object, any air bubbles sticking to the surface of the object when it is submerged cause systematic errors. Will this error give an experimental density that is too high or too low? Explain.

4. Suppose that you were given an irregularly shaped object that floats. Describe how you would experimentally determine its volume.

(continued)

5. A thin circular sheet of aluminum has a radius of 20 cm and a thickness of 0.50 mm. Find the mass of the sheet.

6. Archimedes, a famous Greek scientist, was given a problem by King Hieron II of Syracuse (Sicily). The king suspected that his crown, which was supposed to be made of pure gold, contained some silver alloy, and he asked Archimedes to prove or disprove his suspicion. (It turned out that the crown did contain silver.) How would you experimentally determined whether or not the crown was pure gold? (*Hint*: The method came to Archimedes when getting into a full bathtub. See the footnote in Experiment 22 for Archimedes' solution.)

Name ______________________ Section ____________ Date ____________

Lab Partner(s) ______________________

EXPERIMENT 4

Simple Pendulum Parameters (Angle, Mass, Length, and Damping)

GL Experimental Planning

The Simple Pendulum

1. Scientists use models and theories to describe physical phenomena. When a new model is developed, it must be tested to find out if it is an accurate representation. No theory or model of nature is valid unless its predictions are in agreement with experimental results. The laboratory provides an environment where extraneous factors can be minimized and specific predictions can be tested. The process of making, testing, and refining models is usually called the *scientific method* (see Experiment 1).

   An example of this method will be demonstrated in this experiment for a simple pendulum. A "simple" pendulum is one in which a small but substantial mass is suspended on a relatively light string, like the one pictured in Fig. 4.1. If one were to observe the motion of the mass swinging back and forth, which of the following statements do you think would be the most accurate? (It is understood that the motion takes place in a single plane.)

   The time for the mass to swing back and forth (from point A to B, and back to A in Fig. 4.1.)

   (a) changes randomly from one swing to the next.
   (b) gets consistently bigger from one swing to the next.
   (c) gets consistently smaller from one swing to the next.
   (d) stays about the same from one swing to the next.

   (Circle your choice)

2. The time for the mass to swing back and forth is called the **period** (*T*) of the pendulum. If your physics lab has the appropriate equipment available, you could verify that statement (d) above is the most accurate (with negligible friction). Now consider what might affect the pendulum's period. Look at Fig. 4.1 again and list the physical parameters that could be changed.

3. Did you find three things? Let's consider the length (*L*) first. How do you think the pendulum's length might affect the period? If the length of the pendulum were doubled, would the period (*T*) also double (directly proportional)? Or would it be half of what it was before (inversely proportional)? Or could it be larger or smaller by some other proportion? Write down the relationship that you think is most appropriate.

4. The mass (*m*) of the pendulum bob may be varied. The effect this would have on the period might possibly depend on air resistance, so let's suppose there isn't any. If the pendulum were swinging in a vacuum, would the mass make any difference?

(continued)

To verify your response, look at the forces acting on the bob. Draw a free-body diagram (one showing the forces) for the bob when it would be in the position shown in Fig. 4.1. What is the *component* of the weight force (mg) that acts in the direction of motion?

5. Check with one of your fellow students (or your instructor) to see if the results agree. Notice that there are no other forces acting in the direction of motion (remember, no air resistance). Then, use this force component in Newton's second law and solve for a. Does your result for the acceleration of the bob (and ultimately its pattern of motion) include the mass?

6. Finally, you probably listed the initial (release) angle θ as a factor that would affect the period. Your result for the acceleration above should include this factor (in the form of $\sin\theta$). Since the acceleration depends on $\sin\theta$ instead of θ, the situation is more complicated than those usually encountered in this course. Advanced mathematics is needed to derive the theoretical equation for the period of a simple pendulum oscillating in a plane. This equation includes the factors discussed previously, as well as one (constant factor) you probably wouldn't expect.

$$T = 2\pi\sqrt{\frac{L}{g}}\left(1 + \frac{1}{4}\sin^2\frac{\theta}{2} + \frac{9}{64}\sin^4\frac{\theta}{2} + \cdots\right)$$

This equation predicts that the period will be longer if the length is longer and if the angle is larger, but the relation is not directly proportional. Does this agree with your predictions?

A major problem in using this theoretical equation to make predictions that can be tested by experiment is the infinite series $\left(1 + \frac{1}{4}\sin^2\frac{\theta}{2} + \frac{9}{64}\sin^4\frac{\theta}{2} + \cdots\right)$. If we could find an approximation of this equation, it would be more useful. Since $\sin\theta = 0$ if $\theta = 0$, if the angle is small enough, the terms with θ might be negligible. Test this by calculating the resultant sum of the first three terms in the series for an angle of 5°. Is it bigger than 1.0 by very much? At what angle θ would the first three terms add up to 1.05 (a 5% difference)?

Do you think it is reasonable to say that as long as the angle θ is less than a certain value, then to a very good approximation, $T = 2\pi\sqrt{\frac{L}{g}}$? That is, $\left(\frac{1}{4}\sin^2\frac{\theta}{2} + \frac{9}{64}\sin^4\frac{\theta}{2} + \cdots << 1\right)$. Why or why not?

Name ______________________ Section ______________ Date ______________

Lab Partner(s) ____

EXPERIMENT 4

Simple Pendulum Parameters (Angle, Mass, Length, and Damping)

TI Advance Study Assignment

Read the experiment and answer the following questions.

1. Describe what is meant by the scientific method and how it is applied.

2. What are the physical parameters in the investigation of a simple pendulum?

3. A period is an interval of time. How is this applied to a pendulum?

(continued)

4. How does the period of a pendulum vary theoretically with (a) length, (b) mass of bob, and (c) angular displacement?

5. How will you experimentally check the theoretical predictions in the preceding question?

6. What is meant by a *small-angle approximation*?

7. How can the parabolic form $y = ax^2$ be plotted as a straight line on Cartesian graph paper?

8. What is meant by damped harmonic motion, and what causes it?

PRE-LAB DEMONSTRATION | Problem in Math Analysis

In this experiment, there is some mathematical analysis. Keep in mind that an equation is a balanced statement, numerically equal on each side of the equal sign. And, as long as the same operation is done on each side, for example addition or multiplication, the equation is still balanced or remains an equality.

Your instructor will carry out some simple algebraic operations on an equation but will end up with 2 = 1 (which is not too balanced!). Follow the operations closely—there must be a mistake.

E X P E R I M E N T 4

Simple Pendulum Parameters (Angle, Mass, Length, and Damping)

INTRODUCTION AND OBJECTIVES

The laboratory is a place for the investigation of physical phenomena and principles. In the process, new discoveries may be made and technology advanced. In some instances, while trying to invent things in the laboratory, scientists make various investigations at random. This might be called the *trial-and-error approach.*

Edison's invention of the lightbulb is an example. He kept trying until he found something that worked—a carbonized thread for a filament. Today, the physics laboratory is used in general to apply what is called the *scientific method*: No theory or model of nature is valid unless its predictions are in agreement with experimental results. (See Experiment 1.)

To illustrate the scientific method, in this experiment a theoretical expression or equation that describes the behavior of a simple pendulum is given. The validity of this relationship will then be tested experimentally. In the process, you will learn what variables influence the period of a simple pendulum and how the physical relationship and experimental data can be used to find other useful information (for example, the value of the acceleration due to gravity).

After performing this experiment and analyzing the data, you should be able to do the following:

1. Apply the scientific method to theoretical predictions to check their validity.
2. Understand how physical parameters are varied so as to investigate theoretical predictions.
3. Appreciate the use of approximations to facilitate experimental investigations and analyses.
4. Explain the parameters of damped harmonic motion.

EQUIPMENT NEEDED

- Meterstick
- Laboratory timer(s) or stopwatch(es)
- Protractor
- String
- Three or more pendulum bobs of different masses
- Pendulum clamp (if available)
- 2 sheets of Cartesian graph paper

THEORY

A **simple pendulum** consists of a "bob" (a mass) attached to a string that is fastened such that the pendulum assembly can swing or oscillate freely in a plane (● Fig. 4.1). For a simple or ideal pendulum, all the mass is considered to be concentrated at a point at the center (of mass) of the bob.

Some of the physical properties or parameters of a simple pendulum are (1) the length L of the pendulum, (2) the mass m of the pendulum bob, (3) the angular distance θ through which the pendulum swings, and (4) the period T of the pendulum, which is the time it takes for the pendulum to swing through one complete oscillation (for example, from A to B and back to A in Fig. 4.1).

From experience or preliminary investigation, it is found that the period of a pendulum depends on its length (the longer the length, the greater its period). How do you think the other parameters (m and θ) affect the period?

From physical principles and advanced mathematics, the theoretical expression for the period of a simple pendulum oscillating in a plane is

$$T = 2\pi\sqrt{\frac{L}{g}}\left(1 + \frac{1}{4}\sin^2\frac{\theta}{2} + \frac{9}{64}\sin^4\frac{\theta}{2} + \cdots\right) \quad \textbf{(4.1)}$$

where g is the acceleration due to gravity and the terms in parentheses are part of an infinite series. In calculating T for a given angular distance θ, the more terms of the

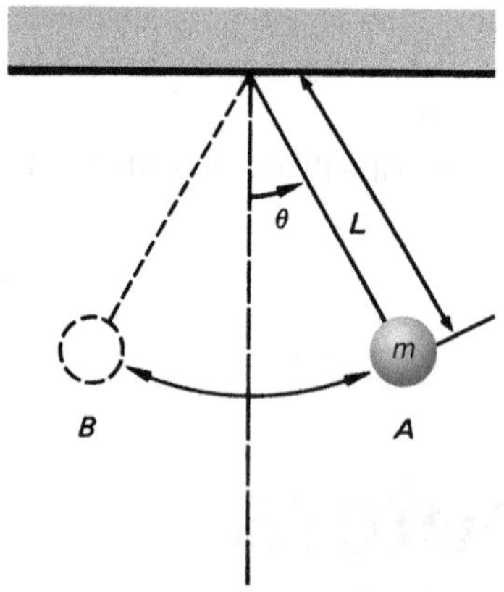

Figure 4.1 The simple pendulum. The physical parameters of a simple pendulum are its length L, the mass m of the bob, and the angle of swing θ. The period T of a pendulum is the time it takes for one completed oscillation—for example, the time it takes to swing from A to B and back to A.

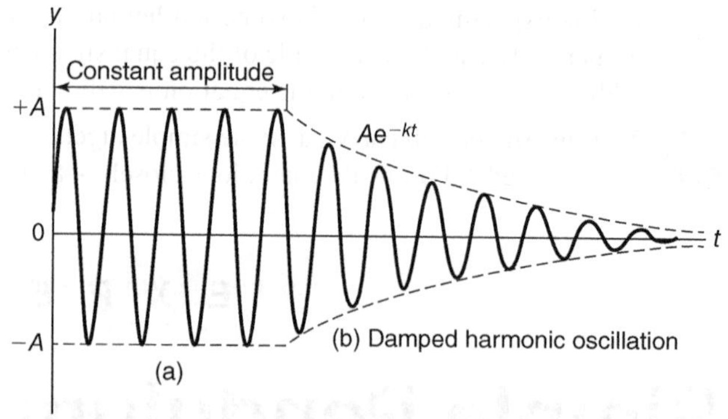

Figure 4.2 Damped Harmonic Motion. Starting with an initial amplitude of A_o, the graph depicts the decay of pendulum amplitude with time. The dashed line described by $A_o e^{-kt}$ is called a decaying exponential function.

series that are evaluated, the greater the accuracy of the theoretical result.

For small angles ($\theta \lesssim 20°$), the θ terms in the series are small compared to unity, that is, ($\frac{1}{4}\sin^2\frac{\theta}{2} + \frac{9}{64}\sin^4\frac{\theta}{2} + \cdots << 1$), and in this case, to a good approximation:

$$\boxed{T = 2\pi\sqrt{\frac{L}{g}}} \tag{4.2}$$

(This is called a **first-order approximation**. If the second term in the series is retained, the approximation is to second order, and so on.)

Notice that even without an approximation [Eq. (4.1)], the period is theoretically independent of the mass of the pendulum bob. Also, within the limits of the small-angle approximation [Eq. (4.2)], the period is independent of the displacement angle.

DAMPED HARMONIC MOTION

The small angle approximation allows a pendulum to *approximate* simple harmonic motion. (Harmonic, because it can be described by harmonic functions—sines and cosines.)*

In an ideal system, the oscillations of a pendulum would occur continuously with a constant amplitude (● Fig. 4.2a). However, in the real world, the amplitude of a oscillating pendulum decreases with time until the oscillations stop altogether. This process is called **damped harmonic motion** (Fig. 4.2b). The dashed e^{-kt} curve is called an *exponential function* with a *damping coefficient k*, which depends on the damping source.†

The damping for a pendulum is generally due to air resistance, whereas in mechanical systems such as springs it is primarily due to internal mechanical forces. In some instances, damping is promoted, for example in automobile shock absorbers, so the auto does not bounce (oscillate) for a long period of time. Similarly, mechanical bathroom scales are damped so you do not have to wait for the needle oscillations to cease to get a good reading. The damping of a pendulum will be investigated in the Procedure section.

EXPERIMENTAL PROCEDURE

1. Set up a simple pendulum arrangement. If a pendulum clamp is not available, the string may be tied around something such as a lab stand arm. Make sure that the string is secure and does not slip on the arm.

2. Experimentally investigate the small-angle approximation [Eq. (4.2)] and the theoretical prediction [Eq. (4.1)] that the period increases with larger angles. Do this by determining the pendulum period for the several angles listed in Data Table 1, keeping the length and mass of the pendulum constant. Measure the angles with a protractor. (Note: θ is the initial angular distance of the bob before release.)

   Rather than timing only one oscillation, time several (four or five) and determine the average period. Timing is generally more accurate if you start the pendulum oscillating before the timing begins. Also, it is usually best to take the timing reference point as the lowest point of the swing.

   Measure and record the pendulum length. The length should be measured to the center of the pendulum bob. (Why?)

   Compute the percent error of the period for each angle θ, using Eq. (4.2) to calculate the theoretical value. (In this case, do not use the absolute difference, so that each percent error will have a sign, + or −.) Further analysis will be done in the Questions section. Proceed to the next step.

*See Experiment 16 for more on simple harmonic motion.

†Theoretically, the damping coefficient is mass dependent, $k = \frac{b}{2m}$.

3. Experimentally investigate whether the period is independent of the mass of the pendulum bob. Using the three masses provided, determine the periods of a pendulum with each mass as the bob (keeping length L and the small angle of oscillation constant). Record your results in Data Table 2, and draw a conclusion from the data.

4. Experimentally investigate the relationship between the length and period of the pendulum. Using four different lengths (such as, 0.20 m, 0.40 m, 0.60 m, and 0.80 m), determine the average period of a pendulum of each length (keeping mass and the small angle of oscillation constant). Record the data in Data Table 3.

5. Compute the theoretical period for each pendulum length [Eq. (4.2)], and enter the results in Data Table 3 ($g = 9.80\ \text{m/s}^2 = 980\ \text{cm/s}^2$).

6. Compute the percent error between the experimental and the theoretical values of the period for each pendulum length, and record in Data Table 3. Draw conclusions about the validity or applicability of Eq. (4.2).

7. The object of the preceding experimental procedures was to determine the validity or applicability of Eq. (4.2)—that is, whether the experimental results agree with the theoretical predictions as required by the scientific method. Once found acceptable, a theoretical expression can then be used to determine experimentally other quantities occurring in the expression.

For example, Eq. (4.2) provides a means for experimentally determining g, the acceleration due to gravity, by measuring the pendulum parameters of length and period, as was done previously.

Squaring both sides of Eq. (4.2),

$$\boxed{T^2 = \frac{4\pi^2}{g} L} \qquad \textbf{(4.3)}$$

or

$$L = \frac{g}{4\pi^2} T^2$$

Hence, the equation has the form $y = ax^2$, that of a parabola. This can be plotted as a straight line with the general form $y = ax$ by letting $L = y$ and $x = T^2$; that is, plotting T^2 on the X-axis. The line will have a slope of $a = g/4\pi^2$.

8. Plot L versus T^2 for the best experimental data (lowest percent error) in Data Table 3, and determine the slope of the graph. Compute the experimental value of g. Record this in the Laboratory Report and compute the percent error of the result.

DAMPED OSCILLATIONS*

9. Set up a pendulum arrangement with an appropriate length (about 75 cm to 1 m) so the complete oscillations are slow enough to be counted. Lay a meterstick on a line under the pendulum with the zero end at the center of the pendulum arc so the amplitudes (horizontal distance from the zero line may be read.

10. In this procedure, one person will count the number of complete oscillations and a lab partner will read the amplitude when the counter indicates a certain number of oscillations. Start the pendulum oscillations with a convenient amplitude (about 15 – 20 cm). Record in Data Table 4. The counter then counts say 25 complete oscillations, and on nearing the end of the final oscillation indicate to the partner to take an amplitude reading at the end of the swing. (The initial amplitude and oscillation group may be varied to best fit your setup.)

With the pendulum continuing to swing, repeat the preceding procedure for another 25 oscillations. Record the cumulative oscillations (25, 50, 75, ...) and the corresponding end amplitudes in the data table.

11. On a sheet of graph paper, plot amplitude versus oscillations. Does the graph describe a decaying exponential function? (See Question 7.)

*This procedure was suggested by Professor Fred B. Otto, author of the Instructor's Resource Manual.

Name ____________________ Section ____________ Date ____________

Lab Partner(s) ____________________

EXPERIMENT 4

Simple Pendulum Parameters (Angle, Mass, Length, and Damping)

TI *Laboratory Report*

DATA TABLE 1

Purpose: To investigate the small-angle approximation.

Mass, m ____________ Pendulum length, L ____________

Angle θ	Period T ( )		Percent error
	Experimental	Theoretical	
5°			
10°			
20°			
30°			
45°			

Conclusion:

DATA TABLE 2

Purpose: To investigate period dependence on mass.

θ ____________ L ____________

m ( )	T ( )		Percent error
	Experimental	Theoretical	

Conclusion:

Don't forget units

(continued)

DATA TABLE 3

Purpose: To investigate period dependence on length.

θ ____________________ m ____________________

L ( )	T ( )		Percent error	T^2 ( )
	Experimental	Theoretical		

Conclusion:

Value of g from experimental data (slope of graph) ____________________ (units)

Percent error ____________________

DATA TABLE 4

Purpose: To investigate damped harmonic motion.

Initial amplitude A ____________________

Number of oscillations	Amplitude ( )

Name ______________________ Section ______________ Date ______________

Lab Partner(s) ____

EXPERIMENT 4 **Simple Pendulum Parameters** *Laboratory Report*

TI QUESTIONS

1. It was suggested that you measure the time for several periods and determine the average period, rather than timing only one period.

   (a) What are the advantages of this method?

   (b) How and why would the result be affected if a very large number of periods were timed?

2. In general, the results of Procedure 2 may not have shown clear-cut evidence that the period increases as dramatically with the angle as Eq. (4.1) might suggest. To understand why, write Eq. (4.1) as

$$T = T_1\left(1 + \frac{1}{4}\sin^2\frac{\theta}{2} + \frac{9}{64}\sin^4\frac{\theta}{2}\right)$$

   and compute T in terms of T_1 $\left(\text{where } T_1 = 2\pi\sqrt{\frac{L}{g}}\right)$ for angles of 5°, 20°, and 60°.

   Comment on the theoretical predictions and experimental accuracy in relation to your results in Data Table 1.

3. Is air resistance or friction a systematic or a random source of error? (See Experiment 2 for error descriptions.) Would it cause the period to be larger or smaller than the theoretical value? (*Hint*: Consider what would happen if the air resistance were much greater—for example, as though the pendulum were swinging in a liquid.)

(continued)

4. Thomas Jefferson once suggested that the period of a simple pendulum be used to define the standard unit of length. What would be the period of a pendulum with a length of 1.0 m?

5. Suppose the 1.0 m pendulum were operated on the Moon. What would its period be there? ($g_M = g/6$)

6. Suppose in the damped equation had $e^{-k_1 t}$ and $e^{-k_2 t}$, with $k_2 > k_1$. How would the graphs differ and why?

7. In Figure 4.2, amplitude was plotted versus time to show exponential damping. How does amplitude versus oscillations show this?

Name ______________________ Section ______________ Date ______________

Lab Partner(s) ____

EXPERIMENT 5

Uniformly Accelerated Motion: Measurement of *g*

GL *Experimental Planning*

GL Figure 5.1 A daring experimenter. See Experimental Planning text for description.

A. Object in Free Fall

An object in free fall falls under the influence of gravity only; air resistance is neglected. A good approximation to free fall in the laboratory is when dense objects fall for relatively short distances. Use the following equipment to determine the acceleration due to gravity (g).

(continued)

EQUIPMENT NEEDED

- 3 objects of different masses (for example, steel balls and/or lead weights)
- Meterstick
- Laboratory timer or stopwatch
- One sheet of graph paper

Review the definition of acceleration. Which of the quantities involved can be directly measured with the given equipment?

The defining equation is not practical to use in many cases. However, the units for acceleration are distance and time (m/s^2), and these quantities can be measured with the equipment given. Can you give a kinematic equation for an object that is dropped (no initial velocity) that only involves length, time, and acceleration? (Use y for length and g for acceleration.)

Hence, we have a simple equation that involves only length, time, and acceleration, along with the equipment to measure length and time for a falling object to find g. Solve the equation for t, which is the time of fall that will be measured.

Answer the following questions.

1. What effect might the distance of fall have on your experimental measurements and results? (Consider the following cases.)
   (a) How long would it take the object to reach the floor if you dropped it from a height of 0.50 m? Could you measure this accurately with a stopwatch? Explain.

   (b) What if an object were dropped from a height of 10 m? Could you measure this distance accurately with a meterstick? Would the acceleration remain constant? Explain.

2. From the preceding calculation, it should be obvious that to experimentally time the distance of fall for a dropped object is critical. To gain an appreciation of how the distance of fall varies with time, consider the daring experimenter shown in ● GL Fig. 5.1. Jo-Jo will illustrate the time-distance relationship of free fall by stepping off a high, vertical cliff with a timer in one hand and a marker in the other. For each second of fall, he makes a mark on the cliff face.

   But wait. Jo-Jo wants you to determine how far he would fall during each second for the first 5 seconds. He requests you plot the results on a distance versus time graph for a visual display.

   Oh, one other thing. He wants to open his parachute when reaching 60 mi/h. At what time, or between which seconds, should he do this?

3. Given three objects with same size and shape, but different masses are dropped simultaneously from a high altitude hot-air balloon. Would the heaviest fall the fastest? If so, would this mean that the acceleration due to gravity depends on mass? Or could there be another factor involved? (*Hint*: Take a look at the opening sentence of this experimental planning.)

4. Suppose that the initial height of the object were measured from the top of the object at the release point to the floor. How would this affect your experimental result for g, that is, would it be too high or too low? Is this a random or a systematic error? (See Experiment 2 on errors.)

TI Advance Study Assignment

Read the experiment and answer the following questions.

B. Linear Air Track

1. How is the acceleration of a car traveling on an elevated air track related to (a) the angle of elevation; (b) the height of elevation?

(continued)

2. What is the equation describing the instantaneous velocity of a car on an elevated air track, and what is the shape of the graph of the instantaneous velocity versus time?

3. Will the graph of instantaneous velocity versus time have a *y*-axis intercept of zero? Explain.

4. Describe how the instantaneous velocity of a car traveling on an elevated air track can be calculated from displacement and time data.

CI Advance Study Assignment

1. What precautions need to be taken when working with a fan-propelled car?

2. For an object moving with constant acceleration, what will be the shape of a graph of position versus time? What will be the shape of a graph of velocity versus time?

PRE-LAB DEMONSTRATION | Apparent Weightlessness (where's the water?)

Water flows freely from two holes in a cup filled with water (● Fig. D2). What happens when the cup is dropped? (This demonstrates acceleration due to gravity in action.)

Figure D2 Apparent Weightlessness

PRE-LAB DEMONSTRATION | Different Distances of Fall (listen to the sound)

The purpose of this demonstration is to illustrate how uniformly accelerated motion gives rise to different distances of fall. This is done by listening to the sounds of dropped, spaced weights hitting a pan. You will be asked to explain the time differences in the sounds in terms of the motions of the weights.

EXPERIMENT 5

Uniformly Accelerated Motion: Measurement of *g*

OVERVIEW (TI, CI)

Experiment 5 examines uniformly accelerated motion using complementary TI and CI approaches. The TI procedures investigate the accelerations of (1) an object in free fall, and (2) a car on a linear air track for both horizontal and inclined motions.

The CI procedures extend the investigation by considering not only the linear relationship for uniformly accelerated motion, $v = at$, but also the parabolic relationship, $x = \frac{1}{2}at^2$. This is done using a fan car and a rotary motion sensor.

INTRODUCTION AND OBJECTIVES (TI, CI)

An important case in kinematics is that of an object in uniformly accelerated motion—one having a uniform or *constant* acceleration. Probably the most common example is a falling object near the surface of the Earth. An object falling solely under the influence of gravity is said to be in *free fall,* and that object falls with an acceleration g (the acceleration due to gravity). Near the Earth's surface, the acceleration due to gravity is approximately constant, with a common value of

$$g = 9.80 \text{ m/s}^2 = 980 \text{ cm/s}^2 = 32.2 \text{ ft/s}^2$$

Of course, air resistance affects the acceleration of a falling object. But for relatively dense objects over short distances of fall, the effects of air resistance are negligible, and objects fall with an acceleration of g.*

In this experiment, the acceleration due to gravity is used to investigate an object undergoing uniformly accelerated motion to see how its velocity and displacement change with time. Conversely, with displacement and time measurements, the value of g can be determined. The experimental data and their analyses will yield a better understanding of the kinetic equations describing the motion.

* The effect of air resistance on falling light objects was demonstrated in Experiment 1B: Muffin Cup Drop

TI OBJECTIVES

After performing this experiment and analyzing the data, you should be able to do the following:

1. Clearly distinguish between average and instantaneous velocity.
2. Express how the velocity of a uniformly accelerated object changes with time.
3. Express how the distance traveled by a uniformly accelerated object changes with time.
4. Explain how the uniform acceleration of an object may be determined from distance and time measurements.

CI OBJECTIVES

1. Analyze the motion of an object that moves with constant acceleration.
2. Understand what it means to say that the position varies with the square of the time.

GENERAL THEORY (TI, CI)

When an object moves with a uniform or constant acceleration, the position x of the object at a time t is given by

$$x = x_o + v_o t + \tfrac{1}{2}at^2 \qquad \textbf{(5.1)}$$

where v_o is the initial velocity and a is the constant acceleration. For an initial position arbitrarily chosen to be $x_o = 0$, and for an object starting from rest ($v_o = 0$), the position at any time reduces to

$$\boxed{x = \tfrac{1}{2}at^2} \tag{5.2}$$

Or for an object in free fall, $y = \frac{1}{2}gt^2$, where y is taken as the vertical direction (downward taken as positive to avoid minus signs). Hence by measuring the time t it takes for an object to fall a distance y, the acceleration due to gravity g can be easily calculated.

Note that for any case where the acceleration is constant, the relationship between position and time is not linear: The position is proportional to the square of the time (t^2), not just to the time (t). A graph of x versus t will be a parabola, not a straight line.

On the other hand, if the object has constant acceleration, then the velocity is changing at a steady rate. The velocity of the object at any time after it starts from rest ($v_o = 0$), is given by

$$\boxed{v = at} \tag{5.3}$$

which is a linear function of time. A graph of v versus t will be a straight line.

The motion of an object undergoing constant acceleration is analyzed to better understand what it means to say that the position varies with the square of the time. This is compared to the velocity function, which is directly proportional to the time. The results apply to any type of uniformly accelerated motion.

TI B. Linear Air Track

Types of linear air tracks are shown in ● TI Fig. 5.2. Air is supplied to the interior of the hollow track and emerges through a series of small holes in the track. This provides a cushion of air on which a car or glider travels along the track with very little friction (an example of the use of a gaseous lubricant).

To have the car move under the influence of gravity, one end of the air track is elevated on a block. The acceleration of the car along the air track is then due to a component of the force due to gravity, $F = ma = mg\sin\theta$ (● TI Fig. 5.3). The acceleration a of the glider along the air track is

$$\boxed{a = g\sin\theta} \tag{TI 5.1}$$

and from the geometry, $\sin\theta = h/L$ (side opposite the angle over the hypotenuse). Hence,

$$\boxed{a = \frac{gh}{L}} \tag{TI 5.2}$$

The magnitude of the instantaneous velocity v of the uniformly accelerating glider at a time t is given theoretically by

$$\boxed{v = v_o + at} \tag{TI 5.3}$$

Hence, a graph of v versus t is a straight line ($y = mx + b$) with a slope $m = a = \Delta v/\Delta t$ and an intercept $b = v_o$. If the car starts from rest, the initial velocity v_o is zero, and

$$\boxed{v = at} \tag{TI 5.4}$$

TI EXPERIMENT 5

Uniformly Accelerated Motion: Measurement of *g*

TI EQUIPMENT NEEDED

A. Object in Free Fall

(see TI Experimental Planning at the beginning of the experiment.)

B. Linear Air Track

- Linear air track
- Several laboratory timers or stopwatches
- Wooden blocks of two different heights
- 1 sheet of Cartesian graph paper

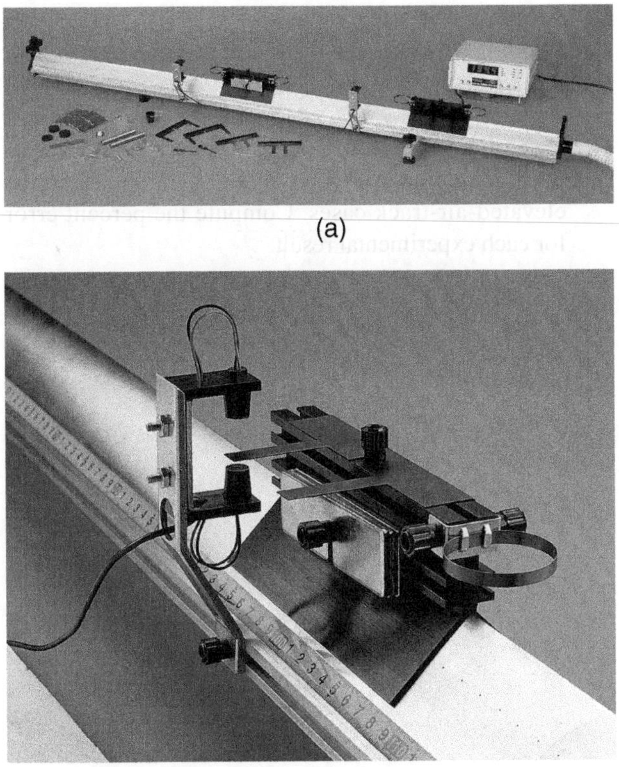

TI Figure 5.2 Air tracks. (a) A blower supplies air to the track through the hose on the right. The cars or gliders travel on a thin cushion of air, which greatly reduces friction. (b) An air track may be equipped with photogates for automatic timing. (Photos Courtesy of Sargent-Welch.)

It can be shown that the instantaneous velocities of the car can be found from the experimental data of the measured displacements x_i of the glider along the air track at times t_i (with $v_o = 0$) by

$$v_i = \frac{2x_i}{t_i} \qquad \textbf{(TI 5.5)}$$

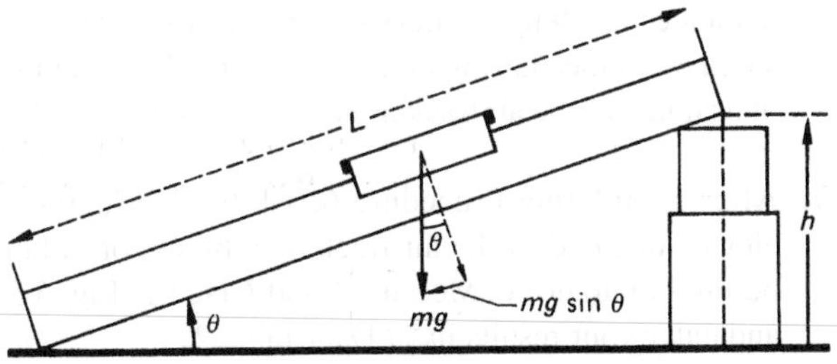

TI Figure 5.3 Accelerating car on air track. When one end of an air track is elevated, the acceleration of the car is due to the component of the (weight) force mg, and $a = g \sin\theta$.

TI EXPERIMENTAL PROCEDURE

A. Object in Free Fall

1. One person should drop the object and do the timing. Lab partners should alternate.

2. Distinguish the objects as m_1, m_2, and m_3. Drop one of them from a fixed height y above the floor and measure its time of fall. Drop it with the arm held horizontally or held upward. (Depending on your height, it may be advantageous to stand on a small step stool. (Why?) Do a couple of practice trials to become familiar with the procedure. Record the data for four trials in TI Data Table 1. Repeat this procedure for the other two objects.

3. Compute the acceleration due to gravity g, using the times of fall. Find the average (mean) value. *Note*: The results obtained by this procedure may have very poor accuracy and precision. (Why?)

B. Linear Air Track

4. The air track should be set up and leveled by the instructor or laboratory assistant. *Do not* attempt to make any adjustments to the air track. Ask your instructor for assistance if you need it.

5. Turn on the air supply, and place the car in motion by applying a small force on the car in a direction parallel to the air track. *Do not* attempt to move the car on the air track if the air supply is not turned on. Use the same small force for each trial—for example, by compressing a spring attached to the car.

6. Using laboratory timers or stopwatches, determine the times required for the car to travel several convenient distances, such as 0.20 m, 0.40 m, 0.50 m, 0.75 m, and so on. Record the times and distances in TI Data Table 2.*

Several students should work together, each with a timer, taking a time reading as the car passes his or her assigned distance mark. Make several practice trials before taking actual data. (Remember that the distances are length intervals and need not be measured from the starting end of the air track. Make use of as much of the air track as is conveniently possible.)

7. After completing Procedure 6, ask the instructor to elevate one end of the air track on a block, or obtain permission to do so. Measure h and L (see TI Fig. 5.3) and enter your results in TI Data Table 2.

8. Start the car from rest near the elevated end of the air track. (To minimize error, it is better to put a block or pencil in front of the car and pull this away smoothly rather than releasing the car by hand.) Measure and record the times required for the car to travel the distances listed, and record your results in TI Data Table 2. Use the experimental method described in Procedure 6.

9. Have the end of the air track elevated to a different height, and repeat the time measurements for this height.

10. Using Eq. (TI 5.5), compute the instantaneous velocity of the car for each of the times in the three experimental sets of data in TI Data Table 2.

11. Plot v versus t for each case on the same graph and determine the slope of each line.

12. Using Eq. (TI 5.2), compute an experimental value of the acceleration due to gravity ($a = g$) for each of the elevated-air-track cases. Compute the percent error for each experimental result.

* If electronic photogate timers are available, your instructor will give you instruction in their use. Electronic timing greatly improves the accuracy and precision of the results. (Why?)

Name ________________ Section ________ Date ________

Lab Partner(s) ________________

TI QUESTIONS

A. Object in Free Fall

1. Objects of different mass were used to see whether the acceleration due to gravity depends on the mass of a falling object. In other words, does a heavier object fall faster than a lighter object? What do your experimental results show?

2. What is probably the greatest source of error in the experimental procedure?

B. Linear Air Track

3. What are the major sources of error in this procedure?

4. What would be the shapes of the curves for a graph of y versus t of the data in each experimental case? How would you determine the value of the car's acceleration from a graph using only y and t values (that is, not computing v_i)?

(continued)

5. What is the physical significance of the slope of the graph for the case of the level air track?

6. What is the maximum possible value of the slope of a v-versus-t curve for a car released from rest on an air track elevated at one end? Describe the experimental setup in this case.

CI EXPERIMENT 5

Uniformly Accelerated Motion: Measurement of *g*

CI EQUIPMENT NEEDED

- 1 collision (or plunger) cart   Pasco Collision Cart ME-9454 (or ME-9430) (Any of the classic carts or the Pascars will work fine.)
- 1 fan accessory   Pasco ME-9491
- 1 dynamics track
- 1 rotary motion sensor (RMS)   CI-6538
- Brackets and pulley mounts:
  - 1 cart-string bracket   CI-6569
  - 1 dynamics track mount accessory   CI-6692 (to mount the RMS to the track)
  - 1 RMS/IDS adapter   ME-6569 (track pulley bracket)
- String
- Optional:
  - Track end-stop

CI THEORY (See TI, CI General Theory at the beginning of the experiment.)

EQUIPMENT SETUP

1. The cart-string bracket and the fan accessory are mounted on top of the cart.
2. The rotary motion sensor (RMS) is mounted to one side of the track, with the small pulley of the RMS/IDS adapter mounted on the opposite end of the same side of the track. ● CI Fig. 5.1 is a diagram of the setup.
3. The string makes a full loop connecting the cart-string bracket with the large pulley of the RMS sensor and the small pulley on the opposite bracket. That string should be tense, but not tight.
4. Adjust the height of the string so that the fan blade clears the string as it spins. The RMS and the small pulley can be moved as far down as needed for the blades to clear the string.

SETTING UP DATA STUDIO

1. Open Data Studio and choose Create Experiment.
2. The Experiment Setup window will open and you will see a picture of the Science Workshop interface. There are seven channels to choose from. (Digital channels 1, 2, 3, and 4 are the small buttons on the left; analog channels A, B, and C are the larger buttons on the right, as shown in ● CI Fig. 5.2.)
3. Click on the Channel 1 button in the picture. A window with a list of sensors will open.
4. Choose the Rotary Motion Sensor from the list and press OK.
5. The diagram now shows you the properties of the RMS sensor directly under the picture of the interface. (See CI Fig. 5.2.)
6. Connect the sensor to the interface, as shown on the computer screen, to channels 1 and 2.
7. Adjust the properties of the RMS as follows:
   First Measurements tab: select Position, Ch 1&2. Deselect all others.

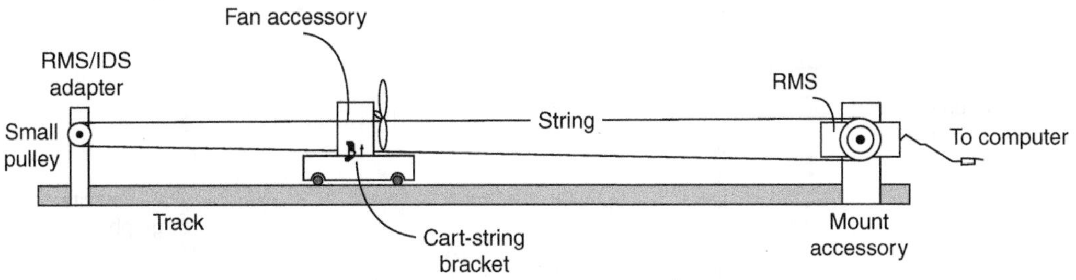

CI Figure 5.1 Rotary motion sensor and cart setup. A string makes a full loop from the cart-string bracket, to the RMS pulley, to the small pulley on the opposite end of the track, and back to the cart. The height of the RMS and the pulley must be adjusted so that the fan blades do not touch the string as they spin.

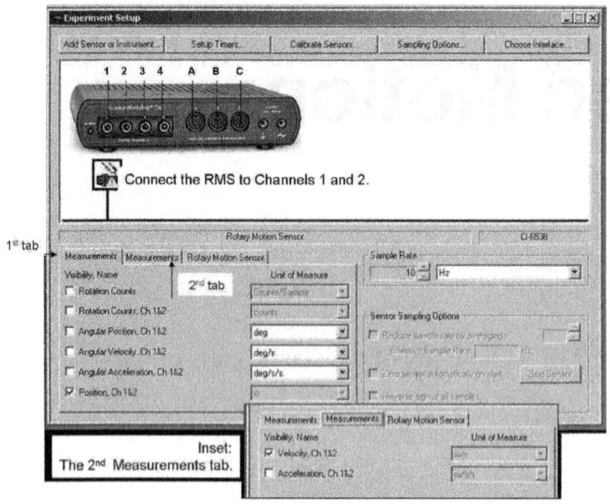

CI Figure 5.2 The Experiment Setup Window. Top of the screen: the Science Workshop interface and the seven available channels. Once a sensor is chosen, an icon for the sensor appears under the appropriate channel. Here, for example, is the RMS icon directly under Channels 1 and 2. Bottom of the window: The properties of the selected sensor can be adjusted as needed. (Reprinted courtesy of PASCO Scientific.)

Second Measurements tab: select Velocity, Ch 1&2. Deselect all others.

The Data list on the left of the screen should now have two icons: one for the position data, the other for the velocity data.

8. Create a graph by dragging the position data icon from the data list and dropping it on top of the graph icon of the displays list. A graph of position versus time will open in a window called "Graph 1."
9. Now drag the velocity data icon and drop it somewhere in the middle of the graph. The graph display will split into two graphs: one of position, the other of velocity, as shown in ● CI Fig. 5.3.

CI EXPERIMENTAL PROCEDURE

WARNING! Be careful not to touch the fan blades while they are spinning.

1. Turn the fan on, but hold on to the car so that it does not yet move.
2. Have a partner press the START button. Let go of the car. Your partner must press the STOP button before the car reaches the end of the track to prevent data being taken of collisions and rebounds with pulleys or end-stops. You may need to do a few practice runs to become familiar with the procedure.

   Note: If the graphs show negative values, reverse the fan so that it is facing in the opposite direction, and start again from the opposite side of the track.

3. Carefully turn off the fan.
4. You should have two graphs on the screen. Use the Scale-to-fit button on the graph toolbar to display the data clearly. Notice that the graph of position versus time is a smooth parabola. The graph of velocity versus time is a straight line.
5. Click anywhere on the position-versus-time graph to make it active. Use the Smart Tool (a button on the graph toolbar, labeled "xy") to choose a data point that is close to the beginning of the motion but for which the position is not zero. Record the position and the time of this point in CI Data Table 1 as the first data points, x_1 and t_1.
6. Find the position at a time $t_2 = 2t_1$. That is, where was the car when the previous time doubled? Record x_2.
7. Repeat for times $t_3 = 3t_1$, $t_4 = 4t_1$, . . . as many multiples of t_1 as you can get from the graph. (The longer the track you use, the more you can get.)

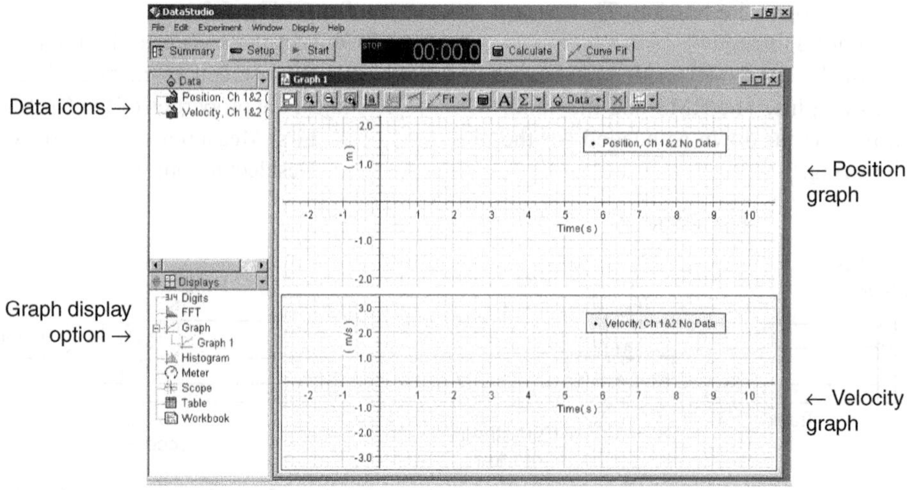

CI Figure 5.3 Graph displays. The graph display in this picture has been maximized to occupy most of the screen. (Reprinted courtesy of PASCO Scientific.)

8. Determine by what factor the distance traveled at time t_2 is greater than the distance at time t_1. Then determine by what factor the distance traveled at time t_3 is greater than the distance at time t_1. Continue until CI Data Table 1 is complete.
9. Now click anywhere on the velocity-versus-time graph to activate it. Use the Smart Tool to find the velocity of the cart at each of the times $t_1, t_2, \ldots, t_n$. Record the velocities in CI Data Table 2, and calculate by how much the velocity increases as the times double, triple, etc.
10. Use the Fit Tool to determine the slope of the velocity graph. (The fit tool is on the graph toolbar; it is a drop menu called "Fit.") Choose a "Linear Fit" for your graph. Report the slope in CI Data Table 2. What is the slope of a velocity-versus-time plot measuring? (*Hint*: Think of the units!)

Name ______________________ Section ______________ Date ______________

Lab Partner(s) ____

CI EXPERIMENT 5

Uniformly Accelerated Motion

Laboratory Report

DATA TABLE 1

Purpose: To investigate a position function that is proportional to the square of the time.

Time		Position		How many times larger is x_n than x_1? x_n/x_1
t_1		x_1		
t_2		x_2		
t_3		x_3		
t_4		x_4		

DATA TABLE 2

Purpose: To investigate a velocity function that is proportional to the time.

Time		Velocity		How many times larger is v_n than v_1? v_n/v_1
t_1		v_1		
t_2		v_2		
t_3		v_3		
t_4		v_4		

Slope of the graph: ______________
(units)

Don't forget units

(continued)

CI QUESTIONS

1. In CI Data Table 1, you measured the position of the car at different times. When the time doubled, did the distance from the origin double also? When the time tripled, did the distance from the origin triple also? Can you see the pattern?

2. Discuss what it means to say that the position function is not directly proportional to the time (t), but to the time squared (t^2).

3. Judging on the basis of the observed pattern, and without using theoretical equations, predict the position of the car when the time is $10t_1$. What will the position be at $20t_1$?

4. In CI Data Table 2, you repeated the procedure for the velocities. What is the pattern now?

5. On the basis of the observed pattern, predict the velocity of the car for times $10t_1$ and $20t_1$.

6. A graph of x versus t is a parabola, because $x \propto t^2$. But if you plot x versus t^2, the resulting graph will be a straight line, with slope $\frac{1}{2}a$, as shown below. Make a graph with your values of x_n on the vertical axis and your *times squared* on the horizontal. Determine the slope, and use it to find the acceleration of the car. (Attach the graph to the Laboratory Report.)

$$\begin{array}{ccccc} x & = & \frac{1}{2}a & t^2 \\ \downarrow & & \downarrow & \downarrow \\ y & = & m & x \end{array}$$

7. By determining a percent difference, compare the acceleration of the car determined from your graph to that measured as the slope of the velocity graph.

Name ______________________ Section ______________ Date ______________

Lab Partner(s) ____

EXPERIMENT 6

The Addition and Resolution of Vectors: The Force Table

TI *Advance Study Assignment*

Read the experiment and answer the following questions.

1. Distinguish between scalar and vector quantities, and give an example of each.

2. How are vectors represented graphically, and how are scalars and vector quantities distinguished when written as symbols?

3. What is meant by drawing a vector to scale? Give a numerical example.

4. Why is the triangle method called the head-to-tail (or tip-to-tail) method?

(continued)

5. How may the resultant of two vectors be computed analytically from a vector triangle?

6. How many vectors may be added by the polygon method? Are other methods of vector addition limited to the number of vectors that can be added? Explain.

7. What is meant by resolving a vector into components? Give an example.

8. Briefly describe the steps in the component method of vector addition.

9. On a force table, what is the difference between the equilibrant and the resultant? Why is only one of these actually determined experimentally?

PRE-LAB DEMONSTRATION | Problem in Kinematics

A student trying to solve a problem with constant acceleration wants to find the velocity v but is not given the acceleration a. In applying some kinematic equations to find this, the surprising result of $1 = 2$ is obtained. Something is wrong.

EXPERIMENT 6

The Addition and Resolution of Vectors: The Force Table

INTRODUCTION AND OBJECTIVES

Physical quantities are generally classified as either scalar or vector quantities. The distinction is simple. A **scalar** quantity (or *scalar*) is one with magnitude only (including units)—for example, speed (15 m/s) and temperature (20 °C). A **vector** quantity (or *vector*), on the other hand, has both magnitude *and* direction. Such quantities include displacement, velocity, acceleration, and force—for example, a velocity of 15 m/s north or a force of 10 N along the $+x$-axis.

Vectors are generally written using a boldface capital letter with an over arrow, for example, $\vec{\mathbf{F}}$. A nonboldface letter, F, indicates a magnitude (or a scalar).*

Because vectors have the property of direction, the common method of addition, that is scalar addition, is not applicable to vector quantities. To find the **resultant** or **vector sum** of two or more vectors, special methods of vector addition are used, which may be graphical and/or analytical. The chief methods of these will be described, and the addition of force vectors will be investigated. The results of graphical and analytical methods will be compared with the experimental results obtained from a force table. The experimental arrangements of forces (vectors) will physically illustrate the principles of the methods of vector addition.

After performing this experiment and analyzing the data, you should be able to do the following:

1. Add a set of vectors graphically to find the resultant.
2. Add a set of vectors analytically to find the resultant.
3. Appreciate the difference in convenience between using graphical and using analytical methods of vector addition.

EQUIPMENT NEEDED

- Force table with four pulleys
- Four weight hangers
- Set of slotted weights (masses), including three of 50 g and three of 100 g
- String
- Protractor
- Ruler
- Level
- 3 sheets of Cartesian graph paper

THEORY

A. Methods of Vector Addition: Graphical

TRIANGLE METHOD

Vectors are represented graphically by arrows (● Fig. 6.1). The length of a vector arrow is proportional to the magnitude of the vector (drawn to scale on graph paper), and the arrow points in the direction of the vector.

The length scale is arbitrary and is usually selected for convenience and so that the vector graph fits nicely on the graph paper. A typical scale for a force vector might be 1 cm:10 N. That is, each centimeter of vector length represents 10 newtons. The *scaling factor* in this case in terms of force per unit length is 10 N/cm. (Note the similarity with the common food cost factor of price/lb—for example, 10¢/lb.)

When two vectors $\vec{\mathbf{A}} + \vec{\mathbf{B}}$ are added by the *triangle method*, the vectors are placed "head-to-tail" (or "tip-to-tail"), that is, the head (tip) of $\vec{\mathbf{A}}$ and the tail of $\vec{\mathbf{B}}$ (Fig. 6.1a). Vector arrows may be moved around as long as they remain pointed in the same direction. Then, drawing a vector from the tail of $\vec{\mathbf{A}}$ to the head of $\vec{\mathbf{B}}$ gives the vector $\vec{\mathbf{R}}$ and completes the triangle. $\vec{\mathbf{R}}$ is the resultant or vector sum of $\vec{\mathbf{A}} + \vec{\mathbf{B}}$; in other words, by vector addition, $\vec{\mathbf{R}} = \vec{\mathbf{A}} + \vec{\mathbf{B}}$.**

The magnitude of $\vec{\mathbf{R}}$ is proportional to the length of the vector arrow, and the direction of $\vec{\mathbf{R}}$ may be specified as being at an angle θ relative to $\vec{\mathbf{A}}$.

*Vectors may also be written in terms of magnitude and a unit vector, $\vec{\mathrm{F}} = \mathrm{F}\hat{x}$, where $\hat{x}$ is a *unit vector* with a magnitude of one (unity) and in the positive x direction. A unit vector with a magnitude of one does nothing to change the magnitude and only indicates the direction of the vector.

**The method may be extended to a parallelogram method. Forming a parallelogram with the vectors, the resultant is the diagonal.

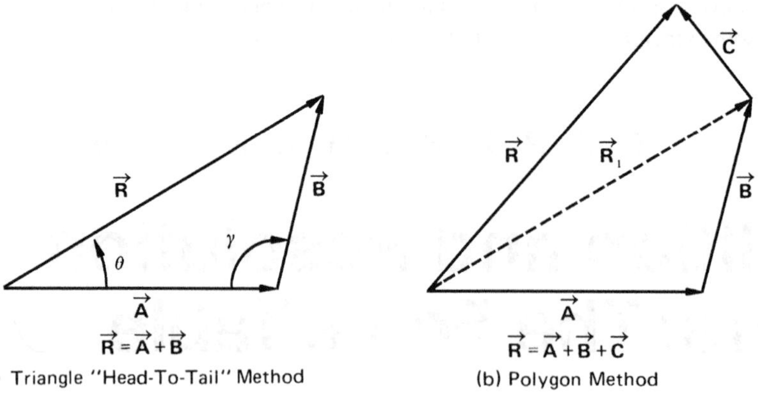

Figure 6.1 Vector addition. Methods of vector addition. Vectors are represented graphically by arrows. See text for description.

Polygon Method

If more than two vectors are added, the head-to-tail method forms a polygon (Fig. 6.1b). For three vectors, the resultant $\vec{\mathbf{R}} = \vec{\mathbf{A}} + \vec{\mathbf{B}} + \vec{\mathbf{C}}$ is the vector arrow from the tail of the $\vec{\mathbf{A}}$ arrow to the head of the $\vec{\mathbf{C}}$ vector. The length (magnitude) and the angle of orientation of $\vec{\mathbf{R}}$ can be measured from the vector diagram. Note that this is equivalent to applying the head-to-tail method twice—the head of $\vec{\mathbf{A}}$ to the tail of $\vec{\mathbf{B}}$, and the head of $\vec{\mathbf{B}}$ to the tail of $\vec{\mathbf{C}}$.

The magnitude (length) R and the orientation angle θ of the resultant vector $\vec{\mathbf{R}}$ in a graphical method can be measured directly from the vector diagram using a ruler and a protractor.

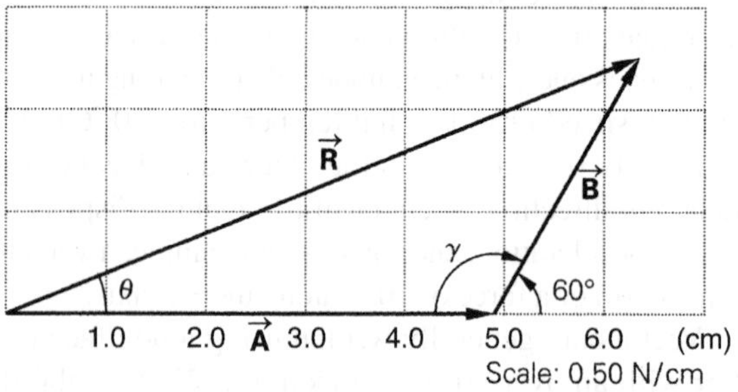

Figure 6.2 Drawing to scale. Figures are often scaled down so as to maintain a convenient size. Here the vector triangle is shown to scale, with a scaling factor of 0.50 N/cm. See text for description.

Example 6.1 To illustrate scaling and the graphical triangle method, let $\vec{\mathbf{A}}$ and $\vec{\mathbf{B}}$ represent forces at angles of 0° and 60°, respectively, with magnitudes of $A = 2.45$ N and $B = 1.47$ N.

Then, choosing a scaling factor (say, 0.50 N/cm), a vector length is found by dividing its magnitude by the scaling factor (magnitude/scaling factor). Note the unit cancellation:

$$A\text{: } 2.45\ \text{N}/(0.50\ \text{N/cm}) = 4.9\ \text{cm}$$

$$B\text{: } 1.47\ \text{N}/(0.50\ \text{N/cm}) = 2.9\ \text{cm}$$

Here, the 0.50-N/cm scaling factor was chosen so as to keep ● Fig. 6.2 an appropriate size. In drawing your vector diagrams, you should choose a scaling factor that will use most of the allotted space on the graph paper—much as in plotting a graph in Experiment 2. Also, a factor with two significant figures was chosen because graph paper grids are usually not fine enough to plot more digits accurately.

The triangle has been drawn in Fig. 6.2, where $\vec{\mathbf{R}} = \vec{\mathbf{A}} + \vec{\mathbf{B}}$. The $\vec{\mathbf{R}}$ vector is measured (with ruler and protractor) to have a length of 6.8 cm and a direction angle of $\theta = 22°$ relative to the $\vec{\mathbf{A}}$ vector. The magnitude of $\vec{\mathbf{R}}$ in newtons is found using the scaling factor:

$$\begin{aligned} R &= (\text{scaling factor})(\text{measured length}) \\ &= (0.50\ \text{N/cm})(6.8\ \text{cm}) = 3.4\ \text{N} \end{aligned}$$

B. Methods of Vector Addition: Analytical

Triangle Method

A resultant vector $\vec{\mathbf{R}}$ is determined by using the head-to-tail method as shown in Fig. 6.2. When the vectors $\vec{\mathbf{A}}$ and B do not form a right triangle, the magnitude of $\vec{\mathbf{R}}$ can be computed from the law of cosines if the angle γ (the angle opposite $\vec{\mathbf{R}}$) is known:

$$\boxed{R^2 = A^2 + B^2 - 2AB\cos\gamma} \qquad \textbf{(6.1)}$$

The angle θ (between $\vec{\mathbf{R}}$ and $\vec{\mathbf{A}}$) can then be computed using the law of sines with the magnitudes of sides $\vec{\mathbf{B}}$ and $\vec{\mathbf{R}}$ known:

$$\boxed{\frac{B}{\sin\theta} = \frac{R}{\sin\gamma}} \qquad \textbf{(6.2)}$$

From Example 6.1, the magnitudes of $\vec{\mathbf{A}}$ and $\vec{\mathbf{B}}$ are 2.45 N and 1.47 N, respectively, and, as can be seen directly from Fig. 6.2, $\gamma = 120°$. (Why?) Using the law of cosines [Eq. (6.1)]:

$$R^2 = A^2 + B^2 - 2AB\cos\gamma$$
$$= (2.45\text{ N})^2 + (1.47\text{ N})^2 - 2(2.45\text{ N})(1.47\text{ N})\cos 120°$$
$$= 6.00\text{ N}^2 + 2.16\text{ N}^2 - 2(3.60\text{ N}^2)(-0.500)^*$$
$$= 11.76\text{ N}^2$$

and taking the square root:

$$R = 3.43\text{ N}$$

The directional angle θ may be found using the law of sines [Eq. (6.2)]:

$$\theta = \sin^{-1}\left(\frac{B\sin\gamma}{R}\right)$$
$$= \sin^{-1}\left(\frac{1.47\text{ N}(\sin 120°)}{3.43\text{ N}}\right) = 21.8°$$

Remember that this is the angle between vectors $\vec{\mathbf{R}}$ and $\vec{\mathbf{A}}$. Note that the results are the same as in Example 6.1 to two significant figures.

Component Method

If two vectors $\vec{\mathbf{A}}$ and $\vec{\mathbf{B}}$ are at right (90°) angles (● Fig. 6.3a), then the magnitude of their resultant is given by the *Pythagorean theorem*, $R = \sqrt{A^2 + B^2}$ (the hypotenuse of a right triangle is equal to the square root of the sum of the squares of the legs of the triangle). Notice that the law of cosines reduces to this formula with $\gamma = 90°$ (because $\cos 90° = 0$). The angle of orientation is given by $\tan\theta = B/A$, or $\theta = \tan^{-1}(B/A)$.

By the inverse process, a vector may be resolved into x and y components (Fig. 6.3b). That is, the vector $\vec{\mathbf{R}}$ is the resultant of $\vec{\mathbf{R}}_x$ and $\vec{\mathbf{R}}_y$, and $\vec{\mathbf{R}} = \vec{\mathbf{R}}_x + \vec{\mathbf{R}}_y$, where $R_x = R\cos\theta$ and $R_y = R\sin\theta$. The magnitude of R is given by

$$R = \sqrt{R_x^2 + R_y^2} \qquad \textbf{(6.3)}$$

and

$$\tan\theta = \frac{R_y}{R_x} \qquad \textbf{(6.4)}$$

or

$$\theta = \tan^{-1}\left(\frac{R_y}{R_x}\right)$$

(resultant, magnitude, and angle)

The vector sum of any number of vectors can be obtained by using the component method. This is conveniently done by having all the vectors originate from the origin and resolving each into x and y components, as shown in ● Fig. 6.4 for $\vec{\mathbf{R}} = \vec{\mathbf{A}} + \vec{\mathbf{B}} + \vec{\mathbf{C}}$.

*Value obtained by calculator or from trig table with $\cos 120° = \cos(180° - 120°) = -\cos 60°$ using the trigonometric identity $\cos(A - B) = \cos A\cos B + \sin A\sin B$.

The procedure is to add vectorially all of the x components together and all of the y components together. The

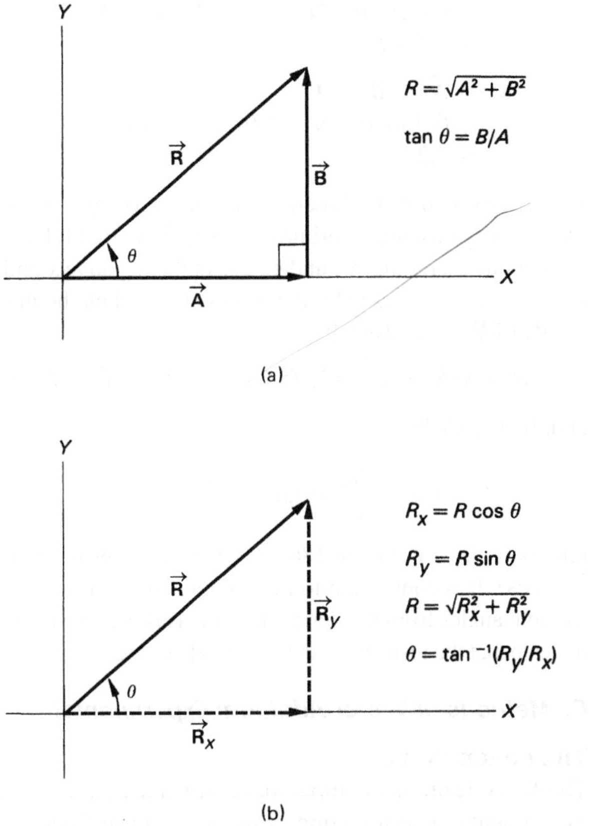

Figure 6.3 Vector resultant and components. (a) The vector addition of $\vec{\mathbf{A}}$ and $\vec{\mathbf{B}}$ gives the resultant $\vec{\mathbf{R}}$. (b) A vector, such as $\vec{\mathbf{R}}$, can be resolved into x and y (rectangular) components: $\vec{\mathbf{R}}_x$ and $\vec{\mathbf{R}}_y$, respectively.

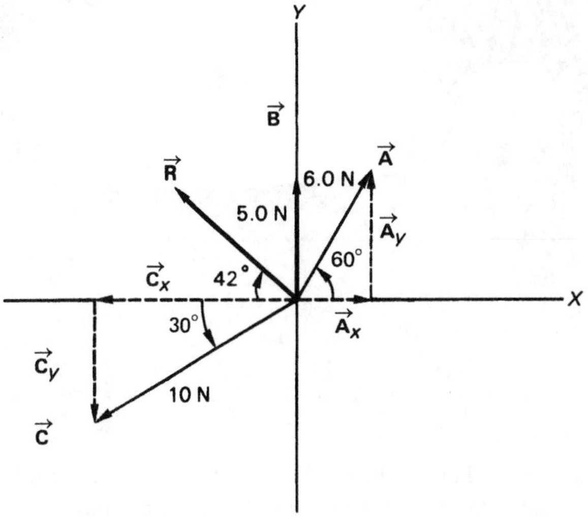

Figure 6.4 Component method. Rather than using the head-to-tail method of vector addition, it is generally more convenient to use the component method, in which all vectors are drawn originating from the origin and resolved into components. See text for description.

$\vec{\mathbf{R}}_x$ and $\vec{\mathbf{R}}_y$ resultants are then added together to get the total resultant $\vec{\mathbf{R}}$. To illustrate this for the vectors in Fig. 6.4,

$$\begin{aligned}\vec{\mathbf{R}}_x &= \vec{\mathbf{A}}_x + \vec{\mathbf{B}}_x + \vec{\mathbf{C}}_x \\ &= 6.0 \cos 60° \text{ N} + 0 - 10 \cos 30° \text{ N} \\ &= -5.7 \text{ N}\end{aligned}$$

$$\begin{aligned}\vec{\mathbf{R}}_y &= \vec{\mathbf{A}}_y + \vec{\mathbf{B}}_y + \vec{\mathbf{C}}_y \\ &= 6.0 \sin 60° \text{ N} + 5.0 \text{ N} - 10 \sin 30° \text{ N} \\ &= 5.2 \text{ N}\end{aligned}$$

where the component directions are indicated by the positive and negative signs (arbitrary units). Note that $\vec{\mathbf{B}}$ has no x component and that $\vec{\mathbf{C}}_x$ and $\vec{\mathbf{C}}_y$ are in the negative x and y directions, as indicated by the minus signs. Then the magnitude of $\vec{\mathbf{R}}$ is [Eq. (6.3)]:

$$R = \sqrt{R_x^2 + R_y^2} = \sqrt{(5.7 \text{ N})^2 + (5.2 \text{ N})^2} = 7.7 \text{ N}$$

and, by Eq. (6.4),

$$\theta = \tan^{-1} \frac{|R_y|}{|R_x|} = \tan^{-1}\left(\frac{5.2 \text{ N}}{5.7 \text{ N}}\right) = 42°$$

relative to the $-x$-axis (or $180° - 42° = 138°$ relative to the $+x$-axis). It is convenient to measure all component angles as acute angles from the x-axis. The minus R_x and positive R_y indicate that the resultant is in the second quadrant.*

C. Methods of Vector Addition: Experimental

The Force Table

The **force table** is an apparatus that makes possible the experimental determination of the resultant of force vectors (● Fig. 6.5). The rim of the circular table is calibrated

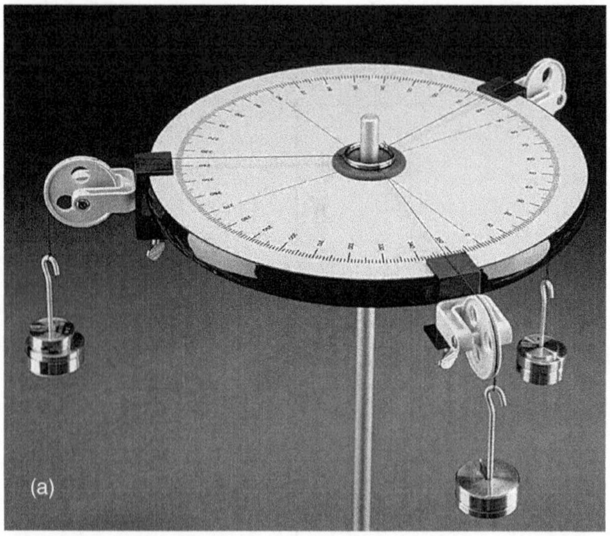

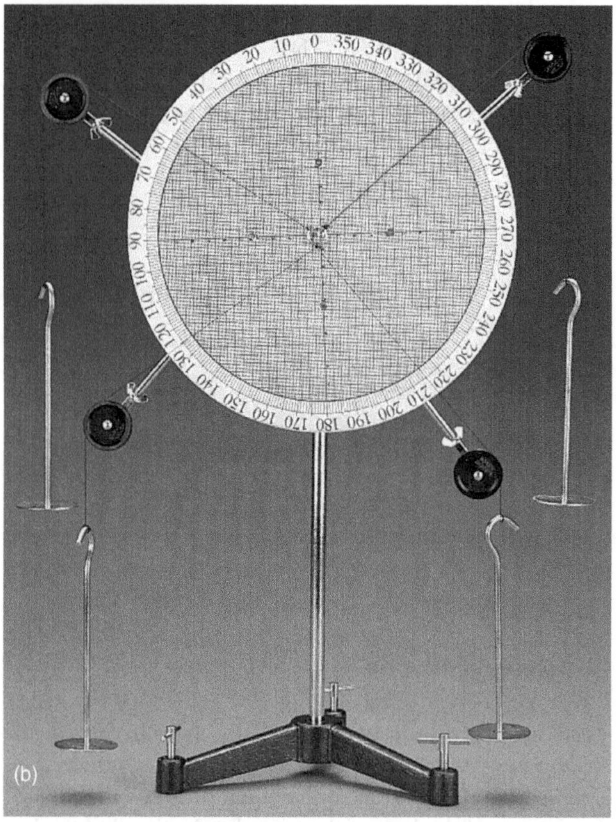

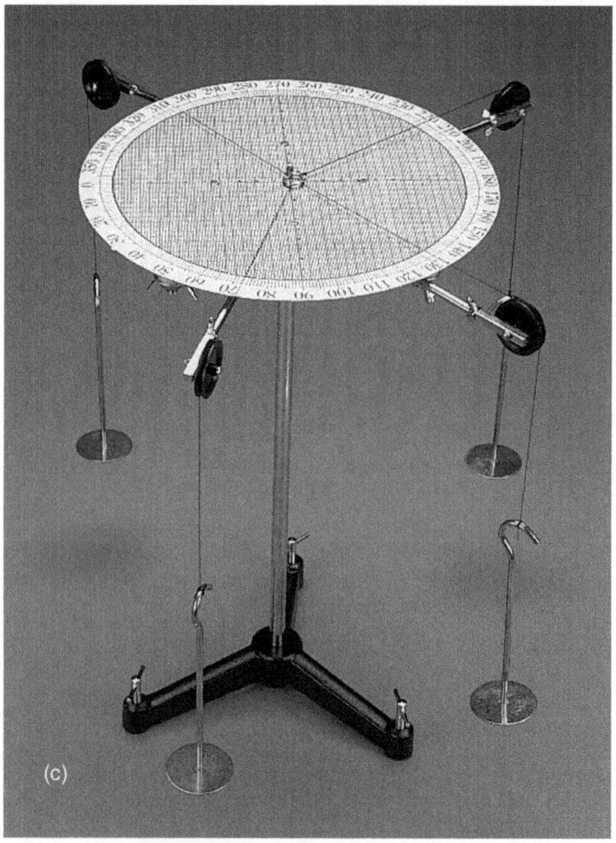

Figure 6.5 Force tables. Various types of force tables. The table in (c) may be used vertically for demonstration (b), or horizontally in the laboratory. (Photos Courtesy of Sargent-Welch.)

*Although it is customary to measure angles counterclockwise from the positive x-axis, this procedure of measuring angles from 60° from the $+x$ axis and 42° from the $-x$ axis in Fig. 6.4 the *nearest x-axis* is convenient in eliminating the need for double-angle equations.

in degrees. Forces are applied to a central ring by means of strings running over pulleys and attached to weight hangers. The magnitude (mg) of a force (vector) is varied by adding or removing slotted weights (masses), and the direction is varied by moving the pulley.

The resultant of two or more forces (vectors) is found by balancing the forces with another force (weights on a hanger) so that the ring is centered around the central pin. The balancing force is *not* the resultant $\vec{\mathbf{R}}$ but rather the *equilibrant* $\vec{\mathbf{E}}$, or the force that balances the other forces and holds the ring in equilibrium.

The equilibrant is the vector force of equal magnitude, but in the *opposite direction*, to the resultant (that is, $\vec{\mathbf{R}} = -\vec{\mathbf{E}}$). See ● Fig. 6.6. For example, if an equilibrant has a magnitude of $(0.30)g$ N in a direction of 225° on the circular scale, the resultant of the forces has a magnitude of $(0.30)g$ N in the opposite direction, 225° − 180° = 45°. It should be evident that the resultant cannot be determined directly from the force table. (Why?)*

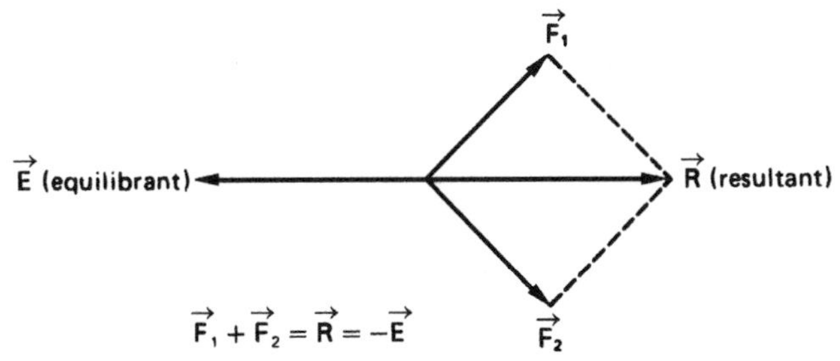

Figure 6.6 Resultant and equilibrant. On a force table, the magnitude and direction of the equilibrant $\vec{\mathbf{E}}$ are measured, rather than those of the resultant $\vec{\mathbf{R}}$, and $\vec{\mathbf{R}} = -\vec{\mathbf{E}}$.

EXPERIMENTAL PROCEDURE

1. Set up the force table with strings and suspended weights, and perform the following cases of vector addition.

2. *Vector addition I.* Given two vectors with magnitudes $F_1 = (0.200)g$ N and $F_2 = (0.200)g$ N at 30° and 120°, respectively, find their vector sum or resultant $\vec{\mathbf{F}} = \vec{\mathbf{F}}_1 + \vec{\mathbf{F}}_2$ by each of the following procedures. (*Note*: Orientation angles of vectors are given relative to the 0° reference line or positive x-axis.)
   (a) *Graphical.* Using the triangle method of vector addition, draw a vector diagram to scale. Use a scale such that the finished vector diagram fills about half a sheet of graph paper. Measure the magnitude and direction of the resultant (with ruler and protractor), and record the results in the data table. Save your graphical sheets to attach to the Laboratory Report.
   (b) *Analytical.* Compute the magnitude of the resultant force. Also, compute the angle of orientation from the relationship $\tan\theta = F_2/F_1$. (Why can you use $\tan\theta$? Remember that θ is the angle between $\vec{\mathbf{F}}$ and $\vec{\mathbf{F}}_1$.) Record the results in the data table.
   (c) *Experimental.* On the force table, clamp pulleys at 30° and 120° and add enough weights to each weight hanger to total 0.200 kg, so as to give weight forces of $F_1 = F_2 = (0.200)$g N in these directions. (The weight hangers usually have masses of 50 g, or 0.050 kg.)

   Using a third pulley and weights, determine the magnitude and direction of the equilibrant force that maintains the central ring centered in equilibrium around the center pin. Record the magnitude and direction of the resultant of the two forces in the data table. Remember, the resultant has the same magnitude as the equilibrant but is in the opposite direction.

   (*Note*: The string knots on the central ring should be of a nontightening variety so that the strings will slip freely on the ring and allow the strings to pull directly away from the center. Pulling the center ring straight up a short distance and releasing it helps adjust the friction in the pulleys as the ring vibrates up and down so that it can settle into an equilibrium position involving only the applied forces. When the forces are balanced, the pin may be carefully removed to see whether the ring is centered on the central hole.)

3. *Vector addition II.* Repeat Procedure 2 for $F_1 = (0.200)g$ N at 20° and $F_2 = (0.150)g$ N at 80°. Use the other half of the sheet of graph paper used in Procedure 2(a) for the graphical analysis. Be careful in the analytical analysis. Can you use $\tan\theta = F_2/F_1$ in this case?

4. *Vector addition III.* Repeat Procedure 2 with $F_1 = F_x = (0.200)g$ N at 0° and $F_2 = F_y = (0.150)g$ N at 90°. In this case, $\vec{\mathbf{F}} = \vec{\mathbf{F}}_x + \vec{\mathbf{F}}_y$, where $\vec{\mathbf{F}}_x$ and $\vec{\mathbf{F}}_y$ are the x and y components of $\vec{\mathbf{F}}$, respectively. That is, the resultant can be resolved into these components. Use half of another sheet of graph paper for the graphical method.

5. *Vector resolution.* Given a force vector of $F = (0.300)g$ N at 60°, resolve the vector into its x and y components and find the magnitudes of $\vec{\mathbf{F}}_x$ and $\vec{\mathbf{F}}_y$ by the following procedures:
   (a) *Graphical.* Draw a vector diagram to scale (on the other half of the sheet of graph paper used in Procedure 4) with the component vectors

*The magnitude of the (weight) force vectors is in general given in the form $R = mg = (0.150)g$ N, for example, where it is understood that the mass is in kilograms and g is the acceleration due to gravity. It is convenient to leave g in symbolic form so as to avoid numerical calculations until necessary. This is similar to carrying along π in symbolic form in equations. Also, note that the masses of the laboratory "weights" usually have values stamped in grams. Don't forget to change grams to kilograms when working in the SI: for example, 150 g = 0.150 kg.

(see Fig. 6.3b), and measure the magnitudes of $\vec{\mathbf{F}}_x$ and $\vec{\mathbf{F}}_y$. Record the results in the data table.

(b) *Analytical.* Compute the magnitudes of $\vec{\mathbf{F}}_x$ and $\vec{\mathbf{F}}_y$ (see the Theory section). Record the results in the data table.

(c) *Experimental.* Clamp pulleys at 240°, 90°, and 0° on the force table. Place a *total* of 0.300 kg on the 240° pulley string using a weight hanger. This force is then the equilibrant of $F = (0.300)g$ N at 60° (since 60° + 180° = 240°), which must be used on the force table rather than the force itself. Add weights to the 0° and 90° hangers until the system is in equilibrium. The 0° and 90° forces are then the $\vec{\mathbf{F}}_x$ and $\vec{\mathbf{F}}_y$ components, respectively, of $\vec{\mathbf{F}}$. Record their magnitudes in the data table.

6. *Vector addition IV.* Given the force vectors $F_1 = (0.100)g$ N at 30°, $F_2 = (0.200)g$ N at 90°, and $F_3 = (0.30)g$ N at 225°, find the magnitude and direction of their resultant $\vec{\mathbf{F}} = \vec{\mathbf{F}}_1 + \vec{\mathbf{F}}_2 + \vec{\mathbf{F}}_3$ by the following procedures:

(a) *Graphical.* Use the polygon method.

(b) *Analytical.* Use the component method.

(c) *Experimental.* Use the force table. Record the results in the data table.

7. *Vector addition V.* Instructor's choice (optional). Your instructor will give you a set of vectors to add. Record the results in the data table as you did for previous procedures.

Name ______________________ Section ______________ Date ______________

Lab Partner(s) ____

EXPERIMENT 7

Newton's Second Law: The Atwood Machine

GL Experimental Planning

Newton's second law expresses a relationship between the net force acting on a specific mass and the resulting acceleration of that mass. This relation is often written in magnitude form, $F_{net} = ma$ (force = mass × acceleration).

EQUIPMENT NEEDED

- Pulley (preferably low-inertia, precision ball bearing type, with support)
- Two weight hangers and weights (masses)
- String
- Laboratory timer or stopwatch
- Meterstick

Take a look at the apparatus shown in Fig. 7.1a. Consider how you could use it to experimentally investigate the validity of Newton's second law. Note from $F_{net} = ma$ that the motion (oscillation) can be affected by varying the net force or mass, $a = F_{net}/m$. In order to find out how one of these variables affects the motion, the other is held constant.

From the figure and given equipment, how could the acceleration a of the mass hangers be determined from direct measurements? (*Hint*: Think of a kinematic equation that includes the three variables.)

In case you missed it, the kinematic equation that applies to this situation is $y = v_o t + \frac{1}{2}at^2$. In this equation, there is a condition on the acceleration. What is it?

The apparatus shown in Fig. 7.1a is called an "Atwood machine," and it allows for all of the variables of $F_{net} = ma$ to be controlled. If the system with loaded masses is released from rest, what quantity in the kinematic equation could be eliminated? For this case, solve the kinematic equation for the acceleration.

(continued)

Did you get $2y/t^2$? Here the meterstick and stopwatch come into play, and the distance traveled (y) and the elapsed time (t) of fall are measured variables for determining acceleration.

Now consider the other two variables in Newton's second law, F and m. For the Atwood machine, can you think of how the net force be expressed in terms of both masses (weights) on the hangers? Remember, the m in $F_{net} = ma$ represents the *total* mass of the system that moves with acceleration a. A free-body diagram may be helpful here.

Substitute your expression for the net force in Newton's second law ($F_{net} = ma$), and solve for a.

Did you get: $a = \dfrac{(m_2 - m_1)g}{m_1 + m_2}$?

Note for the Atwood machine how both hanging masses can be varied. How could you (1) vary the total mass while keeping the (net) force F constant, and (2) vary the net force while keeping the total mass constant? Think about it.

Name ______ Section ______ Date ______

Lab Partner(s) ______

EXPERIMENT 7

Newton's Second Law: The Atwood Machine

TI Advance Study Assignment

Read the experiment and answer the following questions.

1. Write Newton's second law in mathematical form, and describe how the acceleration of an object or system varies with a net force and mass of the system.

2. What are F and m in Newton's second law in terms of the Atwood machine?

3. Explain how F and m are individually varied while the other is held constant. Why is this done?

(continued)

4. How can the frictional force be experimentally determined, and how is it used in the calculations?

5. What is measured in the experiment, and how is this used to compute the acceleration of the system?

CI Advance Study Assignment

Read the experiment and answer the following questions.

1. When the Atwood machine is moving, what is the shape of a velocity-versus-time plot for the motion? Why?

2. The photogate will measure the tangential speed of the pulley. Why is this speed the same as the speed of the ascending and descending masses?

PRE-LAB DEMONSTRATION | Which Way Does the Bubble Go? (mass and inertia)

Mass (m) is an important factor in Newton's second law ($F = ma$ or $m = F/a$). Let's take a look at mass in terms of Newton's first law, which you have probably studied. Mass is a measure of *inertia*, which is defined as the natural tendency of an object to remain in a state of motion or in uniform motion in a straight line (Newton's first law or the law of inertia).

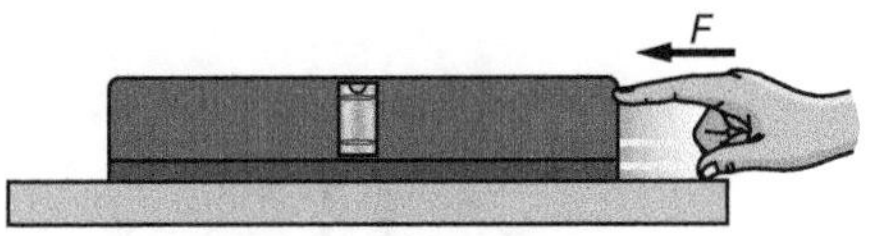

Figure D3 Which Way Does the Bubble Go?

From Newton's second law ($a = F/m$), it can be seen that the greater the mass (or inertia), the smaller the acceleration (or change in velocity) for a given applied force. This can be demonstrated using a small level. When the level is given a push (force), which way will the bubble go? (● Fig. D3)

EXPERIMENT 7

Newton's Second Law: The Atwood Machine

OVERVIEW

Experiment 7 examines Newton's second law using the Atwood machine by TI procedures and/or CI procedures. Both procedures apply the second law by (1) varying the total mass while keeping the unbalanced force constant and (2) varying the unbalanced force while keeping the total mass constant.

The TI procedure determines the accelerations of the system using distance-time measurements. In the CI procedure, speed-time measurements are used by electronically observing the motion of the pulley.

INTRODUCTION AND OBJECTIVES

Newton's second law of motion states that the acceleration, $\vec{a}$, of an object or system is directly proportional to the vector sum of the forces acting on the object, the unbalanced or net force $\vec{F}_{net} = \Sigma\vec{F}_i$, and inversely proportional to the total mass, m, of the system ($\vec{a} \propto \vec{F}_{net}/m$). In vector equation form with standard units, $\vec{a} = \vec{F}_{net}/m$ or, more commonly, $\vec{F}_{net} = m\vec{a}$.*

This relationship will be investigated using an Atwood machine, which consists of two masses connected by a string looped over a pulley (● TI Fig. 7.1a). The Atwood machine is named after the British scientist George Atwood (1746–1807), who used the arrangement to study motion and measure the value of g, the acceleration due to gravity.

In this experiment, the relatively slow, uniform acceleration of the masses will be used to investigate Newton's second law. Since the acceleration a of the system depends on two variables (F_{net} and m, where $a = F_{net}/m$), one of the variables will be held constant while the other is varied. This is common experimental procedure. By varying the net (weight) force and the total mass of the system, the resulting accelerations can be experimentally determined from distance and time measurements and compared with the predictions of Newton's second law.

TI OBJECTIVES

After performing this experiment and analyzing the data, you should be able to do the following:

1. Tell how the acceleration of a system varies with changes in the net force or mass—in particular, for
   a. mass variations with a constant net force, and
   b. force variations with constant mass.
2. Articulate the precise meanings of the variables (F, m, and a) in Newton's second law.
3. Explain how the acceleration of the masses of an Atwood machine may be determined experimentally.

CI OBJECTIVES

Experimentally verify Newton's second law of motion in two ways:

1. By keeping the net force on a system constant and varying the mass, and
2. By keeping the mass of the system constant and varying the net force.

*$\vec{F}_{net} = m\vec{a}$ is often written simply $\vec{F} = m\vec{a}$, where it is understood that $\vec{F}$ is the net force. If $\vec{a} = 0$, there is no net force.

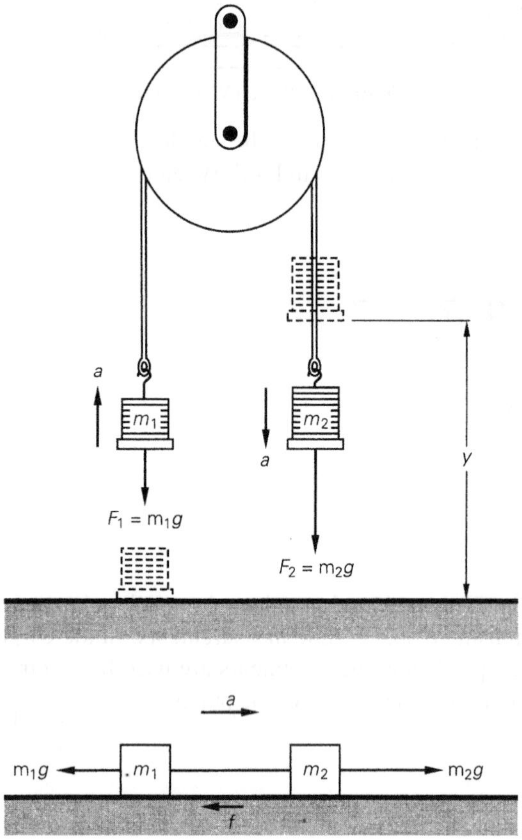

(a)

(b)

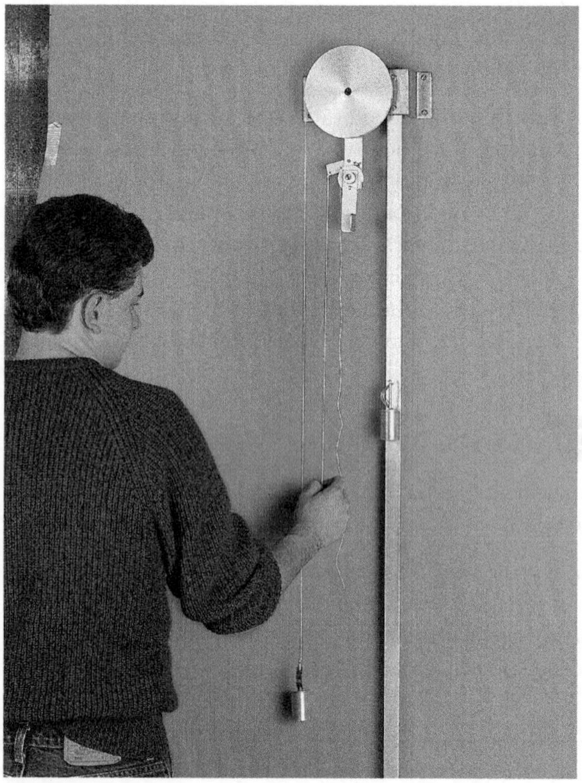

(c)

TI Figure 7.1 The Atwood machine. (a) A single (or double) pulley system is simply a "direction changer," and it is sometimes convenient to draw a horizontal diagram for analysis. (b) A double-pulley system eliminates the possibility of the passing weights hitting each other, which may occur with a single pulley of small diameter. (c) A wall-mounted precision Atwood machine. A trip platform supports the upper weight before the start of each run and is released and reset by control cords. (Photos courtesy of Sargent-Welch.)

TI EXPERIMENT 7

Newton's Second Law: The Atwood Machine

TI EQUIPMENT NEEDED

- Pulley (preferably low-inertia, precision ball bearing type)
- Clamps and support rods
- Two weight hangers
- Set of slotted weights (masses), including small 5-g, 2-g, and 1-g masses
- Paper clips
- String
- Laboratory timer or stopwatch
- Meterstick
- 2 sheets of Cartesian graph paper

TI THEORY

In Fig. TI 7.1a, the light string is considered to be of negligible mass. Masses m_1 and m_2 are taken as the ascending and descending sides of the system, respectively (Fig. TI 7.1). Taking the more massive hanger (m_2) to be moving in the positive direction, the unbalanced or net force is

$$F_{net} = m_2 g - m_1 g = (m_2 - m_1)g \qquad \textbf{(TI 7.1)}$$

where the friction and inertia of the pulley are neglected. By Newton's second law,

$$F_{net} = ma = (m_1 + m_2)a \qquad \textbf{(TI 7.2)}$$

where $m = m_1 + m_2$ is the total mass of the moving system. Then, equating Eqs. (TI 7.1) and (TI 7.2),

$$(m_2 - m_1)g = (m_2 + m_1)a$$

and solving for a:

$$a_t = \frac{F_{net}}{M_{total}} = \frac{(m_2 - m_1)g}{(m_1 + m_2)} \qquad \textbf{(TI 7.3)}$$

(acceleration, theoretical)

where a_t is used to distinguish the theoretical acceleration from the experimentally measured acceleration a_m.

(*Optional*) In the experimental arrangement, there may be an appreciable frictional force f associated with the pulley that opposes the motion. Also, the pulley has inertia. In an attempt to take this inertia into account, an equivalent mass m_{eq} may be added to the total mass in calculations (not physically added in the experiment). Hence, for better accuracy, the equation for the acceleration of the system should be modified as follows:

$$F_{net} = ma$$

$$F - f = (m_1 + m_2 + m_{eq})a$$

$$(m_2 + m_1)g - f = (m_1 + m_2 + m_{eq})a$$

or

$$a = \frac{(m_2 - m_1)g - f}{m_1 + m_2 + m_{eq}} \qquad \textbf{(TI 7.4)}$$

If the masses of the Atwood machine move with a constant speed, the magnitude a of the acceleration of the system is zero, and

$$a = 0 = \frac{(m_2 - m_1)g - f}{m_1 + m_2 + m_{eq}}$$

Solving for f, and equating to $m_f g$ (with $f = m_f g$),

$$f = (m_2 - m_1)g = m_f g \qquad \textbf{(TI 7.5)}$$

(uniform speed)

which provides a method for determining the magnitude of the frictional force of the pulley, or the mass m_f needed to provide the weight to balance the frictional force.

Hence, the expression for the theoretical acceleration of the system [Eq. (TI 7.4)] may be written

$$a_t = \frac{(m_2 - m_1 - m_f)g}{m_1 + m_2 + m_{eq}} \qquad \textbf{(TI 7.6)}$$

(acceleration, theoretical)

Thus, part of the weight of m_2 goes into balancing or canceling the frictional force of the pulley. In the experimental acceleration trials, the m_f determined in each case is left on the descending hanger as part of m_2 to compensate for the opposing frictional force.

To determine the acceleration of the system experimentally so that it may be compared to that predicted by theory, the time t for the descending mass to fall a given distance y is measured. Then, using the kinematic equation,

$$y = v_o t + \frac{1}{2}at^2$$

with the mass starting from rest, $v_o = 0$ (and $y_o = 0$, $t_o = 0$),

$$y = \frac{1}{2}at^2$$

or

$$\boxed{a_m = \frac{2y}{t^2}} \qquad \text{(TI 7.7)}$$

(acceleration, measured)

where a_m is the experimentally measured acceleration.

When a_m is determined experimentally using distance and time measurements, friction and pulley inertia are involved. These are taken into account in the theoretical expression [Eq. (TI 7.6)] so that the experimental and theoretical values of a will be more comparable. Even so, keep in mind that these are approximations and the percent differences may be large. The main purpose of the experiment is to demonstrate how the acceleration of a system depends on the net force and total mass.

TI EXPERIMENTAL PROCEDURE*

1. Set up the Atwood machine as shown in Fig. TI 7.1. Use enough string so that the distance of travel (y) is slightly less than 1 m for convenient measuring. (To measure y, hold one hanger against the floor and measure from the floor to the bottom of the other hanger.) Measure and record y in TI Data Table 1.

A. Varying the Total Mass (Net Force Constant)

Procedure without corrections

(If using inertia and friction corrections, go to Procedure 2a below.)

2. Begin by placing a 10-g mass on the descending hanger so as to create an unbalanced or net force that should cause the system to accelerate from rest. Make a trial run to see if the system moves at an acceleration suitable for timing. If not, adjust the mass accordingly. (See Suggestions 1–3 in the "Comments on Experimental Technique" at the end of the Procedure section.)

   Taking the descending mass as m_2, record m_1 and m_2 in TI Data Table 1 as Trial 1. (Ignore the columns headed with asterisks and the m_{eq} and m_f symbols.)

3. Make three independent measurements of the time it takes for m_2 to travel the distance y from rest. Record the time in TI Data Table 1.†

4. Add 100 g to each hanger. Repeat Procedure 3 (measurement of time with a 10-g mass imbalance). Record the data in the Trial 2 column. *Note*: The distance y should be remeasured for each trial. The length of the string (and y distance) may vary noticeably because of stretching.

5. Repeat Procedure 3 for two more trials with another 100 g being added on each hanger for each trial.

(Procedure using inertia and friction corrections.)

2a. As noted in the Theory section, the pulley contributes to the inertia of the system as though an "equivalent mass" m_{eq} were part of the total mass being accelerated. For better results, a m_{eq} will be added in the calculations. The instructor will provide the value of m_{eq} or tell you how to measure it (*Instructor's Resource Manual*). Record the value of m_{eq} in the data tables.

3a. Begin with the descending mass (m_2) and the ascending mass (m_1), each equal to 50 g (that is, the masses of the hangers alone). With $m_1 = m_2$, the system is in equilibrium—equal forces, $m_1 g = m_2 g$. In the absence of friction, a slight tap or momentary force applied to m_2 should set the system in uniform motion (constant speed). (Why?) However, because of the opposing frictional force, the motion will not persist.

4a. Add small masses to m_2 until a downward push causes m_2 to descend with a uniform (constant) velocity. (See Comment 4 in the "Comments on Experimental Technique" at the end of the Procedure section.) Apply a sufficient push so the masses move at a reasonable speed; they should not move too slowly. You may find it easier to recognize uniform motion by observing the rotating pulley rather than the masses.

Record m_1 and m_2 in TI Data Table 1 in the first column marked with an asterisk. These values are used to calculate the frictional mass, $m_f = m_2 - m_1$, needed in the theoretical calculation of the acceleration of the system [Eq. (TI 7.6)].

5a. (i) Add 10 g to m_2, leaving m_f in place. This creates an unbalanced force that should cause the system to accelerate from rest. Measure the distance y. Record y, m_1, and the new value of m_2 in TI Data Table 1, Trial 1. (See "Comments on

*Refinements in the Experimental Procedure section were developed by Professor I. L. Fischer, Bergen Community College, New Jersey.

† The data tables are arranged to facilitate data taking and analysis. The upper (seven) rows include all the experimental measurements, and the lower (six) rows are for calculations based on these measurements.

Experimental Technique" at the end of the Procedure section.)

(ii) Make three independent measurements of the time it takes for m_2 to travel the distance y from rest. Record the data as Trial 1.

(iii) Remove m_f and the 10-g mass before proceeding to the next trial.

6a. **(i)** Add 100 g to each hanger for a total of 150 g each.

(ii) Repeat Procedure 4a (measurement of frictional mass), and record data in the next asterisked column in TI Data Table 1.

(iii) Repeat timing measurements, Procedure 5a (measurement of acceleration with a net 10-g mass imbalance). Record the data in the Trial 2 column. The calculations for Trial 2 should utilize the value of m_f obtained for the immediately preceding asterisked column. *Note*: The values of m_f and y should be remeasured for each of the trials in TI Data Table 1. As the total mass is changed, the friction will change likewise. The length of the string (y distance) may vary noticeably because of stretching.

7a. Repeat Procedures 4a and 5a for two more trials with another 100 g being added on each hanger for each trial.

B. Varying the Unbalanced Force (Total Mass Constant)

(If using inertia and friction corrections, go to Procedure 1b below.)

1. Begin with an ascending mass of 260 g (50-g hanger + 200 + 5 + 2 + 2 + 1-g masses) and a similar descending mass m_2 = 260 g (50-g hanger + 200 + 10-g masses).*

2. Transfer 1 g from m_1 to m_2 in order to create an unbalanced force without affecting the total mass. Make three measurements of the travel time as done previously in Procedure A3. Record the data as Trial 5 in TI Data Table 2.

3. Leaving the previously transferred 1-g mass in place,
(a) transfer an additional 2 g for Trial 6,
(b) transfer an additional 2 g for Trial 7,
(c) transfer an additional 5 g for Trial 8,
(Procedure using inertia and friction corrections).

1b. Begin with an ascending mass m_1 = 260 g (50-g hanger + 200 + 5 + 2 + 2 + 1-g masses) and a similar descending mass m_2 = 260 g (50-g hanger + 200 + 10-g masses).*

* Mass increments larger than 1 g and 2 g may have to be used, depending on the pulley friction. Friction may not be uniform, so a greater mass difference may be needed to initiate motion.

2b. Measure the frictional mass as done previously in Procedure A4a. Record the data in the asterisked column in TI Data Table 2. The value of m_f from these data may be used in the calculations for all trials in TI Data Table 2, since the total mass (and presumably the friction) will now be constant.

3b. Leaving m_f in place, transfer 1 g from m_1 to m_2 in order to create a net unbalanced force without affecting the total mass. Make three measurements of the travel time as in Procedure 5a. Record all pertinent data in the Trial 5 column.

4b. Leaving m_f and the previously transferred 1-g mass in place,
(a) transfer an additional 2 g for Trial 6,
(b) transfer an additional 2 g for Trial 7,
(c) transfer an additional 5 g for Trial 8.

C. Comments on Experimental Technique

1. The masses must start from rest during the acceleration trials. A good technique is as follows:
(a) Hold m_1 down against the floor.
(b) Simultaneously release m_1 and start the timer.
(c) Stop the timer at the instant m_2 strikes the floor.
The best results are obtained when the same person releases m_1 and operates the timer. (Why?)

2. Some of the masses may be jolted off the hangers by the impact on hitting the floor. It may be helpful to place a shock-absorbing pad on the floor. Also, one lab partner should attend to the upper weight to prevent it, or some of it, from falling. Don't let the upper mass strike the pulley.

3. Take turns at each task.

4. Measure the frictional mass to a precision of $\pm\frac{1}{2}$ g. Fine adjustment of the descending mass may be made by using small "custom" masses (paper clips) as needed. These paper clips can be attached to the cord just above the m_2 hanger. Good precision is necessary for good results because the frictional force is comparable in magnitude to the accelerating force. Small errors in the frictional masses may create large experimental errors.

Name ______________________ Section ______________ Date ______________

Lab Partner(s) ____

TI EXPERIMENT 7

Newton's Second Law: The Atwood Machine

TI *Laboratory Report*

TI DATA TABLE 1

Purpose: To investigate $a = F/m$ by holding F constant. (If not considering pulley inertia and friction, ignore (*) columns and m_{eq} and m_f symbols.)

m_{eq} ________ ( )		Trial							
		*	1	*	2	*	3	*	4
Descending mass m_2 ( )									
Ascending mass m_1 ( )									
Distance of travel y ( )									
Time of travel t ( )	Run 1								
	Run 2								
	Run 3								
	Average								
Measured acceleration $a_m = 2y/t^2$ ( )									
Total mass $= m_1 + m_2 + m_{eq}$ ( )									
Measured frictional mass $m_f = m_2 - m_1$									
Net force $= (m_2 - m_1 = m_f)g$ ( )									
Theoretical acceleration $a_t = \frac{\text{net force}}{\text{total mass}}$									
Percent different between a_m and a_t									

*Measurement of frictional mass m_f. Masses move with constant velocities when given an initial push.

Calculations
(show work)

Don't forget units

(continued)

TI DATA TABLE 2

Purpose: To investigate $a = F/m$ by holding m constant. (If not considering pulley inertia and friction, ignore (*) columns and m_{eq} and m_f symbols.)

m_{eq} ________ ( )		Trial				
		*	5	6	7	8
Descending mass m_2 ( )						
Ascending mass m_1 ( )						
Distance of travel y ( )		X				
Time of travel t ( )	Run 1	X				
	Run 2	X				
	Run 3	X				
	Average	X				
Measured acceleration $a_m = 2y/t^2$ ( )		X				
Total mass $= m_1 + m_2 + m_{eq}$ ( )		X				
Measured frictional mass $m_f = m_2 - m_1$			X	X	X	X
Net force $= (m_2 - m_1 - m_f)g$ ( )		X				
Theoretical acceleration $a_t = \frac{\text{net force}}{\text{total mass}}$		X				
Percent different between a_m and a_t		X				

* Measurement of frictional mass m_f. Masses move with constant velocities when given an initial push.

Name ____________________ Section ____________ Date ____________

Lab Partner(s) ____________________

EXPERIMENT 7 **Newton's Second Law: The Atwood Machine** ***Laboratory Report***

Calculations
(show work)

TI QUESTIONS

1. In the experiment, should the mass of the string be added to the total mass moved by the unbalanced force for better accuracy? Explain.

2. Complete the following sentences:
   (a) When the unbalanced force increases (total mass remaining constant), the acceleration

   of the system ____________________.
   (b) When the total mass that is accelerating increases (unbalanced force remaining

   constant), the acceleration of the system ____________________.
3. How can the value of g, the acceleration due to gravity, be determined using an Atwood machine?

4. Using the data in TI Data Table 2 (constant total mass), plot a_m versus $(m_2 - m_1)$ for each Trial, and draw a straight line that best fits the data. Find the slope and intercept of the line, and enter the values below.

   Rewrite Eq. (TI 7.6) in slope-intercept form ($y = mx + b$), and, using the data in Trial 6, compute the slope and intercept. (Show calculations.) Compare and comment on your results.

	From graph	From Eq. (TI 7.6)
Slope	____________ (units)	____________ (units)
Intercept	____________ (units)	____________ (units)

CI EXPERIMENT 7

Newton's Second Law: The Atwood Machine

EQUIPMENT NEEDED

- Photogate/Pulley System (PASCO ME-6838) (Smart Pulley)
- Mass set that includes 1-g, 2-g, and 5-g weights (Suggested: PASCO ME-8967)
- 2 mass hangers
- Clamps and support rods
- Graph paper

THEORY

When an Atwood machine is unbalanced, the masses move, one ascending and the other descending. (See Fig. TI 7.1.) As the masses move, the string causes the pulley to rotate. With m_1 and m_2 as the ascending and descending masses, respectively, the magnitude of their acceleration is given by

$$a = \frac{F_{net}}{M_{total}} = \frac{(m_2 - m_1)g}{m_2 + m_1} \qquad \textbf{(CI 7.1)*}$$

where the friction and inertia of the pulley have been ignored.

In this part of the experiment, the motion of the ascending and descending masses is analyzed by using a motion sensor to look at the motion of the pulley. The main idea is that all the objects in the system—the ascending mass, the descending mass, and the pulley—must be moving with the same linear speed at any moment. The linear speed of the pulley is measured as the speed of a point on the rim. The sensor detects how many revolutions per second the pulley is making (the angular speed). For a known radius of pulley, the linear speed on the rim is easily determined:

$$v = r\omega$$

where ω is the angular speed.

The sensor performs this calculation automatically. Notice that by measuring the linear speed of the pulley, the ascending speed of mass m_1 and the descending speed of mass m_2 are also measured.

The measured speeds will then be plotted as a function of time. Because the acceleration of the system is constant, the plot of speed versus time will be a straight line with slope equal to the acceleration of the system. The experimental acceleration of the system will be determined by finding the slope of the graph. It will then be compared to the theoretical value predicted by Eq. (CI 7.1).

*See Eq. (TI 7.3) development.

SETTING UP DATA STUDIO

1. Open Data Studio and choose Create Experiment.
2. The Experiment Setup window will open, and you will see a picture of the Science Workshop interface. There are seven channels to choose from. (Digital channels 1, 2, 3 and 4 are the small buttons on the left; analog channels A, B, and C are the larger buttons on the right, as shown in ● CI Fig. 7.1.)
3. Click on the Channel 1 button in the picture. A window with a list of sensors will open.
4. Choose the Smart Pulley from the list, and press OK.
5. Connect the sensor to the interface as shown on the computer screen.
6. The Data list on the left of the screen should now have one icon for velocity.
7. Create a graph by dragging the velocity data icon from the data list and dropping it on top of the graph icon of the displays list. A graph of velocity versus time will open. The window will be called Graph 1.
8. ● CI Fig. 7.2 shows what the screen should look like once the setup is complete.

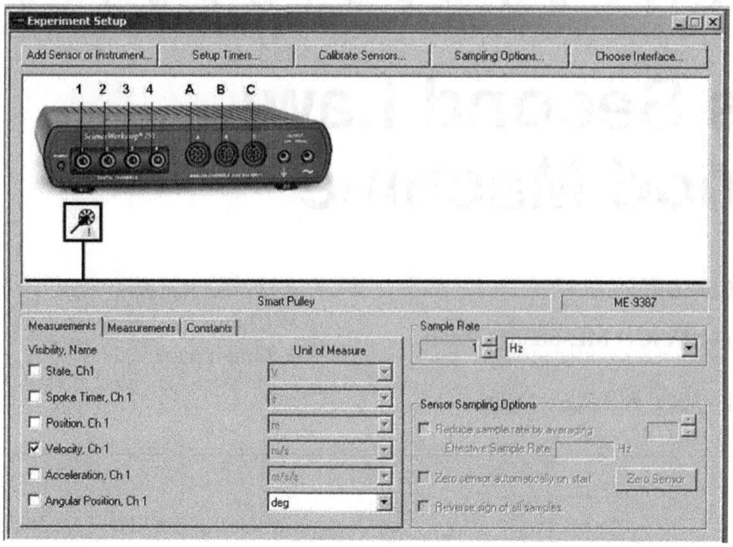

CI Figure 7.1 The Experiment Setup window. The seven available channels are numbered 1 through 4 and A, B, or C. The Smart Pulley is connected to Channel 1 of the Science Workshop Interface. (Reprinted courtesy of PASCO Scientific.)

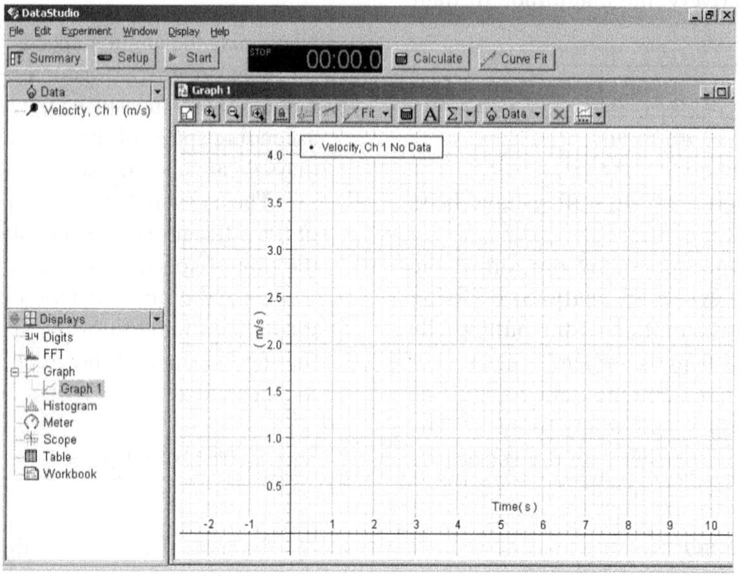

CI Figure 7.2 Data Studio setup. A graph of velocity versus time was created by dragging the "Velocity" icon from the data list and dropping it on the "Graph" icon in the displays list below. In this picture, the graph window has been resized to occupy most of the screen. (Reprinted courtesy of PASCO Scientific.)

CI EXPERIMENTAL PROCEDURE

A. Varying the Total Mass (Unbalanced Force Constant)

1. Set up the Atwood machine using the Photogate/Pulley System (Smart Pulley) instead of a conventional pulley. The ascending mass should begin close to, but not touching, the floor. The descending mass will start at the top ● CI Fig. 7.3 shows the experimental setup. Make the string long enough, and install the pulley high enough, so that the masses can move at least half a meter.

2. If using the PASCO ME-8967 mass and hanger set, begin by placing 50 g on each hanger. This added weight will prevent the system from moving too fast, and data collection will be easier. If you are using a conventional mass and hanger set, a 50-g hanger will work fine with no added mass. For all data collection and calculations, keep track of the total ascending and descending masses, including the mass of the hanger.

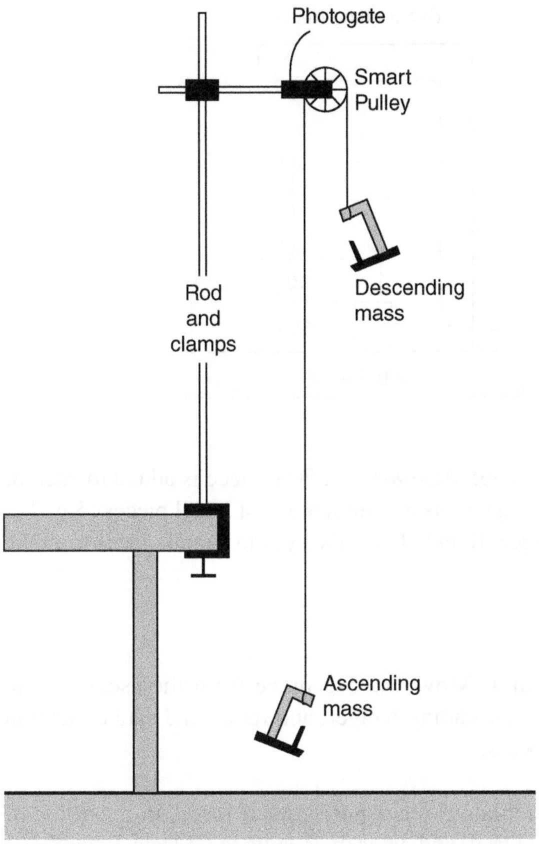

CI Figure 7.3 Experimental setup. A Photogate/Pulley System (Smart Pulley) is used instead of a conventional pulley to set up the Atwood machine. The ascending mass starts near the bottom, close to, but not touching, the floor. The descending mass starts from rest at the top. The Smart Pulley measures the speed of the system as it moves.

3. When the ascending and descending masses are equal, the system should not move. If it does, check that the pulley is level.

4. Trial 1: Add a 5-g piece to the descending mass to unbalance the system. Make a note of the ascending and descending masses in CI Data Table 1. Do not forget to account for the mass of the hangers.

5. Place the ascending mass at the bottom and the descending mass at the top, as shown in CI Fig. 7.3. Gently hold the pulley to prevent the system from moving.

6. Let the system start from rest by letting go of the pulley. Once it starts moving, press the START button. Keep your eyes on the system, and press the STOP button before the masses reach the end of their line and bounce. If the hangers collide while passing each other, try making the strings longer and pressing STOP just before they collide.

7. A straight-line graph should have appeared on the screen. To see it better, press the Scale-to-Fit button on the graph toolbar. (It is the leftmost button of the toolbar.)

8. On the graph toolbar, there is also a drop menu called Fit. Choose to do a "Linear Fit" for the graph. A box will pop up with information about the fit. Make a note of the slope of the line. This is the measured, experimental acceleration. Enter it in CI Data Table 1.

9. Clear the fit information by going to the Fit menu and deselecting the linear fit.

10. Trial 2: Add 10 g to each hanger. The descending mass should still have the 5-g unbalance. Note that this increases the overall total mass of the system but keeps the unbalanced force the same. Repeat the data collection process, and enter the data in CI Data Table 1.

11. Trials 3 and 4: Repeat two more times, each time adding an extra 10 g to each hanger.

12. Clear the graph window of all fit information and then print the graph. Label each of the plots with the total mass of the system corresponding to each trial. Paste the graph to the Laboratory Report.

13. Calculate the net unbalanced force, in newtons.

14. Calculate the theoretical acceleration for each trial, using Eq. (CI 7.1). Compare the theoretical value with the experimental value by taking a percent error.

B. Varying the Unbalanced Force (Total Mass Constant)

1. Erase all previous data by going to the main menu and, under "Experiment," choosing "Delete all data runs."

2. Place the following mass pieces on the ascending hanger: 5 g, 2 g, 2 g, 1 g. If you are using the PASCO mass and hanger set, the hangers should also have a 50-g piece, as discussed previously. If you are using a conventional 50-g hanger, no extra weight is needed. ● CI Fig. 7.4 shows the ascending and descending masses for the PASCO mass and hanger set.

3. Place a 10-g piece on the descending hanger. Again, with the PASCO mass and hanger set, the hanger should also have a 50-g piece, but with a conventional 50-g hanger, no extra weight is needed.

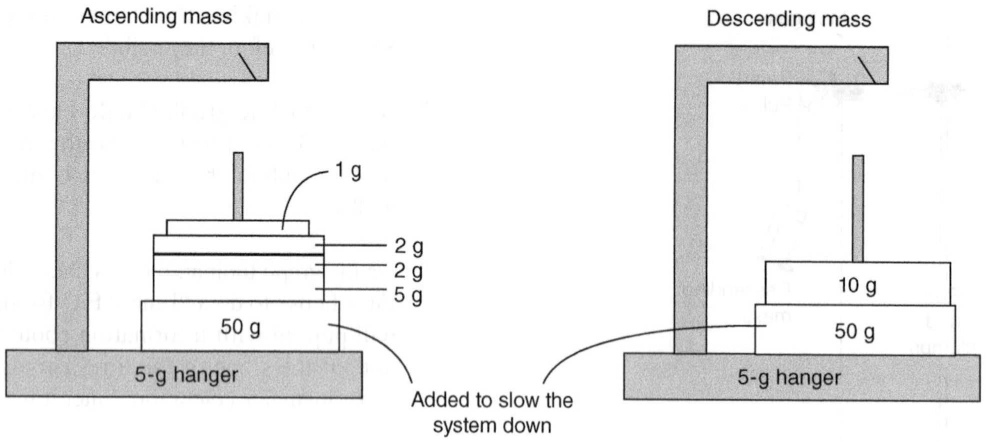

CI Figure 7.4 Ascending and descending masses using PASCO mass and hanger set ME-8967. A 50-g piece is added to each of the small 5-g hangers to prevent them from moving too fast. The ascending mass has a combination of small pieces (5 g, 2 g, 2 g, 1 g) that add to 10 g. A 10-g piece is placed in the descending mass hanger. To unbalance the system, small pieces from the ascending hanger are moved to the descending hanger.

4. Trial 1: Unbalance the system by transferring the 1-g piece from the ascending to the descending hanger. At this time, make a note of the ascending and the descending masses and enter the values in CI Data Table 2. Do not forget to include the mass of the hangers!

5. Collect the data as before and determine the experimental acceleration.

6. Trial 2: Move one of the 2-g pieces from the ascending to the descending hanger, and repeat the data collection process. Note that this changes the amount of unbalanced force without changing the total mass of the system.

7. Trial 3: Move the other 2-g piece from the ascending to the descending hanger, and repeat the data collection process.

8. Trial 4: Move the 5-g piece from the ascending to the descending hanger, and repeat the data collection process.

9. Calculate the net unbalanced force, in newtons, of each trial, and enter the results in CI Data Table 2.

10. Clear the graph window of any fit information and print the graph. Label each of the plots with the unbalanced force corresponding to each trial. Paste the graph to the Laboratory Report.

11. Calculate the theoretical acceleration for each trial, using Eq. (CI 7.1). Compare the theoretical value with the experimental value by taking a percent error.

Name ______________________ Section ______________ Date ______________

Lab Partner(s) ____

CI EXPERIMENT 7

Newton's Second Law: The Atwood Machine

Laboratory Report

DATA TABLE 1

Purpose: To investigate how the acceleration of a system varies as the mass of the system increases, without changing the net applied force.

Trial	Ascending m_1	Descending m_2	Total mass $m_1 + m_2$	Measured acceleration (from graph)	Unbalanced force $(m_2 - m_1)g$	Theoretical acceleration	% error
1							
2							
3							
4							

DATA TABLE 2

Purpose: To investigate how the acceleration of a system varies as the net applied force on the system increases, while the mass remains constant.

Trial	Ascending m_1	Descending m_2	Total mass $m_2 + m_1$	Measured acceleration (from graph)	Unbalanced force $(m_2 - m_1)g$	Theoretical acceleration	% error
1							
2							
3							
4							

Don't forget units

(continued)

CI QUESTIONS

1. What happens to the acceleration of a system when the mass of the system increases but the net force stays constant?

2. What happens to the acceleration of a system when the net applied force increases but the mass of the system does not change?

3. Refer to the data in CI Data Table 2. Make a one-page graph of unbalanced force versus measured acceleration, and draw the best-fitting straight line. Determine the slope of this line. Show the details of the calculation on the graph, and attach the graph to the lab report.

4. What are the units of the slope of your graph?

5. What physical quantity of the system is represented by the slope of the force-versus-acceleration graph? How well does it match the experimental setup?

6. From the results, was there a good agreement between the experimental acceleration and the theoretical (expected) acceleration? What causes the difference? Discuss sources of experimental uncertainty for this experiment.

Name ______________________ Section ______________ Date ______________

Lab Partner(s) ____

EXPERIMENT 8

Conservation of Linear Momentum

TI Advance Study Assignment

Read the experiment and answer the following questions.

1. What is meant when we say that a quantity, such as linear momentum, is conserved?

2. What is the condition for the conservation of linear momentum of a system?

3. Show that Newton's second law can be written in the form $\vec{\mathbf{F}} = \Delta\vec{\mathbf{p}}/\Delta t$.

4. Is the conservation of linear momentum consistent with Newton's first and third laws of motion? Explain.

(continued)

5. In a system of particles for which the total linear momentum is conserved, is the linear momentum of the individual particles constant? Explain.

6. Suppose that a particle of mass m_1 approaches a stationary mass m_2 and that $m_2 >> m_1$. What would you expect to happen on collision?

CI Advance Study Assignment

Read the experiment and answer the following questions.

1. What mechanism will be used to make the collision between the cars an elastic collision?

2. What mechanism will be used to make the collision between the cars an inelastic collision?

PRE-LAB DEMONSTRATION | Newton's Cradle (in and out)

A novelty item, known as Newton's cradle, is shown in ● Fig. D4. It consists of five suspended identical metal balls. When one ball swings in, after multiple collisions, one ball swings out at the other end of the row of balls. When two balls swing in, two swing out; when three swing in, three swing out, and so on—always the same number out as in (even when five swing in).

Question: If two balls swing in with a velocity v, why doesn't one ball swing out with a velocity of $2v$?

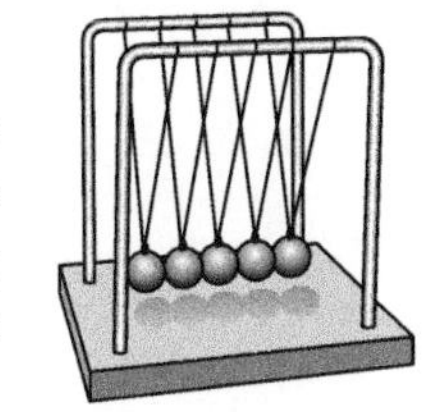

Figure D4 Newton's Cradle

EXPERIMENT 8

Conservation of Linear Momentum

OVERVIEW

Experiment 8 examines the conservation of linear momentum by TI procedures and/or CI procedures. The TI procedure uses distance-time measurements to determine the velocities of air track cars before and after collisions in the investigation of the conservation of linear momentum.

The CI procedure measures the velocities electronically and graphs the data. The velocities, total momentum, and total kinetic energy are obtained from the graphs.

INTRODUCTION AND OBJECTIVES

The conservation of linear momentum ($\vec{\mathbf{p}} = m\vec{\mathbf{v}}$) is an important physical concept. However, the experimental investigation of this concept in an introductory physics laboratory is hampered by ever-present frictional forces.

An air track provides one of the best methods to investigate linear momentum (see TI Fig. 5.2). Aluminum cars or gliders riding on a cushion of air on the track approximate frictionless motion—a necessary condition for the conservation of linear momentum.

In the absence of friction (and other external forces), the total linear momentum of a system of two cars will be conserved during a collision. That is, the total linear momentum of the system should be the same after collision as before collision. By measuring the velocities of the cars before and after collision, the total momentum of a system can be determined and the conservation of linear momentum investigated.

TI OBJECTIVES

After performing this experiment and analyzing the data, you should be able to do the following:

1. Explain when linear momentum is conserved and what this means in terms of force and motion.
2. Apply the conservation of linear momentum to a system.
3. Describe two-body collisions in terms of the conservation of linear momentum.

CI OBJECTIVES

1. Understand that momentum is conserved for both elastic and inelastic collisions.
2. Distinguish between elastic and inelastic collisions in terms of the conservation of kinetic energy.

TI EXPERIMENT 8

Conservation of Linear Momentum

TI EQUIPMENT NEEDED

- Air track
- 3 cars (two of similar mass)
- 4 laboratory timers or stopwatches*
- Laboratory balance
- Masking tape
- Meterstick (if no length scale on air track)
- Velcro (optional)

*If electronic photogates/timers and computer-assisted data analysis are available, your instructor will give you instruction on their use.

TI THEORY

The linear momentum $\vec{\mathbf{p}}$ of a particle or object is defined as

$$\boxed{\vec{\mathbf{p}} = m\vec{\mathbf{v}}} \qquad \textbf{(TI 8.1)}$$

where m is the mass of the object and $\vec{\mathbf{v}}$ its velocity.* Since velocity is a vector quantity, so is linear momentum.

Newton's second law of motion, commonly expressed in the form $\vec{\mathbf{F}} = m\vec{\mathbf{a}}$, can also be written in terms of momentum:

$$\boxed{\vec{\mathbf{F}} = \frac{\Delta\vec{\mathbf{p}}}{\Delta t}} \qquad \textbf{(TI 8.2)}$$

(Recall $\vec{\mathbf{a}} = \Delta\vec{\mathbf{v}}/t$.)

If there is no net or unbalanced external force acting on the object ($F = 0$), then

$$\vec{\mathbf{F}} = \frac{\Delta\vec{\mathbf{p}}}{\Delta t} = 0$$

and

$$\Delta\vec{\mathbf{p}} = 0$$

That is, the change in the momentum is zero, or the momentum is conserved. *Conserved* means that the momentum remains constant (in time). Expanding $\Delta\vec{\mathbf{p}}$,

$$\Delta\vec{\mathbf{p}} = \vec{\mathbf{p}}_f - \vec{\mathbf{p}}_i = 0$$

and

$$\boxed{\vec{\mathbf{p}}_f = \vec{\mathbf{p}}_i} \qquad \textbf{(TI 8.3)}$$

and the "final" momentum $\vec{\mathbf{p}}_f$ at any time t_f is the same as the initial momentum $\vec{\mathbf{p}}_i$ at time t_i. Notice that this is consistent with Newton's first law of motion, since

$$\vec{\mathbf{p}}_f = \vec{\mathbf{p}}_i \quad \text{or} \quad m\vec{\mathbf{v}}_f = m\vec{\mathbf{v}}_i$$

and

$$\vec{\mathbf{v}}_f = \vec{\mathbf{v}}_i$$

That is, an object remains at rest ($\vec{\mathbf{v}}_i = 0$) or in uniform motion ($\vec{\mathbf{v}}_i = \vec{\mathbf{v}}_f$) unless acted on by some external force.

The previous development also applies to the total momentum of a system of particles or objects. For example, the total linear momentum ($\vec{\mathbf{P}}$) of a system of two objects m_1 and m_2 is $\vec{\mathbf{P}} = \vec{\mathbf{p}}_1 + \vec{\mathbf{p}}_2$, and if there is no net external force acting on the system, then

$$\Delta\vec{\mathbf{P}} = 0$$

In the case of a collision between two objects of a system (with only internal forces acting), the initial total momentum before the collision is the same as the final total momentum after the collision. That is,

$$\boxed{\begin{matrix}\text{(before)} & & \text{(after)} \\ \vec{\mathbf{p}}_{1_i} + \vec{\mathbf{p}}_{2_i} & = & \vec{\mathbf{p}}_{1_f} + \vec{\mathbf{p}}_{2_f}\end{matrix}} \qquad \textbf{(TI 8.4)}$$

or

$$m_1\vec{\mathbf{v}}_{1_i} + m_2\vec{\mathbf{v}}_{2_i} = m_1\vec{\mathbf{v}}_{1_f} + m_2\vec{\mathbf{v}}_{2_f}$$

In one dimension, the directions of the velocity and momentum vectors are commonly indicated by plus and minus signs, that is, $+v$ and $-v$.†

The internal forces of a system do not change the total momentum, because, according to Newton's third law, $\vec{\mathbf{F}}_{12} = -\vec{\mathbf{F}}_{21}$ [the force on object 1 due to object 2 is equal to and opposite in direction (minus) to the force on object 2 due to object 1]. Thus the change in momentum for one object will be equal in magnitude and opposite in direction to the change in momentum for the other object, and the total momentum will be unchanged.

*Boldface symbols with over arrows indicate vectors (see Expt. 6).

†In two (or three) dimensions, the momentum is conserved in both (or all) directions. That is, $\vec{\mathbf{p}} = \vec{\mathbf{p}}_x + \vec{\mathbf{p}}_y = 0$, and $\vec{\mathbf{p}}_x = 0$ and $\vec{\mathbf{p}}_y = 0$ (Why?) *Note*: $\vec{\mathbf{p}}_x = \Sigma\vec{\mathbf{p}}_x$ and $\vec{\mathbf{p}}_y = \Sigma\vec{\mathbf{p}}_y$.

TI EXPERIMENTAL PROCEDURES

(Review the operation of the air track in Experiment 5, if necessary.)

1. Determine the mass of each car and record it in the TI Trial Data Table. Let the masses of the two cars of nearly equal mass be m_1 and m_2 and the mass of the third car be m_3.

2. Mark off two equal and convenient lengths (for example, $\frac{1}{2}$ m or 1 m) on both sides of the center position of the air track. Make full use of the length of the track, but leave some space near the ends of the track. Place the four tape reference marks at the lower edges of the track so as not to interfere with the car motion. *Do not* mark the air track surface itself with tape or anything else.

3. *Time trials.* By measuring the time interval Δt it takes a car to move the reference mark length d, one can determine the magnitude of the velocity $v = d/\Delta t$ of the car, where $\Delta t = t_2 - t_1$. The actual timing of the motion of a car moving between the two sets of reference marks is done by either method.

   In Method A (below), involving four observers, each has a timer and is assigned to an individual reference mark. In Method B, involving two observers, each has a timer and is assigned to one *set* of reference marks, as described below. Time trials will be done to determine the better method.*

   In addition to giving timing practice and determining the better method of timing, check out the experimental setup for possible systematic errors. The time intervals for the individual cars to travel the equal distances between the reference marks should be very similar for any one trial. If not, the air track may need leveling, and/or there may be some frictional problem with part of the track. Should this be the case, notify your instructor. *Do not* attempt to level the air track on your own.

   Experimentally carry out each of the following timing methods to determine which is better.

   Method A—Four Timers. Set one of the cars in motion with a *slight* push so that it moves with moderate speed up and down the track. (A few practice starts help.) As the car hits the bumper at one end of the track, all four observers should start their timers.

   As the leading edge of the car passes the assigned reference marks, each respective observer stops his or her timer. (Making a dry run or two to become familiar with the timing sequence is helpful.) Carry out this procedure twice for each of the three cars, and record the data in the TI Trial Data Table.

   Method B—Two Timers. Set the car in motion. The two observers should start and stop their individual timers as the leading edge of the car passes their respective reference mark set. Carry out this procedure twice for each of the three cars, and record the data in the TI Trial Data Table.

4. Compute the Δt's for each trial and calculate the percent difference for each trial set. From the data, decide which timing method should be used on the basis of consistency or precision.

Case 1: Collision Between Two Cars of (Nearly) Equal Mass, With One Initially at Rest

5. With one of the cars (m_2) of nearly equal mass stationary at the center position of the air track, start the other car (m_1) moving toward the stationary car. (See ● TI Fig. 8.1). (It may be more convenient to start m_1 moving away from m_2 and take measurements as m_1 returns from rebounding from the end of the track.) A trial run should show that m_1 remains at rest, or nearly at rest, after collision and that m_2 is in motion.

   Determine the time it takes for m_1 to travel between the reference marks as it approaches m_2 and the time it takes for m_2 to travel between the other set of reference marks after collision. Carry out this procedure three times, and record the data in TI Data Table 1.

   Compute the velocities and the total momentum before and after collision and the percent difference in these values for each trial.

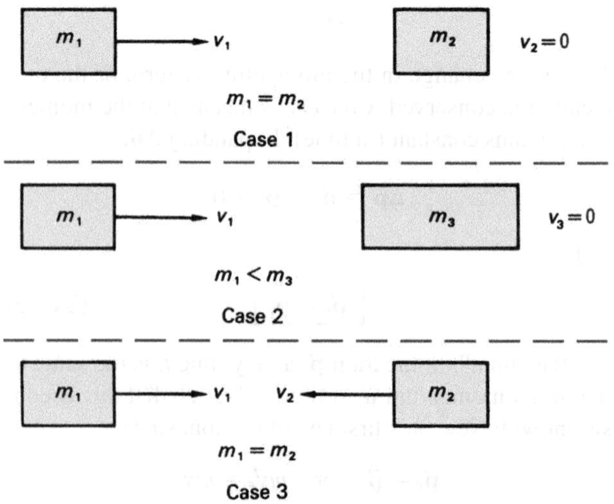

TI Figure 8.1 Experimental collision cases. See text for descriptions.

*If electronic photogate timers are available, your instructor will give you instruction in their use. Electronic timing greatly improves the accuracy and precision of the results. (Why?)

CASE 2: COLLISION BETWEEN TWO CARS OF UNEQUAL MASS, WITH THE MORE MASSIVE CAR INITIALLY AT REST

6. Repeat Procedure 5 with m_2 replaced by m_3 (more massive than m_1). See TI Fig. 8.1. In this case, m_1 will travel in the opposite direction after collision, as a trial run will show. Make appropriate adjustments in the timing procedure to measure the velocity of m_1 before *and* after collision. Record the data and the required calculations in TI Data Table 2. Be careful with the directional signs of the velocities and momenta.

CASE 3: COLLISION BETWEEN TWO CARS OF (NEARLY) EQUAL MASS INITIALLY TRAVELING IN OPPOSITE DIRECTIONS

7. With m_1 and m_2 initially moving toward each other (TI Fig. 8.1), determine the total momentum before and after collision. (*Note*: Speeds do not have to be, and probably won't be, equal.)

   Make appropriate adjustments in the timing procedure to measure the velocities of m_1 and m_2 before and after collision. Carry out the procedure three times, and record the data in TI Data Table 3.

   Compute the percent difference for the total momentum before and after collision for each trial.

(OPTIONAL PROCEDURE)

Another procedure, which may be done at the instructor's option, is as follows:

8. Attach pieces of Velcro to the collision bumpers of both cars, and repeat one or more of the preceding cases as directed by your instructor. Make up a data table, and analyze your results as done previously. (*Hint*: Read in your textbook about elastic and inelastic collisions—in particular, completely inelastic collisions.)

Name ______________________ Section ______________ Date ______________

Lab Partner(s) ____

TI EXPERIMENT 8

Conservation of Linear Momentum

TI Laboratory Report

Distance between marks ______________

TI TRIAL DATA TABLE

Purpose: To determine the better method of timing.

		Method A							Method B		
Car mass ( )		t_1 ( )	t_2 ( )	Δt_{12} ( )	t_3 ( )	t_4 ( )	Δt_{34} ( )	Percent diff.	Δt_{12} ( )	Δt_{34} ( )	Percent diff.
m_1											
m_2											
m_3											

TI DATA TABLE 1

Purpose: To analyze $m_1 = m_2$ case, with $v_{2_i} = 0$.

Trial	Before collision			After collision			
	m_1			m_2			
	Δt_1 ( )	v_{1_i} ( )	p_{1_i} ( )	Δt_2 ( )	v_{2_f} ( )	p_{2_f} ( )	Percent diff.
1							
2							
3							

Don't forget units

(continued)

TI DATA TABLE 2

Purpose: To analyze $m_3 > m_1$ case, with $v_{3_i} = 0$.

Trial	Before collision			After collision							
	m_1			m_1			m_3				
	Δt_{1_i} ( )	v_{1_i} ( )	Total momentum ( )	Δt_{1_f} ( )	v_{1_f} ( )	p_{1_f} ( )	Δt_{3_f} ( )	v_{3_f} ( )	p_{3_f} ( )	Total momentum ( )	Percent diff.
1											
2											
3											

TI DATA TABLE 3

Purpose: To analyze the $m_1 = m_2$ case, initial motions in opposite directions.

Trial	Before collision						
	m_1			m_2			
	Δt_{1_i} ( )	v_{1_i} ( )	p_{1_i} ( )	Δt_{2_i} ( )	v_{2_i} ( )	p_{2_i} ( )	Total momentum ( )
1							
2							
3							

Trial	After collision							
	m_1			m_2				
	Δt_{1_f} ( )	v_{1_f} ( )	p_{1_f} ( )	Δt_{2_f} ( )	v_{2_f} ( )	p_{2_f} ( )	Total momentum ( )	Percent diff. ( )
1								
2								
3								

Name ________________ Section ________________ Date ________________

Lab Partner(s) ________________

EXPERIMENT 8 **Conservation of Linear Momentum** *Laboratory Report*

TI QUESTIONS

1. Do the results of the experiment support the conservation of linear momentum? Consider possible sources of error.

2. Was it necessary to have equal length intervals in the experiment to investigate properly the conservation of momentum? Explain.

3. In Cases 1 and 2, one of the cars was initially at rest, so it must have received an acceleration. Is the car accelerating as it passes between the reference marks? Explain.

4. In each of the three cases, was kinetic energy ($K = \frac{1}{2}mv^2$) conserved? Justify your answers with a sample calculation for a trial from each case. If the kinetic energy is not conserved, where did it go?

CI EXPERIMENT 8

Conservation of Linear Momentum

EQUIPMENT NEEDED

- 2 rotary motion sensors (PASCO CI-6538)
- Brackets and pulley mounts:
  - 2 cart-string brackets (CI-6569)
  - 2 dynamics track mount accessories (CI-6692, to mount the RMS to the track)
  - 2 RMS/IDS adapters (ME-6569, track pulley bracket)
- 2 collision carts (PASCO Classic Cars, ME-9454)
- 1 track
- Clay or Velcro strips
- String
- Optional: track end stop

THEORY

The purpose of this experiment is to investigate the momentum and kinetic energy for elastic and inelastic collisions. The momentum and kinetic energy before the collision of two cars are compared with the momentum and kinetic energy after the collision by looking at a plot of these quantities versus time.

In a collision between two objects, the total momentum at any time is found by adding the momentum of one of the objects to that of the other:

$$\vec{\mathbf{P}}_{Total} = \vec{\mathbf{P}}_1 + \vec{\mathbf{P}}_2 = m_1\vec{\mathbf{v}}_1 + m_2\vec{\mathbf{v}}_2 \qquad \text{(CI 8.1)}$$

This is vector addition, which means the directions of motion of both objects must be taken into account. The sensor used to measure the speeds of the objects will also assign a positive or negative sign, depending on direction. In general, an object moving toward the sensor is assigned a positive velocity, and an object moving away from the sensor is assigned a negative velocity.

An object in motion also has kinetic energy. The total kinetic energy in a system can be determined by adding the kinetic energies of all objects in the system:

$$K_{Total} = K_1 + K_2 = \tfrac{1}{2}m_1v_1^2 + \tfrac{1}{2}m_2v_2^2 \qquad \text{(CI 8.2)}$$

The total momentum and the total kinetic energy just before and just after a collision are determined and compared. First, an elastic collision between two cars is considered. The cars have magnets that make them repel each other when they get close enough. The effect is that the cars bounce off each other (collide) without touching. Next, an inelastic collision is considered. The magnets are replaced by a piece of clay (or Velcro) that will make the cars stick to each other after the collision.

BEFORE YOU BEGIN

1. Install a cart-string bracket on each of the collision carts. The cart-string bracket is mounted on the side of the cart, as shown in ● CI Fig. 8.1.
2. Choose one cart to be Car 1 and measure its mass, in kilograms, including the cart-string bracket. Report the mass of Car 1 in the Laboratory Report.
3. The other cart will be Car 2. Measure its mass and also record that mass in the Laboratory Report.
4. Do not lose track of which is Car 1 and which is Car 2. If needed, put a small tape label on the cars so that you will not confuse them later.

This information will be needed during the setup of Data Studio.

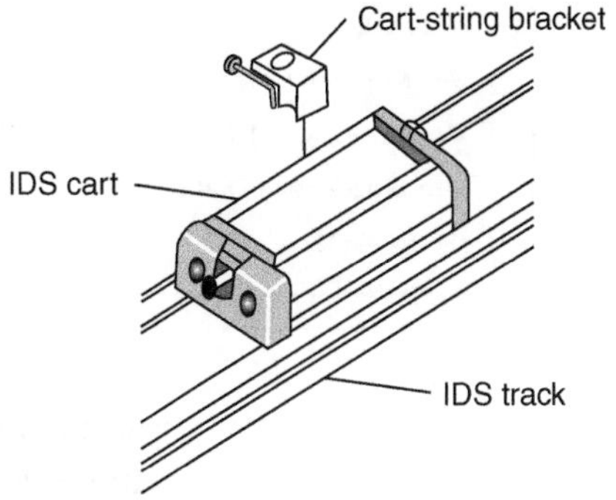

CI Figure 8.1 Installing cart-string brackets. The cart-string brackets are installed on top of the collision carts, secured with a side screw. The top screw is used to tie a string. When measuring the mass of the car, include the cart-string bracket.

SETTING UP DATA STUDIO

1. Open Data Studio and choose "Create Experiment."
2. The Experiment Setup window will open, and you will see a picture of the Science Workshop interface. There are seven channels to choose from. (Digital channels 1, 2, 3, and 4 are the small buttons on the left; analog channels A, B, and C are the larger buttons on the right, as shown in CI Fig. 8.2.)
3. Click on the Channel 1 button in the picture. A window with a list of sensors will open.
4. Choose the Rotary Motion Sensor from the list, and press OK.
5. Click on the Channel 3 button in the picture, and again choose a Rotary Motion Sensor from the list and press OK.
6. Connect the sensors to the interface as shown on the computer screen: One goes to Channels 1 and 2, the other goes to Channels 3 and 4.
7. The properties of each RMS sensor are shown directly under the picture of the interface. (See CI Fig. 8.2.)
8. Click on the icon of the first sensor and adjust the properties as follows:
   First Measurements tab: deselect all options.
   Second Measurements tab: select Velocity and deselect all others.
   Rotary Motion Sensor tab: set the Resolution to high (1440 divisions/rotations); and set the Linear Scale to Large Pulley (Groove).
   Set the Sample Rate to 100 Hz.
9. Click on the icon of the second sensor, and repeat the process of adjusting the properties, as done in step 8.
10. Open the program's calculator by clicking on the Calculate button, on the top main menu. Usually, a small version of the calculator opens, as shown in ● CI Fig. 8.3. Expand the calculator window by clicking on the button marked "Experiment Constants."
11. The expanded window (shown in ● CI Fig. 8.4) is used to establish values of parameters that will remain constant throughout the experiment. In this case, these are the masses m_1 and m_2 of the carts, which have already been measured. This is how to do it:
    a. Click on the lower New button (within the "Experiment Constants" section of the calculator window) and enter the name of the constant as m1, the value as the mass of Car 1 measured before, and the units as kg.

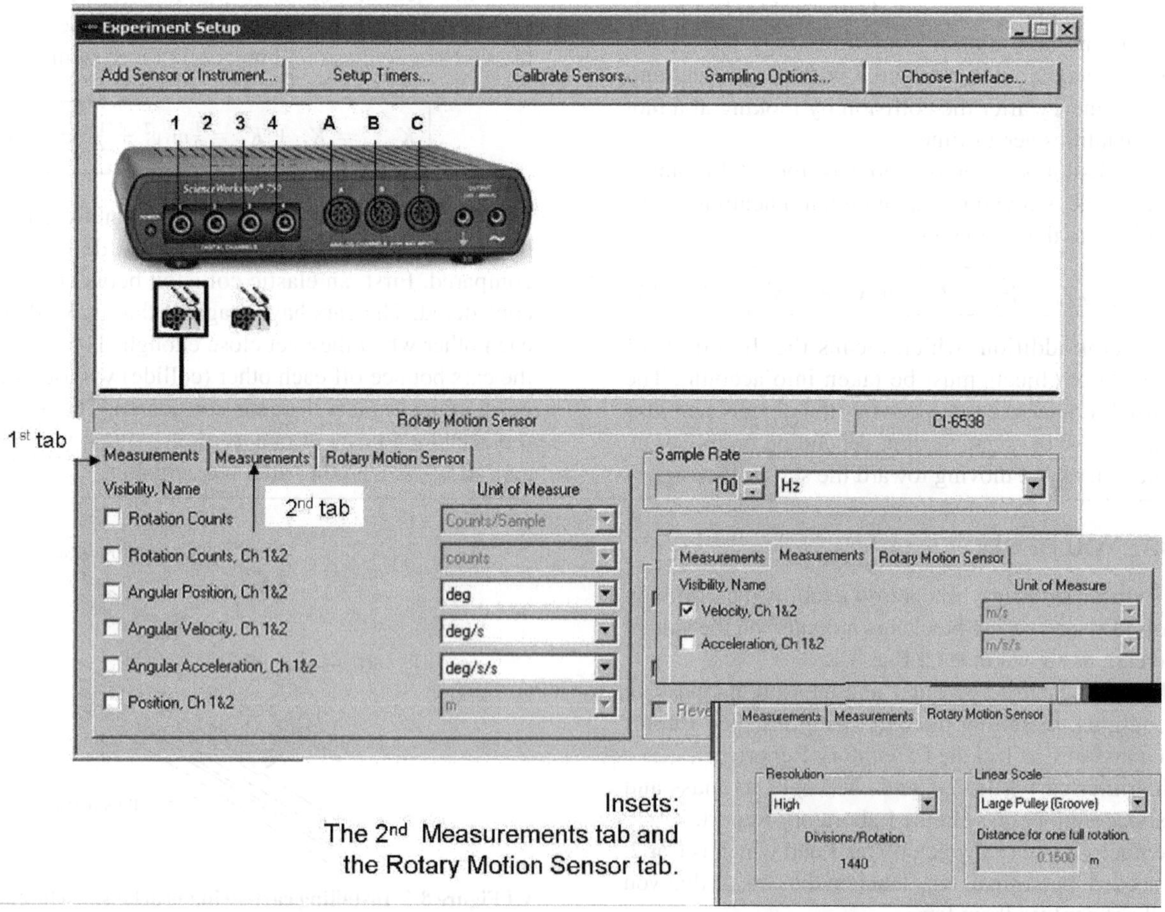

CI Figure 8.2 The Experiment Setup window. The seven available channels are numbered 1 through 4 and A, B, or C. One rotary motion sensor is connected to Channels 1 and 2, and the other is connected to Channels 3 and 4. The sensor properties are adjusted by selecting appropriate tabs. Make sure the properties of both sensors are adjusted equally. (Reprinted courtesy of PASCO Scientific.)

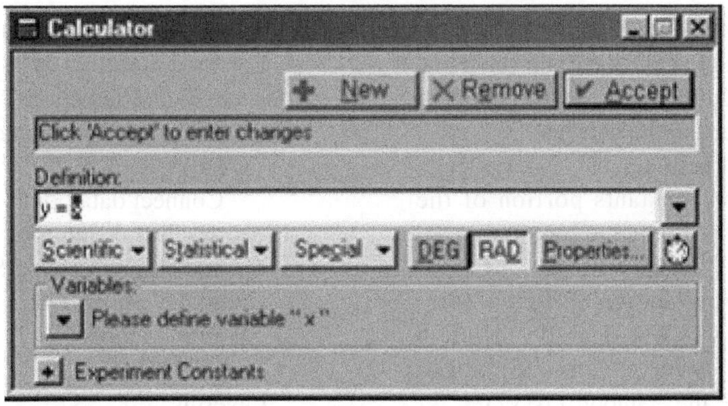

CI Figure 8.3 The calculator window. This small version of the calculator window opens when the Calculate button is pressed. The calculator will be used to enter formulas that handle the values measured by the sensor. The computer will perform the calculations automatically as the sensor takes data. (Reprinted courtesy of PASCO Scientific.)

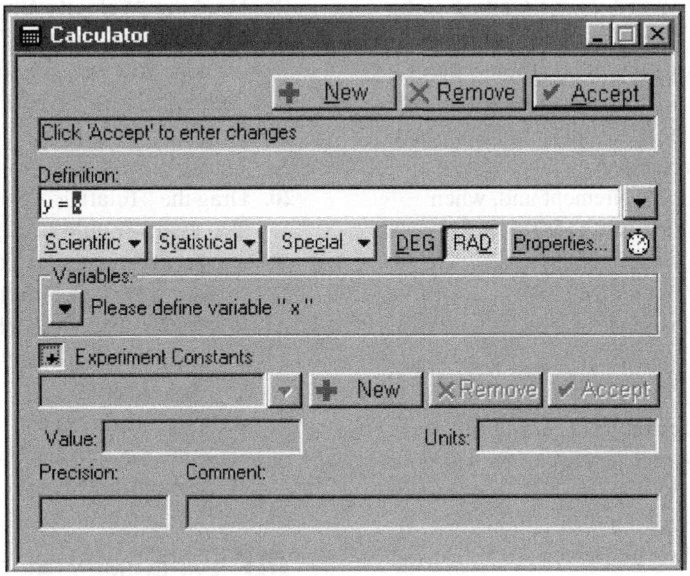

(a)

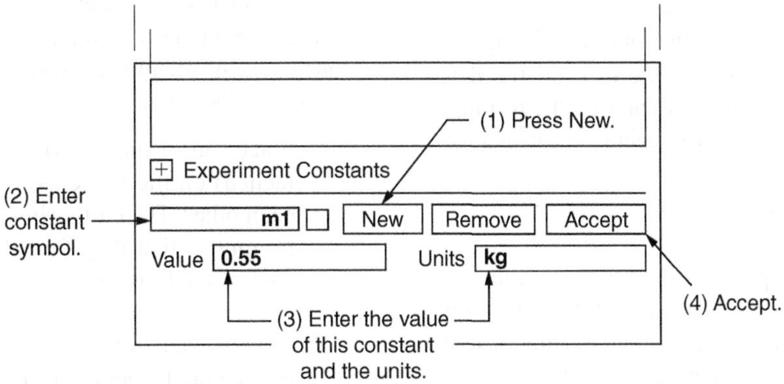

(b)

CI Figure 8.4 The expanded calculator window. (a) After the Experiment Constants button is pressed, the calculator window expands to full size. (b) The "Experiment Constants" section is the lower part of the expanded calculator. This section is used to define parameters that are to remain constant during the experiment. The diagram shows the steps needed to enter experimental constants into the calculator. (Reprinted courtesy of PASCO Scientific.)

b. Click the lower Accept button.
c. Click on the New button again and enter the name of the constant as m2, the value as the mass of Car 2 measured before, and the units as kg.
d. Click the lower Accept button.
e. Close the experiment constants portion of the calculator window by pressing the button marked "Experiment Constants" again.

12. **Calculation of the total momentum of the system:**
a. In the same calculator window, clear the definition box and enter the following equation: TotalP = m1 * smooth(10,v1) + m2 * smooth(10,v2)
This is the calculation of the total momentum, $\vec{\mathbf{P}}_{\text{Total}} = m_1\vec{\mathbf{v}}_1 + m_2\vec{\mathbf{v}}_2$, that we will call TotalP. The smooth function will help produce a cleaner graph.
b. Press the top Accept button after entering the formula. Notice that the variables m1, m2, v1, and v2 will appear in a list. The masses were already assigned values, but v1 and v2 are waiting to be defined.
c. To define variables v1 and v2, do them one at a time by clicking on the drop menu button on the left side of each variable. A list of options appears, asking what type of variable this is.
- Define v1 as a Data Measurement and, when prompted, choose Velocity(Ch1&2).
- Define v2 as a Data Measurement and, when prompted, choose Velocity(Ch3&4).

d. Press the Accept button again.

Please notice that Channels 1 and 2 will keep track of Car 1 and that Channels 3 and 4 will track Car 2. Make sure the equipment is set up accordingly.

13. **Calculation of the total kinetic energy of the system:**
a. Still in the same calculator window, press the top New button again to enter a new equation.
b. Clear the definition box and enter the following equation: TotalKE = 0.5 * m1 * smooth(10,v1)^2 + 0.5 * m2 * smooth(10, v2)^2
This is the calculation of the total kinetic energy, $K_{\text{Total}} = \frac{1}{2}m_1v_1^2 + \frac{1}{2}m_2v_2^2$, that we will call TotalKE.
c. Press the Accept button after entering the formula. Notice that the variables will again appear in a list. Define them exactly as before.
d. Press the Accept button again.

14. Close the calculator window.
15. The data list at the top left of the screen should now have four items: Velocity from Ch1&2, Velocity from Ch3&4, TotalP, and TotalKE. A small calculator icon identifies the quantities that are calculated.
16. Create a graph by dragging the "Velocity Ch1&2" icon from the data list and dropping it on the "Graph" icon on the displays list. A graph of velocity versus time will open in a window titled Graph 1.
17. Double-click anywhere on the graph. The Graph Settings window will open. Make the following changes and selections:
Under the tab Appearance:
Data:
Connect data points in bold;
deselect the buttons marked "Show Data Points" and "Show Legend Symbols"
Under the tab Layout:
Multiple graphs:
Vertical
Layering:
Do not layer
Measurement adding:
Replace matching measurement
Group measurement:
Do not group
Click OK to accept the changes and to exit the graph settings window.
18. Drag the "Velocity Ch3&4" data icon and drop it in the middle of Graph 1. The graph will split in two. At the top, you will see the Velocity Ch1&2 and at the bottom the Velocity Ch 3&4 on a separate y-axis.
19. Drag the "TotalP" icon and drop it on the split graph. The graph will split again, this time into three sections.
20. Drag the "TotalKE" icon and also drop it on the graph. The result should be a graph split into four sections, one section for each of the quantities.
21. Press the "Align Matching X Scales" button on the graph's toolbar. (It is a button with a picture of a padlock.) This will make all graphs aligned to a common $t = 0$ on the x-axis.
22. ● CI Fig. 8.5 shows what the screen should look like after all setup is complete. The size of the graph window can be maximized so that you can observe the plots better.

CI EXPERIMENTAL PROCEDURE

The complete experimental setup is shown in ● CI Fig. 8.6. Each car is connected to its own sensor and pulley system, one on each side of the track. Here are the instructions for setting up the carts.

1. Place Cars 1 and 2 (with the cart-string brackets attached) on the track with the magnetic sides facing each other. The cart-string brackets may need repositioning so that they face the outside of the track, as shown in CI Fig. 8.6.

2. Install a rotary motion sensor (RMS) on each side of the track, with the pulleys facing the inside of the track.

3. On the opposite side of the track, install the RMS/IDS adapters (small pulleys). See ● CI Fig. 8.7 for reference.

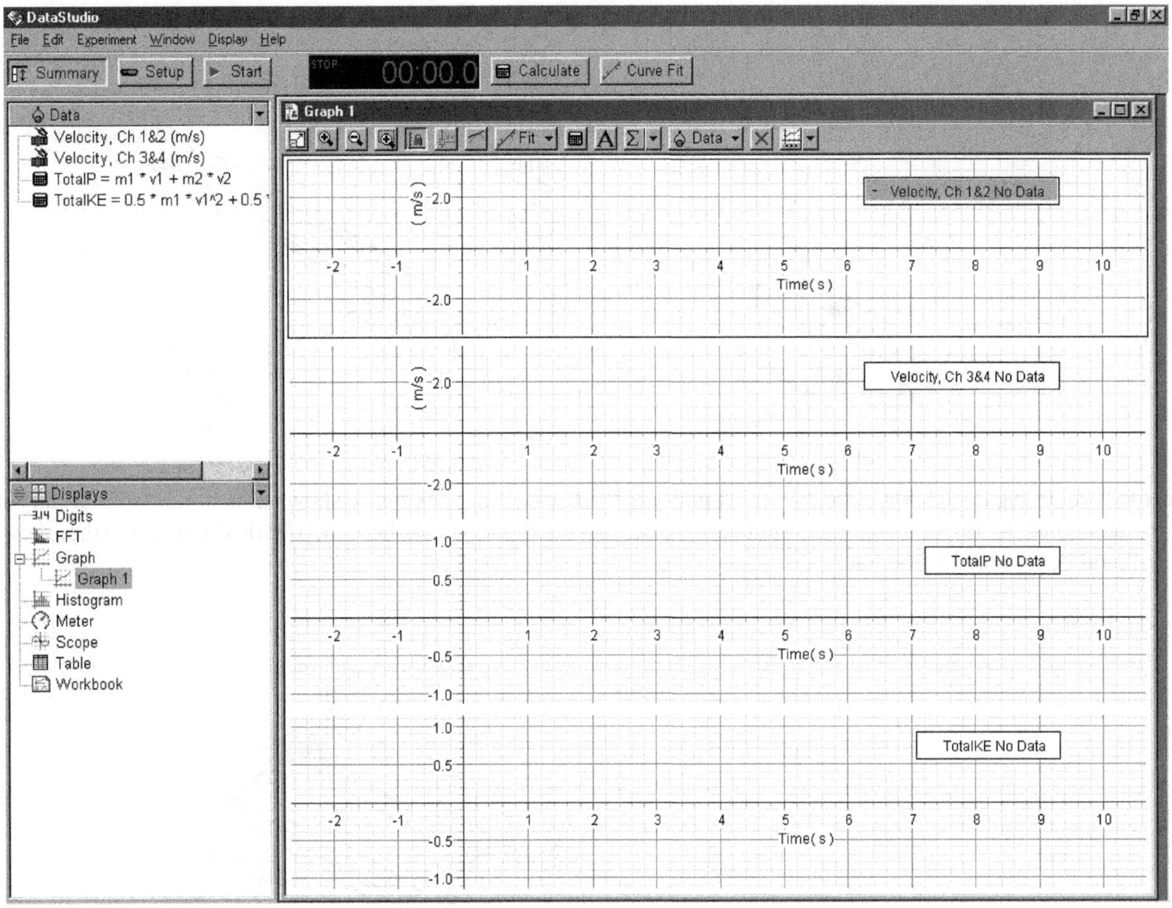

CI Figure 8.5 Data Studio setup. Data for velocity of each car, total momentum, and total kinetic energy will appear simultaneously on four plots, with matching time axes. The graph window may be maximized to occupy the whole screen in order to display the experimental results better. (Reprinted courtesy of PASCO Scientific.)

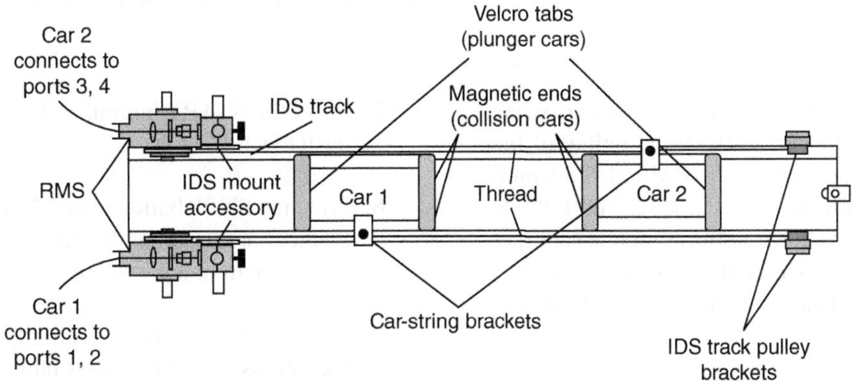

CI Figure 8.6 Experimental setup. Two collision carts are installed on the same track. Each cart is connected to its own rotary motion sensor on one side and to its own IDS-RMS adapter (track pulley bracket) on the other side. An elastic collision can be performed by having the magnetic ends of the cars face each other. An inelastic collision can be performed by having the nonmagnetic sides face each other and putting clay or Velcro on the ends of the cars.

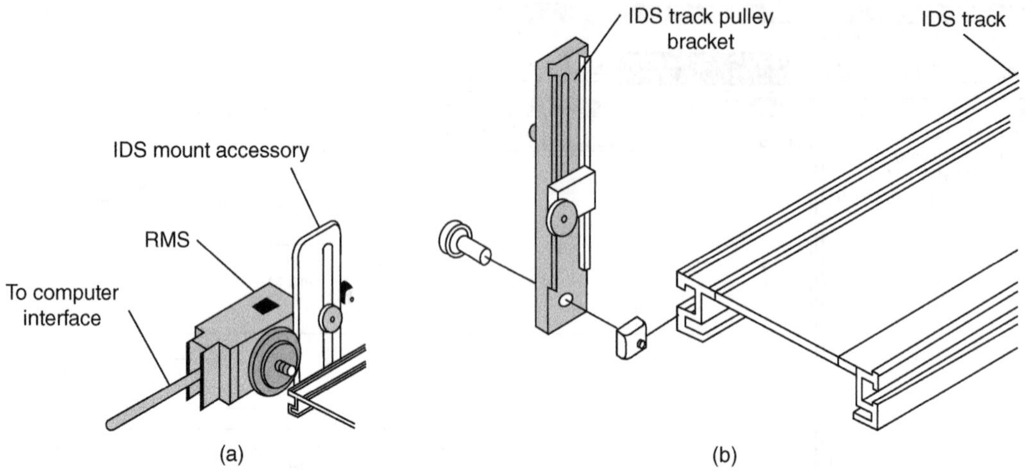

CI Figure 8.7 Mounting the RMS and the IDS track pulley to the track. (a) This figure shows how to mount the rotary motion sensor to one end of the track, using the mount accessory. (b) This figure shows how the IDS/RMS adapter (the track pulley bracket) should be mounted to the track.

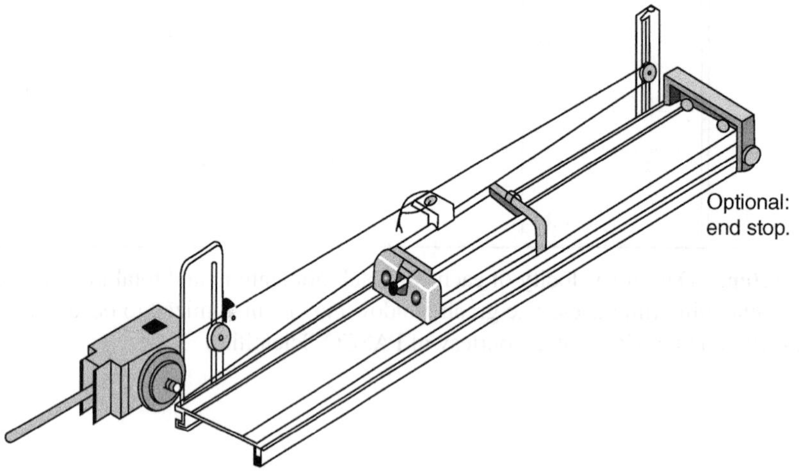

CI Figure 8.8 Example of one side of the experimental setup. This diagram illustrates one of the carts completely set up. Notice the string connecting the pulleys to the cart-bracket is to have tension but not be so tight that the cart cannot move freely.

4. A string will make a loop starting from the cart-string bracket on top of the car, to the large pulley of the RMS, to the small pulley of the RMS/IDS adapter, and back to the cart, as shown in ● CI Fig. 8.8. Do this for both cars, as shown in the complete set up of CI Fig. 8.6. Adjust the height of the pulleys so that the strings are tense, not sagging, but the cars are able to move freely.

Case 1: Elastic Collision Between Two Cars of (Nearly) Equal Mass, With One Initially at Rest

5. Set Car 2 somewhere on the middle of the track, at rest.

6. Set Car 1 all the way to the end of the track.

7. Press the START button, and then give Car 1 a good push toward Car 2.

8. Press the STOP button after the collision, before Car 2 bounces at the end of the track. (Several practice runs, and the help of a partner, may be needed.)

9. Click anywhere on the Velocity Ch1&2 graph, and then press the Scale-to-Fit button on the graph toolbar (The Scale-to-Fit button is the leftmost button on the graph toolbar.) This will make the data scale to the length of the graph on the screen. Repeat for the other three graphs.

10. If any of the graphs of velocity is reading negative values, switch the yellow and black cables of the

corresponding rotary motion sensor in the interface so that the yellow cord connects to where the black cord was, and vice versa. Repeat the data collection process, and use the new data in the rest of the analysis.

11. Print the graph. If no printer is available, make a careful drawing of the graph, paying special attention to dips and peaks in the graphs. Attach the graph to the Laboratory Report.

12. Click anywhere on the Velocity Ch1&2 graph, and then press the Smart-Tool button on the graph toolbar. (The Smart-Tool is a button on the graph toolbar labeled XY.) A set of crosshairs will appear on the graph. Repeat for each of the other graphs to get a set of crosshairs on each graph. The crosshairs can be dragged around to determine the exact (x, y) value of any point in the graphs.

13. Use the Smart-Tools to find the time t_o that corresponds to the moment just before the collision. Report the value of t_o in the Laboratory Report. (*Hint*: Use the velocity graphs and think of what the cars were doing just before the collision.)

14. In the graph printout, mark the time t_o in all graphs by drawing a single, vertical line from top to bottom of the page crossing time t_o.

15. Use the Smart-Tools to find the time t_f that corresponds to the moment just after the collision ended. Report the value of t_f in the Laboratory Report (*Hint*: The collision does not end at the same time as when it started—look carefully! Again, think of what the cars were doing right after the collision.)

16. In the graph printout, mark the time t_f in all graphs by drawing a single, vertical line from top to bottom of the page crossing time t_f. The two vertical lines now separate the before-collision from the after-collision moments.

17. Determine how long (in time) the collision lasted.

18. Use the Smart-Tool to determine, at time t_o:
- the velocity of Car 1
- the velocity of Car 2
- the total momentum of the system
- the total kinetic energy of the system

Enter the results in CI Data Table 1.

19. Use the Smart-Tool to determine, at time t_f:
- the velocity of Car 1
- the velocity of Car 2
- the total momentum of the system
- the total kinetic energy of the system

Enter the results in CI Data Table 1.

20. Calculate the change in velocity of each car, the change in momentum of each car, the change in the total momentum of the system, and the change in the total kinetic energy of the system. Enter the results in CI Data Table 1.

Case 2: Inelastic Collision Between Two Cars of (Nearly) Equal Mass, With One Initially at Rest

21. Switch the cars on the track so that their magnetic ends are facing away from each other. The easiest way to do this without altering the strings is to unscrew the cart-string brackets from the carts but not from the strings. The cars can then be switched under the brackets and the brackets reinstalled.*

22. Place a small piece of clay on the colliding end of both cars. (*Note*: Velcro strips and sticky masking tape also work well for this. Some PASCO carts already come with Velcro strips attached.)

23. Set Car 2 somewhere on the middle of the track, at rest.

24. Set Car 1 all the way to the end of the track.

25. Press the START button, and then give Car 1 a good push toward Car 2.

26. Press the STOP button after the collision, before the cars reach the end of the track and bounce. (The cars must stick together after the collision. Several practice runs, and the help of a partner, may be needed to get the hang of it.)

27. Repeat steps 10 to 20, for this set of data, but enter the results in CI Data Table 2.

*Some PASCO carts have magnets on both ends. These won't work. A new set of carts with no magnets (plunger carts) will be needed, which means new masses must be measured and entered in the calculator, if this is the case.

Name ______________________ Section ______________ Date ______________

Lab Partner(s) ____

CI EXPERIMENT 8

Conservation of Linear Momentum

CI Laboratory Report

Case 1: Elastic Collision Between Two Cars of (Nearly) Equal Mass, With One Initially at Rest

Car 1: $m_1 =$ ______________

Car 2: $m_2 =$ ______________

Collision started at $t_o =$ ______________ Total collision time $\Delta t = t_f - t_o =$ ______________

Collision ended at $t_f =$ ______________

CI DATA TABLE 1

Purpose: To analyze an elastic collision between two objects of nearly identical mass.

	Just before the collision	Just after the collision	Changes	
Velocity of Car 1, $\vec{\mathbf{v}}_1$			$\Delta\vec{\mathbf{v}}_1$	
Velocity of Car 2, $\vec{\mathbf{v}}_2$			$\Delta\vec{\mathbf{v}}_2$	
Total momentum, $\vec{\mathbf{P}}_{Total}$			$\Delta\vec{\mathbf{P}}$	
Total kinetic energy, K_{Total}			ΔK	

Don't forget units

(continued)

CASE 2: INELASTIC COLLISION BETWEEN TWO CARS OF (NEARLY) EQUAL MASS, WITH ONE INITIALLY AT REST

Car 1: $m_1 =$ ____________________

Car 2: $m_2 =$ ____________________

Collision started at $t_o =$ ____________________ Total collision time $\Delta t = t_f - t_o =$ ____________________

Collision ended at $t_f =$ ____________________

CI DATA TABLE 2

Purpose: To analyze an inelastic collision between two objects of nearly identical mass.

	Just before the collision	Just after the collision	Changes	
Velocity of Car 1, $\vec{\mathbf{v}}_1$			$\Delta\vec{\mathbf{v}}_1$	
Velocity of Car 2, $\vec{\mathbf{v}}_2$			$\Delta\vec{\mathbf{v}}_2$	
Total momentum, $\vec{\mathbf{P}}_{\text{Total}}$			$\Delta\vec{\mathbf{P}}$	
Total kinetic energy, K_{Total}			ΔK	

Name ______________________ Section ______________ Date ______________

Lab Partner(s) ____

EXPERIMENT 8 **Conservation of Linear Momentum** *Laboratory Report*

CI QUESTIONS

1. How well do the results support the law of conservation of momentum, considering the possible sources of uncertainty?

2. Which collision took a longer time, the elastic or the inelastic collision? Discuss the possible reasons.

3. Was the kinetic energy of the system conserved? Discuss by comparing the results for the elastic collision and the inelastic collision.

4. During the inelastic collision, the kinetic energy was obviously not conserved. What do you think happened to the "lost" energy?

(continued)

5. During the collisions, both cars changed their momentum. Discuss for both collisions: How does the change in momentum of each car compare to that of the other? Does one car change more than the other? What do you think would happen if the cars had different mass? (If time is available, try it.)

6. For an object to undergo a change in its momentum, a net force needs to be applied. The amount of change in momentum produced by the force depends on the length of the time during which the force acts and is called the *impulse*. That is,

$$\text{Impulse} = \Delta\vec{\mathbf{p}} = \vec{\mathbf{F}}\Delta t$$

where the force $\vec{\mathbf{F}}$ is assumed to be constant, or to be an "average force." For each of the collisions, calculate the average force acting on the cars during the collision, and compare them.

7. Suppose a ball falls on your head. What is better for you (less damage)—for the ball to bounce straight back off your head, or for it to stop and stick to you? Justify your answer.

Name ______________________ Section ______________ Date ______________

Lab Partner(s) ____

EXPERIMENT 9

Projectile Motion: The Ballistic Pendulum

GL Experimental Planning

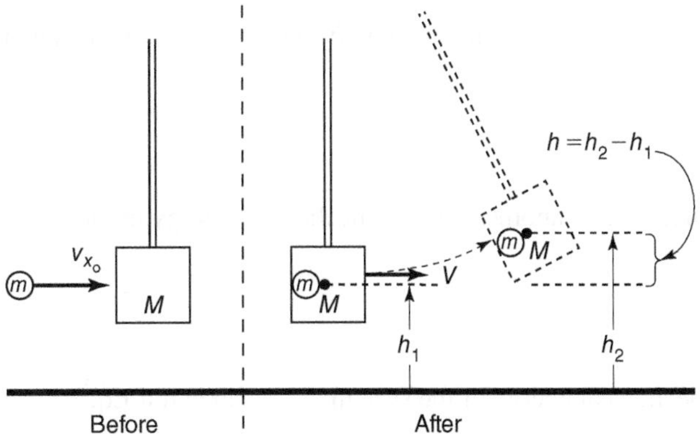

GL Figure 9.1 Ballistic pendulum parameters. See Experimental Planning text for description.

A. The Ballistic Pendulum

The ballistic pendulum allows the experimental determination of the velocity of a projectile that is launched horizontally. This is done using two conservation principles and a few simple measurements.

The parameters of a ballistic pendulum system are shown in GL Fig. 9.1. A projectile of mass *m* is fired with velocity v_o into a stationary pendulum bob of mass *M* and becomes embedded. The (horizontal) momentum of the system can be expressed in terms of the variables given in GL Fig. 9.1. What is the momentum of the system immediately after the projectile is fired (that is, just before it hits the pendulum bob)?

1. In terms of the variables given in GL Fig. 9.1, what is the momentum of the system immediately after the mass *m* becomes embedded in the pendulum bob?

2. If the horizontal momentum is considered to be conserved in the collision, what can you say about the two expressions for momentum that you determined above?

3. Write an equation for the conservation of momentum for this collision. Designate it Eq. 1.

(continued)

4. Verify that your equation has the masses and the velocities before and after the collision. If not, review your result with a classmate or your instructor. Solve the equation for the initial velocity of the projectile, v_o.

Note that to calculate the initial projectile velocity v_o, the velocity V of the block and projectile combination needs to be known. (The values of the masses can be determined with a balance.) So far, only one conservation principle has been used—the conservation of linear momentum. Now consider the mechanical energy of the system *after* the collision. Write an expression for the kinetic energy of the system (the mass and bob combo) immediately after collision, and label it Eq. 2.

As the bob swings upward from h_1 to a maximum height h_2 (GL Fig. 9.1), what is happening to the kinetic energy of the system (neglecting friction)?

If the kinetic energy is decreasing, is there another form of mechanical energy in the system that may be increasing? If so, what is it?

Write an equation for the mechanical energy of the system at h_2, and call it Eq. 3.

5. How are Eqs. 2 and 3 related by the conservation of mechanical energy?

6. If you apply the conservation of mechanical energy for the system after the collision, the expressions in Eqs. 2 and 3 are equal. Set them equal to each other, and call the resulting equation Eq. 4.

7. What is the only variable in this equation that cannot be measured directly? You should recognize that it is the velocity V. (h_1 and h_2 can be measured directly with a meterstick.) Solve Eq. 4 for V.

8. Recall that the velocity V needed to be determined to find v_o in Eq. 1. Your last result gives V in terms of measurable quantities. Substitute your expression for V into Eq. 1, and solve for v_o.

9. Is v_o now expressed in terms of known and measurable quantities? Explain. It should be, and this is the theory of how the projectile initial velocity can be determined experimentally using the ballistic pendulum.

Name ______________________ Section ______________ Date ______________

Lab Partner(s) ____

TI Advance Study Assignment

Read the experiment and answer the following questions.

A. The Ballistic Pendulum

1. In determining the magnitude of the initial velocity of the ballistic pendulum projectile, what conservation laws are involved and in which parts of the procedure?

2. Why is it justified to say that the momentum in the horizontal direction is conserved over the collision interval? Is momentum conserved before and after the collision? Explain.

3. Why are the heights measured to the center of mass of the pendulum-ball system?

(continued)

B. Determination of the Initial Velocity of a Projectile from Range-Fall Measurements

4. After the horizontal projectile leaves the gun, what are the accelerations in the x- and y-directions?

5. How is the location where the ball strikes the floor determined?

6. Besides the range, what else is needed to determine the magnitude of the initial velocity of the ball?

C. Projectile Range Dependence on the Angle of Projection

7. For a given initial velocity, how does the range of a projectile vary with the angle of projection θ?

8. Theoretically, the angle of projection for maximum range is 45°. Does this set a limit on the range? Explain.

PRE-LAB DEMONSTRATION | Pendulum and Peg

A pendulum-peg setup is shown in ● Fig. D5. If the peg were not there, the pendulum bob would rise to the same height on the opposite side of the swing. But, what happens if the string hits the peg?

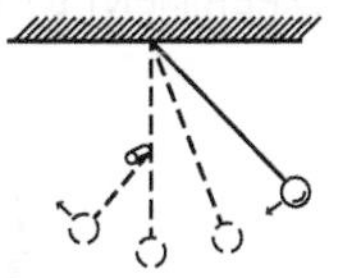

Figure D5 Pendulum-Peg Setup

EXPERIMENT 9

Projectile Motion: The Ballistic Pendulum

INTRODUCTION AND OBJECTIVES

Projectile motion is the motion of an object in a plane (two dimensions) under the influence only of gravity (free fall, air resistance neglected). The kinematic equations of motion describe the components of such motion and may be used to analyze projectile motion. In most textbook cases, the initial velocity of a projectile (speed and angle of projection) is given and other quantities calculated.

However, in this experiment, the unknown initial velocity will be determined from experimental measurements. This will be done (1) through the use of the ballistic pendulum and (2) from range-fall distance measurements. The dependence of the projectile range on the angle of projection will also be investigated so as to obtain an experimental indication of the angle of projection that gives the maximum range.

These procedures will greatly assist in understanding some of the most basic physical principles. After performing the experiment and analyzing the data, you should be able to do the following:

1. Explain the use of conservation laws (linear momentum and mechanical energy) in determining the initial velocity of a projectile using the ballistic pendulum.
2. Describe the components of motion and how they are used in determining the velocity of a projectile with range-fall measurements.
3. Tell how the range of a projectile varies with the angle of projection.

EQUIPMENT NEEDED

- Ballistic pendulum
- Sheets of plain paper (and carbon paper)*
- Meterstick
- Protractor
- Laboratory balance
- Masking tape
- Wooden blocks
- 1 sheet of Cartesian graph paper
- Safety glasses

*Carbon paper may or may not be needed.

THEORY

A. The Ballistic Pendulum

Types of ballistic pendula apparatus are shown in ● Fig 9.1. The ballistic pendulum is used to experimentally determine the initial velocity of a horizontally projected object (a metal ball) fired from a spring gun. The projectile is fired into a stationary, pendulum bob suspended by a rod, and on collision, the pendulum and the embedded projectile swing upward.

A catch mechanism stops the pendulum at its highest position of swing. By measuring the vertical distance that the center of mass of the pendulum-ball system rises, the initial velocity of the projectile can be computed through the use of the conservation of linear momentum and the conservation of mechanical energy (neglecting rotational considerations).

Consider the schematic diagram of a ballistic pendulum shown in ● Fig. 9.2. A projectile of mass m with an initial horizontal velocity of v_{x_o} is fired into and becomes embedded in a stationary pendulum of mass M.

To a good approximation, the horizontal momentum is conserved during collision over the time interval of the collision. Therefore, the horizontal component of total momentum is taken to be the same immediately before and after collision. The velocity of the pendulum bob is initially zero, and the combined system ($m + M$) has a velocity of magnitude V just after collision. Hence, by the conservation of linear momentum for the horizontal direction,

$$\underset{\text{(before)}}{mv_{x_o}} = \underset{\text{(after)}}{(m + M)V} \tag{9.1}$$

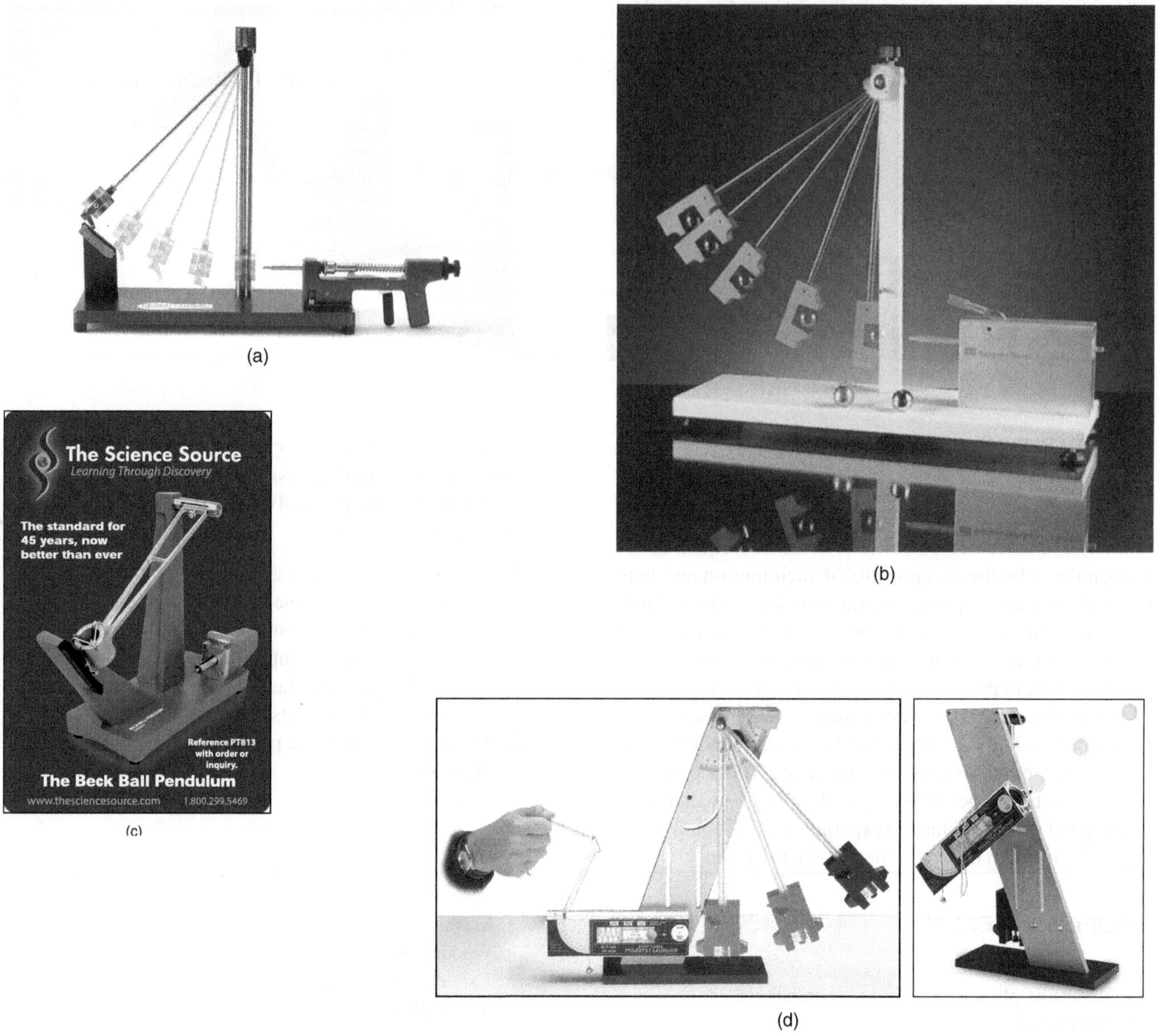

Figure 9.1 Ballistic pendula. Types of ballistic pendula. (Photos Courtesy of (a) and (b) Sargent-Welch, (c) Daedalon, The Science Source Company, and (d) Reprinted courtesy of PASCO Scientific. Note that the ballistic launcher may be mounted on the other side of the stand for projectile range.

After the collision, the pendulum with the embedded projectile swings upward (momentum of the system no longer conserved, why?) and stops. The center of mass of the pendulum-ball system is raised a maximum vertical distance $h = h_2 - h_1$.* By the conservation of mechanical energy, the increase in potential energy is equal to the kinetic energy of the system just after collision (friction of the support is considered negligible). Hence,

$$\boxed{\tfrac{1}{2}(m + M)V^2 \quad = \quad (m + M)gh} \tag{9.2}$$

kinetic energy just after collision = change in potential energy

Solving Eq. (9.2) for V,

$$\boxed{V = \sqrt{2gh}} \tag{9.3}$$

*The center of mass of the system is used because this represents the point where all the mass is considered concentrated.

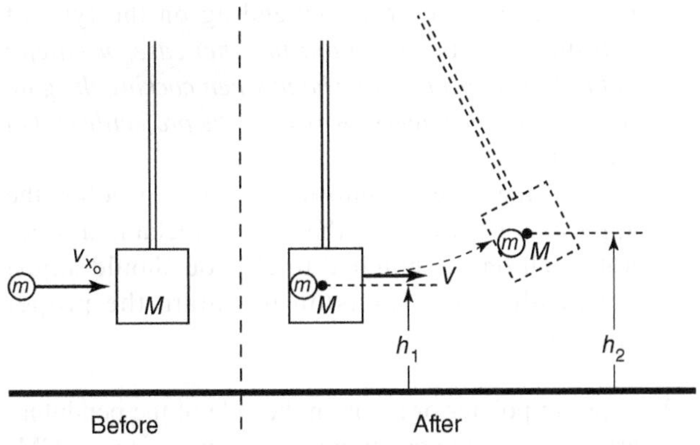

Figure 9.2 Ballistic pendulum action. Ideally, the horizontal linear momentum is conserved during collision. After collision, work is done against gravity, and kinetic energy is converted into potential energy. (Rotational considerations neglected.)

Substituting this expression into Eq. (9.1) and solving for v_{x_o} yields

$$\boxed{v_{x_o} = \left(\frac{m + M}{m}\right)\sqrt{2gh}} \qquad \textbf{(9.4)}$$

(initial speed)

Hence, by measuring *m*, *M*, and *h*, the magnitude of the initial velocity of the projectile can be computed.

B. Determination of the Initial Speed of a Horizontal Projectile from Range-Fall Measurements

If a projectile is projected horizontally with an initial velocity of magnitude v_{x_o} from a height of *y*, it will describe an arc as illustrated in ● Fig. 9.3. The projectile will travel a horizontal distance *x* (called the *range*) while falling a vertical distance *y*.

The initial vertical velocity is zero, $v_{y_o} = 0$, and the acceleration in the *y*-direction is the acceleration due to gravity ($a_y = g$). There is no horizontal acceleration, $a_x = 0$; hence the components of the motion are described by

$$\boxed{x = v_{x_o}t} \qquad \textbf{(9.5)}$$

and

$$\boxed{-y = -\tfrac{1}{2}gt^2} \qquad \textbf{(9.6)}$$

Eliminating *t* from these equations and solving for v_{x_o}, we have (neglecting air resistance):

$$\boxed{v_{x_o} = \sqrt{\frac{gx^2}{2y}} = \left(\frac{g}{2y}\right)^{\frac{1}{2}}x} \qquad \textbf{(9.7)}$$

Hence, by measuring the range *x* and the distance of fall *y*, the initial speed of the projectile can be computed.

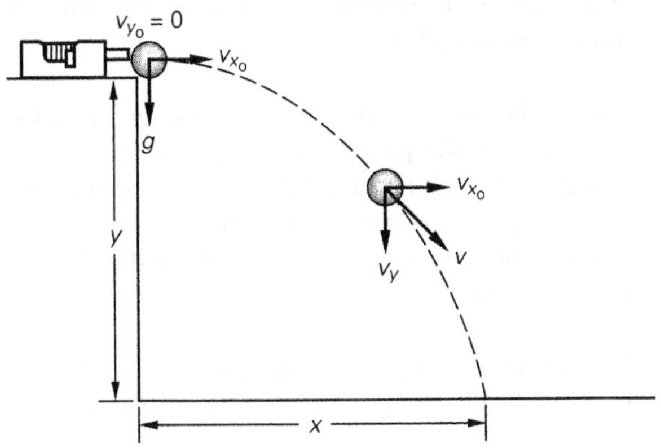

Figure 9.3 Range-fall. The configuration for range-fall measurements. See text for description.

C. Projectile Range Dependence on the Angle of Projection

The projectile path for a general angle of projection θ is shown in ● Fig. 9.4. The components of the initial velocity have magnitudes of

$$\boxed{\begin{aligned} v_{x_o} &= v_o\cos\theta \\ v_{y_o} &= v_o\sin\theta \end{aligned}} \qquad \textbf{(9.8)}$$

At the top of the arc path, $v_y = 0$, and since

$$\begin{aligned} v_y &= v_{y_o} - gt \\ &= v_o\sin\theta - gt \end{aligned}$$

(downward taken as negative), then

$$v_o\sin\theta - gt_m = 0$$

or

$$\boxed{t_m = \frac{v_o\sin\theta}{g}} \qquad \textbf{(9.9)}$$

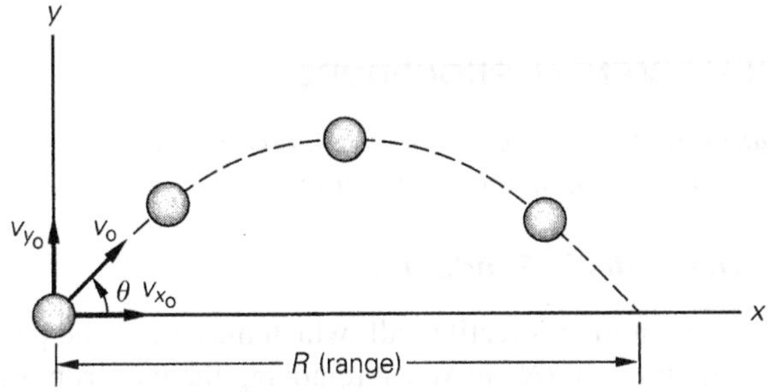

Figure 9.4 Projectile motion. For an arbitrary projection angle above the horizontal, the range *R* of a projectile depends on the initial velocity—that is, on the speed and angle of projection. (*Note*: If the PASCO launcher is used with elevated launch, place a box or horizontal surface at the landing location so the ball will land on the box at the same level height as the muzzle of the launcher.)

where t_m is the time for the projectile to reach the maximum height of y_m.

If the projectile returns to the same elevation as that from which it was fired, then the total time of flight t is

$$t = 2t_m = \frac{2v_o \sin\theta}{g} \qquad (9.10)$$

During the time t, the projectile travels a distance R (range) in the x-direction:

$$R = v_{x_o} t = (v_o \cos\theta)t = \frac{2v_o^2 \sin\theta\cos\theta}{g}$$

where t is from Eq. 9.10.

But using the trigonometric identity $2\sin\theta\cos\theta = \sin 2\theta$, we find that the **range** or maximum distance in the x-direction is

$$R = \frac{v_o^2 \sin 2\theta}{g} \qquad (9.11)$$

(range)

From Eq. 9.11, it can be seen that the range of the projectile depends on the angle of projection θ. The maximum range R_{max} occurs when $\sin 2\theta = 1$. Since $\sin 90° = 1$, by comparison

$$2\theta = 90° \quad \text{or} \quad \theta = 45°$$

Hence, a projectile has a maximum range for $\theta = 45°$, and

$$R_{max} = \frac{v_o^2}{g} \qquad (9.12)$$

(maximum range, $\theta = 45°$)

which provides another convenient method to determine experimentally the initial speed of a projectile.

(*Note*: This development neglects air resistance, but the equations give the range to a good approximation for relatively small initial speeds and short projectile paths. Why?)

EXPERIMENTAL PROCEDURE

Caution: *With projectiles involved, it is recommended that safety glasses be worn during all procedures.*

A. The Ballistic Pendulum

1. Obtain the projectile ball, which may be in the pendulum arm. (*Note*: When removing the ball from the pendulum arm of some types of ballistic pendula, be sure to push up on the spring catch that holds the ball in the pendulum so as not to damage it.)

   Place the projectile ball on the ball rod of the spring gun, and cock the gun by pushing on the ball. Both ball and rod may move backward, or the ball may slip over the rod, depending on the type of ballistic pendulum. ***Caution:*** *In either case, be careful not to bruise or hurt your hand when cocking the gun. Also, keep your fingers away from the projectile end of the gun.**

   Fire the projectile into the pendulum to see how the apparatus operates. If the catch mechanism does not catch on the notched track, you should adjust the pendulum suspension to obtain the proper alignment.

2. A preset pointer or a dot on the side of the pendulum arm indicates the position of the center of mass (CM) of the pendulum-ball system. With the pendulum hanging freely, measure the height h_1 of the pointer above the base surface (Fig. 9.2) and record it in Data Table 1.

3. Shoot the ball into the freely hanging stationary pendulum, and note the notch at which the catch mechanism stops on the curved track. Counting upward on the curved track, record the notch number in Data Table 1. Repeat this procedure four times, and for each trial record the notch number in the data table. Alternatively, the height may be measured each time. See Procedure 4 note. (For the PASCO pendulum, the angle of arm swing is measured by an angle indicator. Record this angle. The height of the swing is then $h = l - l\cos\theta$, where l is the length from the arm pivot to the center of mass.)

4. Determine the average of these observations, which is the average notch position of the pendulum. Place the catch mechanism in the notch corresponding most closely to the average, and measure the height h_2 of the CM dot above the base surface used for the h_1 measurement (Fig. 9.2).

   Note: To minimize frictional losses, the catch mechanism may be disabled by tying it up with thread or using a rubber band. The mechanism then acts as a pointer to indicate the highest notch, which is observed by a lab partner. Holding some reference object, such as a pencil, by the notched track helps to determine the proper notch number.

5. Loosen the screw of the pendulum support and carefully remove the pendulum. Weigh and record the masses of the ball (m) and the pendulum arm (M). Do not attempt to remove the support rod from the bob. Consult your instructor for the procedure if a different model is used.

6. From the data, compute the magnitude of the initial velocity using Eq. (9.4) ($g = 9.80\ \text{m/s}^2 = 980\ \text{cm/s}^2$).

* The PASCO pendulum comes with a ramrod for loading the ball.

B. Determination of the Initial Velocity of a Projectile from Range-Fall Measurements

7. With the pendulum removed or in the upper catch mechanism notch so as not to interfere with the projectile, position the apparatus near one edge of the laboratory table as shown in Fig. 9.3.

   Shoot the ball from the gun, and note where the ball strikes the floor. (The range of the ball is appreciable, so you may have to shoot the ball down an aisle. Be careful not to hit anyone with the ball, particularly the instructor.)

8. Place a sheet of paper where the ball hits the floor. Tape the paper to the floor (or weight it down) so that it will not move. When the ball strikes the paper, the indentation mark will enable you to determine the range of the projectile.* Also mark the position of the apparatus on the table (for example, using a piece of tape as a reference). It is important that the gun be fired from the same position each time.

9. Shoot the ball five times, hitting the paper, and measure the horizontal distance or range x the ball travels for each trial (see Fig. 9.3). [If faint indentation marks cannot be found on the paper, cover it with a sheet of carbon paper (carbon side down). The ball will then make a carbon mark on the paper on impact.]

   Record the measurements in Data Table 2, and find the average range. The height y is measured from the bottom of the ball (as it rests on the gun) to the floor. Measure this distance, and record in the data table.

10. Using Eq. (9.7), compute the magnitude of the initial velocity of the ball ($g = 9.80\ \text{m/s}^2 = 980\ \text{cm/s}^2$). Compare this to the velocity determined in Part A, and compute the percent difference.

* The range will be measured from the floor position directly below the center of the ball just as it leaves the gun to the marks on the paper on the floor. The floor location is determined by putting the ball on the gun without loading the spring.

C. Dependence of Projectile Range on the Angle of Projection

11. With the ballistic pendulum apparatus on the floor (with pendulum removed), elevate the front end so that it can be fired at an angle θ relative to the horizontal. (Your instructor will tell you how to do this.) Aim the projectile down an aisle or hallway, *being careful not to aim at anything or anybody.*

12. Using a protractor to set the angles of projection, fire the projectile at angles of 20°, 30°, 40°, 45°, 50°, 60°, and 70° with two or three trials for each angle. The projectile should be aimed so that it lands as close as possible to the same spot for the trials of a particular angle.

    Station one or more lab partners at a safe distance near where the projectile strikes the floor. They are to judge the average range of the two or three trials. Measure the average range for each angle of projection, and record the data in Data Table 3.

    Suggestion: It is convenient to measure the distance from the gun to the position where the ball lands and to mark this position. The range measurement then can be made relative to this measured mark, instead of from the starting point each time. Also, it is convenient to shoot toward a wall at the end of the hall or aisle or to lay a meterstick on the floor perpendicularly to the line of flight in order to stop the ball from rolling.

13. Plot the range versus the angle of projection, and draw a smooth curve that fits the data best. As might be expected, the points may be scattered widely because of the rather crude experimental procedure. Even so, you should be able to obtain a good idea of the angle for the maximum range. Determine this angle from the graph, and record it in Data Table 3.

Name ______________________ Section ______________ Date ______________

Lab Partner(s) ____

EXPERIMENT 9

Projectile Motion: The Ballistic Pendulum

TI Laboratory Report

A. The Ballistic Pendulum

DATA TABLE 1

(Modify the Data Table if it does not apply to your ballistic pendulum.)

Purpose: To determine the magnitude of initial projectile velocity.

Trials	Notch number of pendulum catch (or angle)	Height h_2 of pointer with pendulum catch in closest-to-average notch number ______________
1		Height h_1 of pointer with pendulum freely suspended
2		______________
3		$h = h_2 - h_3$ ______________ (or $h = l - l\cos\theta$)
4		Mass of ball m
5		______________
		Mass of pendulum M (bob and support)
Average		______________

Calculations
(show work)

Computed v_{x_o} ______________
(units)

Don't forget units

(continued)

B. Determination of the Initial Velocity of a Projectile from Range-Fall Measurements

DATA TABLE 2

Purpose: To determine the magnitude of initial projectile velocity.

Trial	Range
1	
2	
3	
4	
5	
Average	

Vertical distance of fall, y ______________

Computed v_{x_o} ______________
(units)

Percent difference between results of Parts A and B ______________

Calculations (show work)

C. Dependence of Projectile Range on the Angle of Projection

DATA TABLE 3

Purpose: To investigate projection angle from maximum range.

Angle of projection	Average range
20°	
30°	
40°	
45°	
50°	
60°	
70°	

Angle of projection for maximum range from graph ______________

Name ____________ Section ____________ Date ____________

Lab Partner(s) ____________

EXPERIMENT 9 Projectile Motion: The Ballistic Pendulum — *Laboratory Report*

TI QUESTIONS

A. The Ballistic Pendulum

1. Is the collision between the ball and the pendulum elastic or inelastic? Justify your answer by calculating the kinetic energy of the system before collision using the value of v_{x_o} in the experiment and the kinetic energy just after collision using the experimental value of h in Eq. 9.2.

2. Using the results of Question 1 that would apply if the collision were inelastic, find the fractional kinetic energy loss during the collision. Express the "loss" as a percent. What became of the "lost energy"?

3. Expressing the kinetic energy in terms of momentum ($K = \frac{1}{2}mv^2 = p^2/2m$), prove using symbols, not numbers, that the fractional loss during the collision is equal to $M/(m + M)$.

4. Compute the fractional energy loss from the experimental mass values using the equation developed in Question 3, and compare this to the result in Question 2. Explain the difference, if any.

(continued)

5. Is the friction of the pendulum (catch mechanism, support axis, etc.) a random or systematic error? Will this source of error cause your calculated velocity to be less than or greater than the actual velocity? (See Experiment 2 on errors.)

B. Determination of the Initial Velocity of a Projectile from Range-Fall Measurements

6. What effect does the force of gravity have on the horizontal velocity of the projectile? Explain.

7. What effect would air resistance have on the range of the projectile?

C. Dependence of Projectile Range on the Angle of Projection

8. Using experimental data, compute the magnitude of the initial velocity v_o of the projectile from Eq. (9.12), and compare this to the results of Parts A and B of the procedure.

9. If, for a given initial velocity, the maximum range is at a projection angle of 45°, then there must be equal ranges for angles above and below this. Show this explicitly.

Name ______________________ Section ______________ Date ______________

Lab Partner(s) ____

EXPERIMENT 10

Centripetal Force

TI Advance Study Assignment

Read the experiment and answer the following questions.

1. Define centripetal force.

2. What supplies the centripetal force for (a) a satellite in orbit around the Earth and (b) the mass in uniform circular motion in this experiment?

3. An object moving in *uniform* circular motion is accelerating. How can this be, since uniform motion implies constant motion?

4. For an object in uniform circular motion, on what parameters does the experimental determination of the centripetal force depend when using $F = ma$?

(continued)

5. If the centripetal force acting on an object in uniform motion suddenly ceased to act (went to zero), what would happen to the object? That is, what would be its subsequent motion?

6. Suppose that the centripetal force acting on an object in circular motion were increased to a new value, and the object remained in a circular path with the same radius. How would the motion be affected?

7. Explain how the centripetal force is directly determined for the apparatus you will be using in the experiment.

PRE-LAB DEMONSTRATION | Bucket Swing (what keeps the water in the bucket?)

A person swings a bucket of water in a vertical circle. If done properly, the water stays in the bucket at the top of the swing. Why is this? (If not done properly, the person gets wet.)

EXPERIMENT 10

Centripetal Force

INTRODUCTION AND OBJECTIVES

The Earth revolves about the Sun; atomic electrons move around the nucleus. What keeps these objects in orbit? The answer is **centripetal force** (centripetal means "center-seeking"). The centripetal force is supplied by gravitational and electrical interactions, respectively, for each of these cases.

The study of centripetal force in the laboratory is simplified by considering objects in uniform circular motion. An object in uniform circular motion moves with a constant speed (a scalar) but has a changing velocity (a vector) because of the continual change in direction. This change in velocity results from centripetal acceleration due to a centripetal force.

In the experimental situation(s) of this experiment, the centripetal force will be supplied by a spring and can be readily measured. However, the magnitude of the centripetal force can also be determined from other experimental parameters, for example, the frequency of rotation of the object, mass, and radius of orbit. Centripetal force will be experimentally investigated by measuring these parameters and comparing the calculated results with the direct measurement of the spring force, which mechanically supplies the center-seeking centripetal force.

After performing the experiment and analyzing the data, you should be able to do the following:

1. Explain why a centripetal force is necessary for circular motion.
2. Describe how the magnitude of the centripetal force for uniform circular motion may be determined from motional parameters.
3. Summarize what determines the magnitude of the centripetal force necessary to keep an object in uniform circular motion.

EQUIPMENT NEEDED

Manual Centripetal Force Apparatus

- Laboratory timer or stopwatch
- Meterstick
- Weight hanger and slotted weights
- String
- Laboratory balance
- Safety glasses

THEORY

An object in uniform circular motion requires a centripetal, or center-seeking, force to "hold" it in orbit. For example, when one swings a ball on a rope in a horizontal circle around one's head (● Fig. 10.1), the centripetal force, $F_c = ma_c$, is supplied by the person and transmitted to the ball through the rope. In the absence of the centripetal force (for example, if the rope breaks or if the person releases the rope), the ball would no longer be held in orbit and would initially fly off in the direction of its tangential velocity v.

An object in uniform circular motion moves with a constant speed. Even though the object's speed is constant, its velocity is changing because the direction of the motion is continually changing. This change in velocity results from a centripetal acceleration a_c that is due to the applied centripetal force F_c. The direction of the acceleration (and force) is always toward the center of the object's circular path, and it can be shown (see your textbook) that the magnitude of the acceleration is given by

$$a_c = \frac{v^2}{r} \tag{10.1}$$

(centripetal acceleration)

where v is the tangential or orbital speed of the object and r is the radius of the circular orbit. By Newton's second law, $F = ma$, the magnitude of the centripetal force is

$$F_c = ma_c = \frac{mv^2}{r} \tag{10.2}$$

(centripetal force)

Figure 10.1 Centripetal acceleration. An object in uniform circular motion must have a centripetal acceleration with a magnitude of $a_c = v^2/r$ directed toward the center of the circular path. In the case of swinging a ball on a rope around one's head, the centripetal force $F_c = ma_c$ is supplied by the person and transmitted through the rope. (Tony Freeman/PhotoEdit.)

where m is the mass of the object. In terms of distance and time, the orbital speed v is given by $v = 2\pi r/T$, where $2\pi r$ is the circumference of the circular orbit of radius r, and T is the period.

Notice that Eq. 10.2 describes the centripetal force acting on an object in uniform circular motion in terms of the properties of the motion and orbit. It is equal to the expression of a physical force that actually supplies the centripetal action. For example, in the case of a satellite in uniform circular motion around the Earth, the centripetal force is supplied by gravity, which is generally expressed as $F_g = Gm_1m_2/r^2$, and $F_c = F_g$. Similarly, for an object being held in uniform circular motion by the tension force of a string, the tension force (F_t) is equal to Eq. 10.2 (that is, $F_t = mv^2/r$).*

The centripetal force given by Eq. 10.2 can also be expressed in terms of the angular speed ω or frequency f of rotation, using the expressions $v = r\omega$ and $\omega = 2\pi f$:

$$F_c = \frac{mv^2}{r} = \frac{m(r\omega)^2}{r} = mr\omega^2$$

and

$$\boxed{F_c = mr(2\pi f)^2 = 4\pi^2 mrf^2} \qquad (10.3)$$

*Technically, it is the component of F_t directed toward the center of the circular orbit. The rope cannot be exactly horizontal. See Question 4 at the end of the experiment.

where ω is in radians per second and f is in hertz (Hz, 1/s or cycles per second). In this experiment, it is convenient to think of f as being in revolutions per second.

EXPERIMENTAL PROCEDURE

Manual Centripetal Force Apparatus

1. A type of hand-operated centripetal force apparatus is shown in ● Fig. 10.2. By rolling the rotor between the thumb and fingers, the operator sets a suspended mass bob into circular motion, with the centripetal

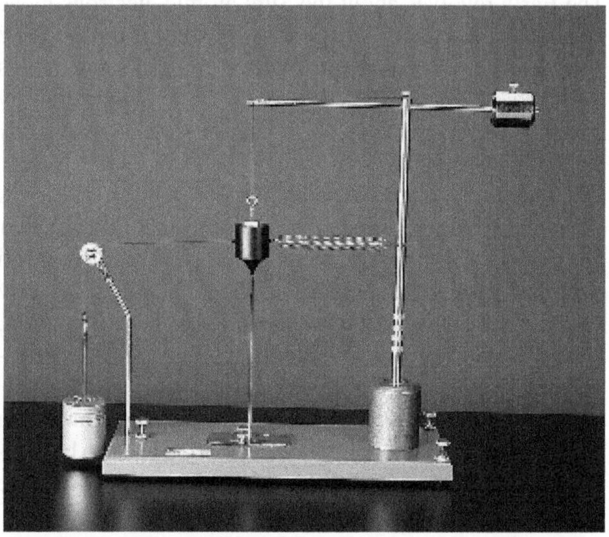

(a)

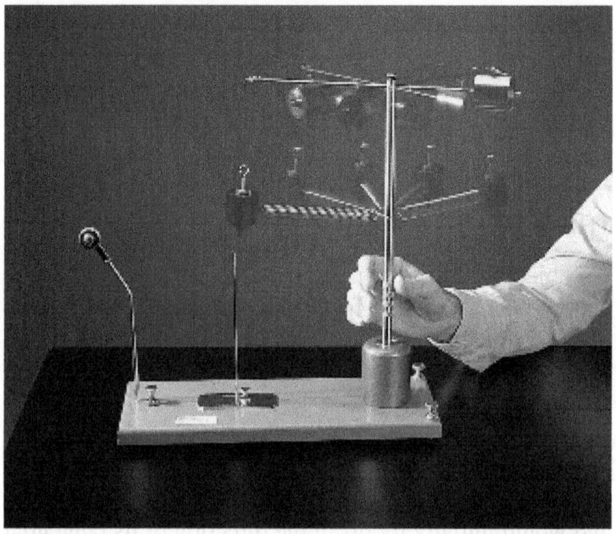

(b)

Figure 10.2 Hand-operated centripetal force apparatus. (a) The suspended weights, used to determine the centripetal force supplied by the spring, are not attached to the bob when the apparatus is operationally rotating. (b) Apparatus in action. See text for description. (Photos Courtesy of Sargent-Welch.)

force being supplied by a spring. The horizontal support arm is counterbalanced for ease of operation; the position of the counterbalance is not critical.

A pulley mounted to the base of the apparatus is used to make direct measurement of the spring tension supplying the centripetal force for uniform circular motion of a particular radius indicated by the distance between the vertical pointer rod P and the axis of rotation.

2. Remove the bob and determine its mass on a laboratory balance. Record the mass value in Data Table 1. Adjust the position of the vertical pointer rod, if possible, to the smallest possible radius (distance between the pointer tip and the center of the vertical rotor shaft). Measure this distance and record.

3. Attach the bob to the string on the horizontal support arm, and with the bob hanging freely (spring unattached), adjust the support arm so that the bob is suspended directly over the pointer. Attach the spring to the bob, and practice rolling the rotor between your thumb and fingers so that the bob revolves in a circular path and passes over the pointer on each revolution in uniform circular motion. (Adjust the position of the counterbalance on the support arm, if necessary, for ease of operation.)

   Make sure the locking screws are tight, and be careful of the rotating counterweight. ***Caution:*** *Safety glasses should be worn. This is always a good practice in a laboratory with moving equipment.*

   While one lab partner operates the rotor, another lab partner with a laboratory timer or stopwatch times the interval for the bob to make about 25 revolutions. The number of counted revolutions may have to be varied depending on the speed of the rotor. Count enough revolutions for an interval of at least 10 s. Record the data in Data Table 1. Practice the procedure before making an actual measurement.

4. Repeat the counting-timing procedure twice. Compute the time per revolution of the bob for each trial, and determine the average time per revolution of the three trials.

   From the data, calculate the average speed of the bob. Recall $v = c/t = 2\pi r/T$, where c is the circumference of the circular orbit, r is the radius of the orbit, and T is the average time per revolution or period. Then, using Eq. 10.2, calculate the centripetal force.

5. Attach a string to the bob opposite the spring, and suspend a weight hanger over the pulley. Add weights to the hanger until the bob is directly over the pointer. Record the weight, Mg, in the data table. (Do not forget to add the mass of the weight hanger.) This weight is a direct measure of the centripetal force supplied by the spring during rotation. Compare this with the calculated value, and compute the percent difference of the two values.

6. *Variation of mass.* Unscrew the nut on the top of the bob, insert a slotted mass of 100 g or more under it, and retighten the nut. Repeat Procedures 3 through 5 for determining the period of rotation and comparing the computed value of the centripetal force with the direct measurement of the spring tension. (*Question*: Does the latter measurement need to be repeated?) Record your findings in Data Table 2.

7. *Variation of radius.* Remove the slotted masses from the bob, and if pointer P is adjustable, move it farther away from the axis of rotation to provide a larger path radius. Measure and record this distance in Data Table 3. Repeat Procedures 3 through 5 for this experimental condition.

8. *Variation of spring tension (optional).* Replace the spring with another spring of different stiffness. Repeat Procedures 3 through 5, recording your findings in Data Table 4.

Name ______________________ Section ______________ Date ______________

Lab Partner(s) ____

EXPERIMENT 10

Centripetal Force

TI Laboratory Report

A. Manual Centripetal Force Apparatus

DATA TABLE 1

Purpose: To determine period of revolution for computation of centripetal force.

	Trial 1	Trial 2	Trial 3
Number of revolutions			
Total time ( )			
Time/revolution ( )			

Mass of bob ______________

Radius of circular path ______________

Average time per revolution ______________

Average speed of bob (v) ______________

Computed value of centripetal force ______________

Direct measurement of centripetal force ______________

Percent difference ______________

Computation of centripetal force (attach additional sheet)

DATA TABLE 2

Purpose: To observe the effect of varying mass.

	Trial 1	Trial 2	Trial 3
Number of revolutions			
Total time ( )			
Time/revolution ( )			

Mass of bob ______________

Radius of circular path ______________

Average time per revolution ______________

Average speed of bob (v) ______________

Computed value of centripetal force ______________

Direct measurement of centripetal force ______________

Percent difference ______________

Computation of centripetal force (attach additional sheet)

(continued)

DATA TABLE 3

Purpose: To observe the effect of varying radius.

	Trial 1	Trial 2	Trial 3
Number of revolutions			
Total time ( )			
Time/revolution ( )			

Mass of bob ____________

Radius of circular path ____________

Average time per revolution ____________

Average speed of bob (v) ____________

Computed value of centripetal force ____________

Direct measurement of centripetal force ____________

Percent difference ____________

Computation of centripetal force
(attach additional sheet)

DATA TABLE 4 (Optional)

Purpose: To observe the effect of varying spring tension.

	Trial 1	Trial 2	Trial 3
Number of revolutions			
Total time ( )			
Time/revolution ( )			

Mass of bob ____________

Radius of circular path ____________

Average time per revolution ____________

Average speed of bob (v) ____________

Computed value of centripetal force ____________

Direct measurement of centripetal force ____________

Percent difference ____________

Computation of centripetal force
(attach additional sheet)

Name ______ Section ______ Date ______
Lab Partner(s) ______

TI QUESTIONS

1. How does the centripetal force vary with the radius of the circular path? Consider (a) constant frequency and (b) constant speed. Was this substantiated by experimental results?

2. If the centripetal force on an object in uniform circular motion is increased, what is the effect on (a) the frequency of rotation f (with r constant) and (b) f and r when both are free to vary?

3. Does the centripetal force acting on an object in uniform circular motion do work on the object? Explain.

(continued)

4. Figure 10.1 shows a student swinging a ball in a circle about his head. Show that the rope cannot be exactly horizontal. (*Hint*: Take the rope's tension force T to be at an angle below the horizontal, and examine the components of T. Use a diagram to illustrate.)

Name ______________________ Section ______________ Date ______________

Lab Partner(s) ____

EXPERIMENT 11

Friction

TI Advance Study Assignment

Read the experiment and answer the following questions.

1. State the three general empirical rules used to describe friction.

2. What is the normal force, and why is it used instead of the load?

3. Why is it important to have the string parallel to the horizontal surface in the procedures where suspended weights are used?

4. What is the coefficient of friction, and in what units is it expressed? Distinguish between μ_s and μ_k. Which is generally greater?

(continued)

5. Explain how graphs of weight versus normal force in Procedures A and B give the coefficients of friction.

CI Advance Study Assignment

Read the experiment and answer the following question.

1. Under what conditions is the tension in the string pulling horizontally on the cart equal in magnitude to the frictional force?

EXPERIMENT 11

Friction

OVERVIEW

Experiment 11 examines friction using complementary TI and CI approaches. The TI procedures are concerned with determination of the coefficients of friction, μ_s and μ_k, with an option of investigating the dependence of μ on various parameters, such as different materials, lubrication, and so on.

The CI procedures extend the investigation by examining the effect of speed on sliding friction.

INTRODUCTION AND OBJECTIVES

In general, the term **friction** refers to the force or resistance to motion between contacting material surfaces. (Internal friction occurs in liquids and gases.) The friction between unlubricated solids is a broad and complicated topic, because it depends on the contacting surfaces and the material properties of the solids. Three general empirical "rules" are often used to describe friction between solid surfaces. These are that the frictional force is

1. independent of the surface area of contact.
2. directly proportional to the *load*, or the contact force that presses the surfaces together.
3. independent of the sliding speed.

Let's take a look at each of these rules:

1. Intuitively, one would think that friction depends on the roughness or irregularities of the surfaces, and the greater the area of contact, the more the friction. This would seem to contradict rule 1.
2. However, the actual contact area of the surfaces should depend on the force that presses the surfaces together, or the load. Increasing this force should increase the amount of contact of the irregularities between the surfaces and, hence, the friction. Rule 2 then seems logical.
3. Is it consistent that the friction between a sliding object and a surface be independent of the sliding speed? It would seem that the rate at which the surface irregularities meet, which is dependent on the sliding speed, should have some effect.

With such thoughts in mind, in this experiment, the validity of the foregoing empirical rules will be investigated. Experimentally, you might find that they are very general and, at best, approximations when applied to different materials and different situations.

TI OBJECTIVES

After performing the experiment and analyzing the data, you should be able to do the following:

1. Comment on the validity of the empirical rules of friction.
2. Describe how coefficients of friction are determined experimentally.
3. Tell why the normal reaction force of a surface on an object is used to determine the frictional force rather than the weight of the object.

CI OBJECTIVES

1. Verify that friction is proportional to the normal force.
2. Indicate whether or not friction is independent of sliding speed.

TI EXPERIMENT 11

Friction

TI EQUIPMENT NEEDED

- Board with attached low-friction pulley
- Rectangular wooden block with hook (for example, a piece of 2 × 4 lumber or commercially available block)
- Weight hanger and set of weights
- String
- Protractor
- Laboratory balance
- Table clamp and support
- Meterstick
- Masking tape
- 2 sheets of Cartesian graph paper

(*Optional*)

- Plastic block
- Aluminum block
- Wheel cart
- Dry lubricating powder (for example, graphite or molybdenum sulfide, MoS_2)

TI THEORY (general, TI and CI)

It is sometimes assumed that the *load*, or the contact force that presses the surfaces together, is simply the weight of the object resting on a surface. Consider the case of a block resting on a horizontal surface as illustrated in ● TI Fig. 11.1a. The force that presses the surfaces together is the downward weight force of the block (magnitude mg), which is the load. However, on an inclined plane, only a component of the weight contributes to the load, the component perpendicular to the surface. (See TI Fig. 11.3, where the magnitude of the load is $mg \cos \theta$.)

In order to take such differences into account, the frictional force f is commonly taken to be directly proportional to the normal force N, which is the force of the surface *on* the block—that is, $f \propto N$ (see TI Fig. 11.1). In the absence of other perpendicular forces, the normal force is equal in magnitude to the load, $N = mg$ in TI Fig. 11.1 and $N = mg \cos \theta$ in TI Fig. 11.3, which avoids any confusion between weight and load.

With $f \propto N$, written in equation form:

$$\boxed{f = \mu N}$$

or **(TI 11.1)**

$$\boxed{\mu = \frac{f}{N}}$$

where the Greek letter mu (μ) is a unitless constant of proportionality called the **coefficient of friction**. (Why doesn't μ have units?)

When a force F is applied to the block parallel to the surface and no motion occurs, then the applied force is balanced by an opposite force of static friction. (TI Fig. 11.1b, $F - f_s = ma = 0$). As the magnitude of the applied force is increased, f_s increases to a *maximum* value given by (see TI Fig. 11.1c)

$$\boxed{f_{s_{max}} = \mu_s N} \quad \textbf{(TI 11.2)}$$

(static friction)

where μ_s is the **coefficient of static friction**.* The maximum force of static friction is experimentally approximated by the smallest force applied parallel to the surface that will just set the block into motion.

At the instant the applied force F becomes greater than $f_{s_{max}} = \mu_s N$, however slightly, the block is set into motion, and the motion is opposed by the force of kinetic (sliding) friction f_k (TI Fig. 11.1d), and

$$\boxed{f_k = \mu_k N} \quad \textbf{(TI 11.3)}$$

(kinetic friction)

where μ_k is the **coefficient of kinetic (sliding) friction**.

The unbalanced force causes the block to accelerate ($F - f_k = ma$). However, if the applied force is reduced so that the block moves with a uniform velocity ($a = 0$), then $F = f_k = \mu_k N$.

Usually, for a given pair of surfaces, $\mu_k < \mu_s$. That is, it takes more force to overcome static friction (get an object moving) than to overcome kinetic friction (keep it moving). Both coefficients may be greater than 1, but they are usually less than 1. The actual values depend on the nature and roughness of the surfaces.

*These conditions on f_s are sometimes written $f_s \leq \mu_s N$; that is f_s is less than or equal to the maximum value of $\mu_s N$. As the applied force is increased, f_s increases but there is no motion until $f_{s_{max}}$ is reached.

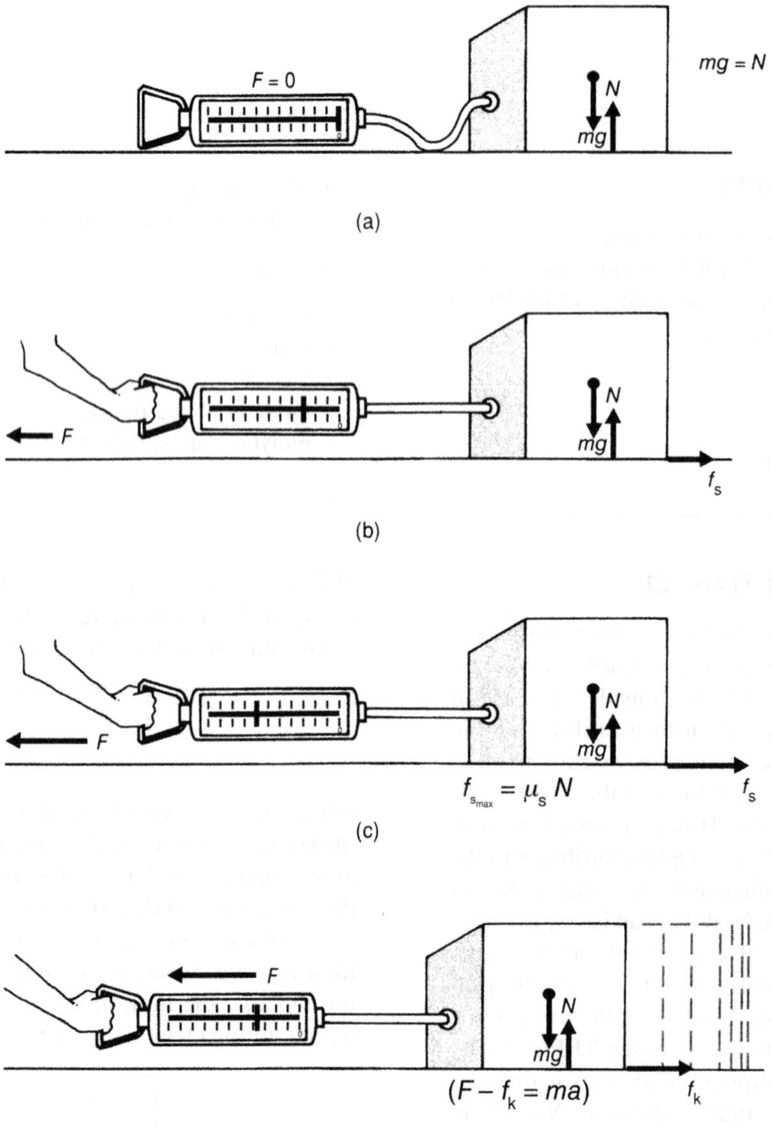

TI Figure 11.1 Friction. The applied force is balanced by the force of static friction f_s(a − c), and $F - f_s = ma = 0$. As the applied force increases, so does the force of static friction, until a maximum value is reached ($f_{s_{max}} = \mu_s N$). A slightly greater force (d) sets the block into motion ($F - f_k = ma$), with the applied force being opposed by the force of kinetic friction, f_k.

TI EXPERIMENTAL PROCEDURE

A. Determination of μ_s

1. Determine the mass of the wooden block on a laboratory balance, and record it in the Laboratory Report.

2. Clean the surfaces of the board and block so they are free from dust and other contaminants. Place the board with the pulley near the edge of the table so that the pulley projects over the table's edge (● TI Fig. 11.2). Attach one end of a length of string to the wooden block and the other end to a weight hanger. Place the block flat on the board, and run the string over the pulley so that the weight hanger is suspended over the end of the table. Be sure that the string is parallel to the board, otherwise there will be a vertical component of the force *F*.

3. With the rectangular block lying on one of its sides of larger area, add weights to the hanger until the block just begins to move. (*Note*: If the 50-g hanger causes the block to move, add some weights to the block and add this mass to the mass of the block, m_b.) Determine the required suspended mass within 1 g. Record the weight force (Mg) required to move the block in TI Data Table 1. This is equal in magnitude to the frictional force, $f_{s_{max}}$. (Friction of the pulley is neglected.)

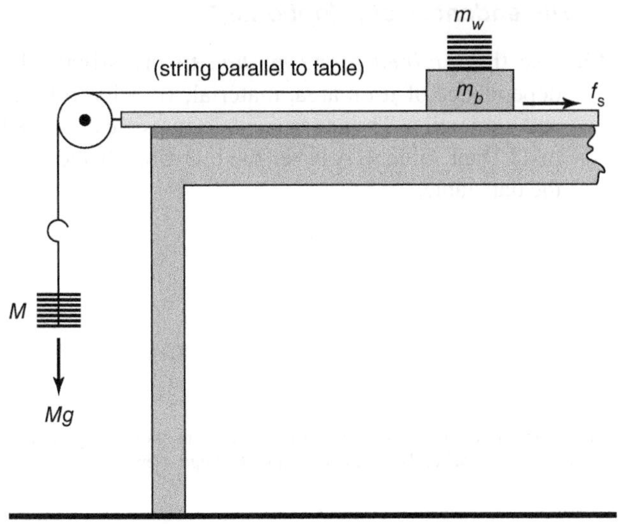

TI Figure 11.2 Coefficient of static friction. Experimental setup to determine μ_s. See text for description.

Suggested experimental technique:
(a) Keep the block in the middle of the plane.
(b) Lift the block, gently lower it onto the plane, *restrain* it from moving for a count of 5 (*do not* press it against the plane), and then release the block. If the block moves, the suspended mass M is too large; if it doesn't move, M is too small; and if the block moves about half the time, M is about right.

4. Repeat Procedure 3 with m_w = 100-, 200-, 300-, 400-, and 500-g masses, respectively, added to the block. Record the results in TI Data Table 1.

5. Plot the weight force just required to move the block (or the maximum force of static friction, $F = f_{s_{max}}$) versus the normal force N of the surface on the block $[N = (m_b + m_w)g]$. Draw a straight line that best fits the data. Include the point (0, 0). (Why?)

Since $f_s = \mu_s N$, the slope of the straight line is μ_s. Determine the slope, and record it in TI Data Table 1.

B. Determination of μ_k

Horizontal Board

6. In the experimental setup in Fig. 11.2, when the block moves with a uniform (constant) speed, its acceleration is zero. The weight force F and the frictional force f_k are then equal and opposite ($F - f_k = ma = 0$, and $F = f_k$).

7. Using the larger side (surface area) of the block and the series of added masses as in Part A. Add mass to the weight hanger until a slight push on the block will cause it to move with a uniform speed. It may be helpful to tape the weights to the block. The required weight force for the motion in each case should be less than that for the corresponding case in Part A. (Why?) Record the data in TI Data Table 2.

Suggested experimental technique:
(a) Begin with the block at one end of the plane, and give it a push so that it slides across the entire plane.
(b) Observe the behavior of the block in the same region as before, namely in the middle of the plane. This is where the block should be observed for constant speed.

8. Plot the weight force (or the force of kinetic friction, $F = f_k$) versus the normal force N for these data on the same graph as for Part A. Draw a straight line that best fits the data.

Since $f_k = \mu_k N$, the slope of the straight line is μ_k. Determine the slope, and record it in TI Data Table 2. Calculate the percent decrease of μ_k from the μ_s value.

Elevated Board (Inclined Plane)

9. Elevate the pulley end of the board on a support to form an inclined plane (● TI Fig. 11.3.) Note that in Fig. 11.3 the magnitude of the normal force (perpendicular to the plane) is equal to a *component* of the weight force.

With the block laying on a *side* of its larger surface area, determine the angle θ of incline that will allow the block to slide down the plane with a constant speed after being given a slight tap. (No suspended weight is used in this case.) *Note*: The maximum angle before slipping *without* tapping gives μ_s, whereas the angle of constant velocity *with* tapping gives μ_k.

10. Using a protractor, measure the angle θ and record in TI Data Table 3. Also, with a meterstick, measure the length L of the base (along the table) and the height h of the inclined plane. Record the ratio h/L in TI Data Table 3.

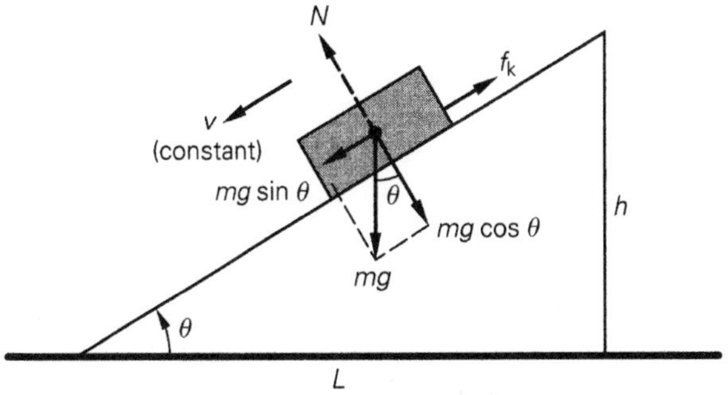

TI Figure 11.3 Coefficient of kinetic friction. Experimental setup to determine μ_k. See text for description.

11. Repeat this procedure for the block with the series of added masses as in the previous procedure for the horizontal board, and record in TI Data Table 3. It may be helpful to tape the masses to the block.

12. Using a calculator, find the tangents of the θ angles, and record them. Compute the average of these values and the average of the ratios *h/L*. These averages should be similar. (Why?)

13. Compare the average value of tan θ with the value of μ_k found in the procedure for the horizontal board. It can be shown theoretically that tan $\theta = \mu_k$ in this case. Compute the percent difference of the experimental values.

*C. Dependences of μ (optional)**

14. Use the *inclined plane method* to investigate the dependence of μ on area, material, velocity, rolling, and lubrication. The experimental setups are described in TI Data Table 4. Answer the questions listed after the data table.

*This experimental procedure and modifications were suggested by Professor I. L. Fischer, Bergen Community College, New Jersey.

Name ______________________ Section ______________ Date ______________

Lab Partner(s) ____

TI EXPERIMENT 11

Friction

TI Laboratory Report

Note: Attach graphs to laboratory report

Mass of block m_b ______________

A. Determination of μ_s

TI DATA TABLE 1

Purpose: To investigate $f_s = \mu_s N$, where N depends on $m_b + m_w$, by measuring μ_s on a level plane (see TI Fig. 11.2).

m_w	0					
$N = (m_b + m_w)g$*						
$f_s = F = Mg$						

* It is convenient to express the force in terms of mg, where g is left in symbol form [e.g., $(0.250)g$ N], even when graphing.

Calculations
(show work)

μ_s ______________
(from graph)

Don't forget units

(continued)

B. DETERMINATION OF μ_k

TI DATA TABLE 2

Purpose: To investigate $f_k = \mu_k N$, where N depends on $m_b = m_w$, by measuring μ_k on a level plane.

m_w	0					
$N = (m_b + m_w)g$						
$f_k = F = Mg$						

Calculations
(show work)

μ_k ________________ (from graph)

Percent decrease of μ_k relative to μ_s ________________

TI DATA TABLE 3

Purpose: To investigate $\mu_k = \tan\theta$, where θ is independent of $m_b + m_w$, by measuring μ_k by the inclined plane method (see TI Fig. 11.3).

m_w	0						
θ							Average
h/L							
$\tan\theta$							

Calculations
(show work)

Percent difference between $\tan\theta = \mu_k$ and μ_k from TI Data Table 2 ________________

Name ______________________ Section ______________ Date ______________

Lab Partner(s) ______________________

EXPERIMENT 11 **Friction** *Laboratory Report*

C. Dependences of μ (optional)

TI DATA TABLE 4

Purpose: To investigate dependences of μ by various measurements using the inclined plane method and other materials (if available).

No.	Conditions	θ	$\mu = \tan\theta$
1	Wooden block on larger area, static (μ_s)		
2	Wooden block on smaller area, static (μ_s)		
3	Wooden block on smaller area, kinetic (μ_k)		
Other materials			
4	Plastic block		
5	Aluminum block, moving slowly		
6	Aluminum block, moving faster		
7	Wheeled cart		
8	Aluminum block with dry lubricating powder		
9	Plastic block with dry lubricating powder		

Answer the following questions on a separate sheet of paper, and attach it to the TI Laboratory Report.

(a) Compare No. 1 with TI Data Table 1: Is the inclined plane method valid for μ_s?
(b) Compare No. 2 with No. 1 and No. 3 with TI Data Table 4: Does μ depend on area?
(c) Compare Nos. 3, 4, and 5: Does μ_k depend on material?
(d) Compare No. 5 with No. 6: Does μ_k depend on velocity?
(e) Compare No. 7 with anything: How does rolling friction compare with other types of friction?
(f) Compare Nos. 8 and 9 with Nos. 5 and 4: What is the effect of adding the lubricant?

TI QUESTIONS

1. Explain why $f_s \leq \mu_s N$; that is, why is f_s less than or equal to $\mu_s N$?

(continued)

2. Speculate, in terms of the microscopic surface irregularities, about why $\mu_k < \mu_s$ and what effect a lubricant has on the coefficient of friction.

3. (a) Prove that tan θ is equal to μ_k when the block slides down the incline with a constant speed. (Use symbols, not numbers.)

   (b) If θ is the maximum angle of incline just before the block moves, what is μ_s in terms of θ?

4. Suppose that the block were made to move up the inclined plane with a uniform speed by suspending masses on a string over the pulley. Derive an equation for the coefficient of kinetic friction for this case in terms of the suspended masses, the mass of the block, and the angle of decline. (Neglect any friction and mass effects of the pulley.)

5. On the basis of your experimental results, draw and justify conclusions about the validity of the empirical rules for friction. What does this tell you about applying general rules to all materials and about the nature of friction?

Friction

CI EQUIPMENT NEEDED

- 1 wooden block (The block used in the TI procedure can be used here also. Another option is the "Friction Block" included in the PASCO Classic Dynamics System.)
- Additional blocks as needed to make the string horizontal when connected to the force sensor (Two PASCO cars, ME–9430 or 9454, stacked upside down on top of each other and on top of the friction block will make a tower of the correct height.)
- 1 straight, smooth track (PASCO dynamics track)
- 1 force sensor (PASCO CI-6537)
- 1 constant-speed motorized car (PASCO ME-9781)
- Extra weights to load the sliding object (200-g or 500-g pieces will work fine) (The PASCO Classic Dynamics System includes mass bars that can be used in this part.)
- Graph paper

CI THEORY

In this experiment, we will study two of the general empirical rules used to describe the friction between solid surfaces. In the first part, we will examine the relationship between friction and the normal force to verify that they are proportional to each other. In the second part, we will examine the effect of the speed of the object on the amount of frictional force. In both cases, a force sensor will be used to measure the frictional force between a sliding wooden block and a track.

● CI Fig. 11.1 illustrates the experimental situation. The sliding object is a wooden block. Other blocks are shown added as needed so that the string is *horizontal* when connected to a force sensor riding on a motorized car. As an alternative, the figure also shows the setup using the suggested PASCO equipment, where a stack of cars is used to make the object the correct height. Other alternatives include using a single 2 × 4 board with a nail that makes it possible to attach the string at the proper height (not pictured).

● CI Fig. 11.2 shows a free-body diagram of a block as it slides with constant speed along a level track. The horizontal forces are F, the tension of the string, and f, the frictional force provided by the track. With the speed constant, there is no acceleration. From Newton's second law, we have

$$\Sigma F_x = F - f = ma = 0$$

or

$$F = f$$

In this experiment, the force sensor will directly measure F, the tension in the string. Notice that as long as the car moves at a constant speed, the magnitude of F is equal to the magnitude of the frictional force acting on the sliding block.

On the other hand, the vertical forces balance each other out, so the magnitude of the normal force N can be determined as the magnitude of the weight of the object: $N = mg$.

SETTING UP DATA STUDIO

Note: The force sensor needs to be calibrated before use. Refer to the user's manual for instructions on how to calibrate the sensor. The procedures described here assume that the force sensor has been properly calibrated.

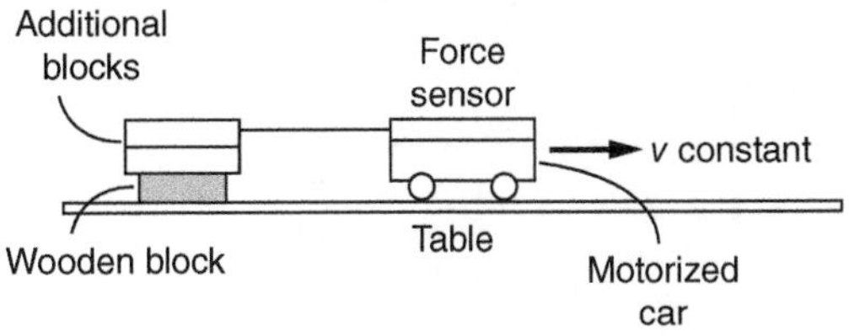

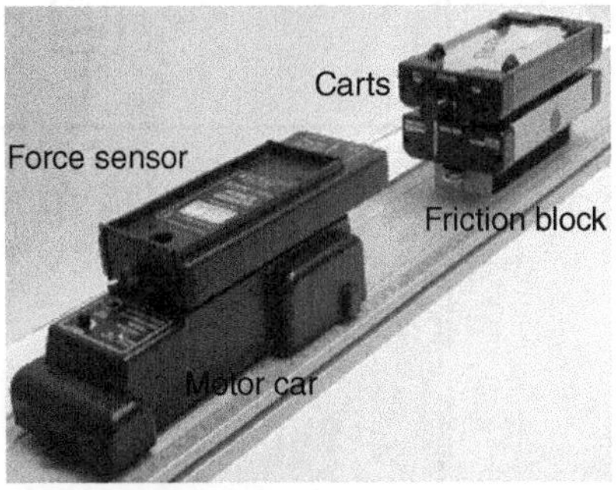

CI Figure 11.1 The experimental setup. A wooden block slides on a flat surface while being pulled by a motorized car that moves at a constant speed. Additional blocks can be added as necessary on top of the wooden block so that the string is horizontal when connected to the force sensor. The force sensor rides on the motorized car. As an alternative, PASCO dynamic cars can be stacked on top of a friction block to achieve the same effect. (Reprinted courtesy of PASCO Scientific.)

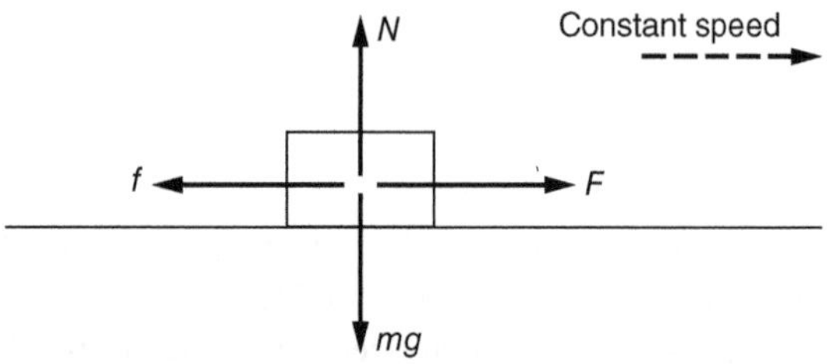

CI Figure 11.2 Free-body diagram of the sliding block. The horizontal forces are F, the tension on the string, and f, the friction from the surface. The force sensor measures F. At constant speed, the horizontal force vectors are equal and opposite, and $F = f$. The force sensor readings can be taken to be the friction as long as the block slides at constant speed.

1. Open Data Studio, and choose "Create Experiment."
2. The Experiment Setup window will open, and you will see a picture of the Science Workshop interface. There are seven channels to choose from. (Digital channels 1, 2, 3, and 4 are the small buttons on the left; analog channels A, B, and C are the larger buttons on the right, as shown in ● CI Fig. 11.3.)
3. Click on the channel A button in the picture. A window with a list of sensors will open.
4. Choose the Force Sensor from the list, and press OK.
5. Connect the sensor to channel A of the interface, as shown on the computer screen.
6. The properties of the force sensor are shown directly under the picture of the interface. Set the sample rate to 200 Hz.
7. Create a digits display by double-clicking on "Digits" in the displays list (lower left of the screen). A display window called Digits 1 will open. It will show the force readings from the sensor when data are collected.
8. Double-click anywhere on the Digits 1 window. The Digits Setting window will open.
9. Select the Statistics button from the Toolbar box, and click OK. There will now be a drop menu with the sigma symbol on the Digits 1 window toolbar.
10. Press the sigma symbol, and choose "Mean." This will show the average of a series of measurements on the display.
11. The size of the display window can be adjusted for easier viewing, if needed. The bigger the screen, the more digits you will be able to see once data are collected. For the purpose of this experiment, keep the size such that only two decimal places are shown. (Wait until data are collected to adjust this. There have to be data on the display before any change can be noticed.)
12. ● CI Fig. 11.4 shows what the screen will look like after the setup is complete and data are taken.

CI EXPERIMENTAL PROCEDURE

A. The Effect of the Load

1. Measure the mass of the wooden block and of any other block or car that will be placed on top of it to add height, as illustrated in CI Fig. 11.1. Record the total mass in Trial 1 of CI Data Table 1.

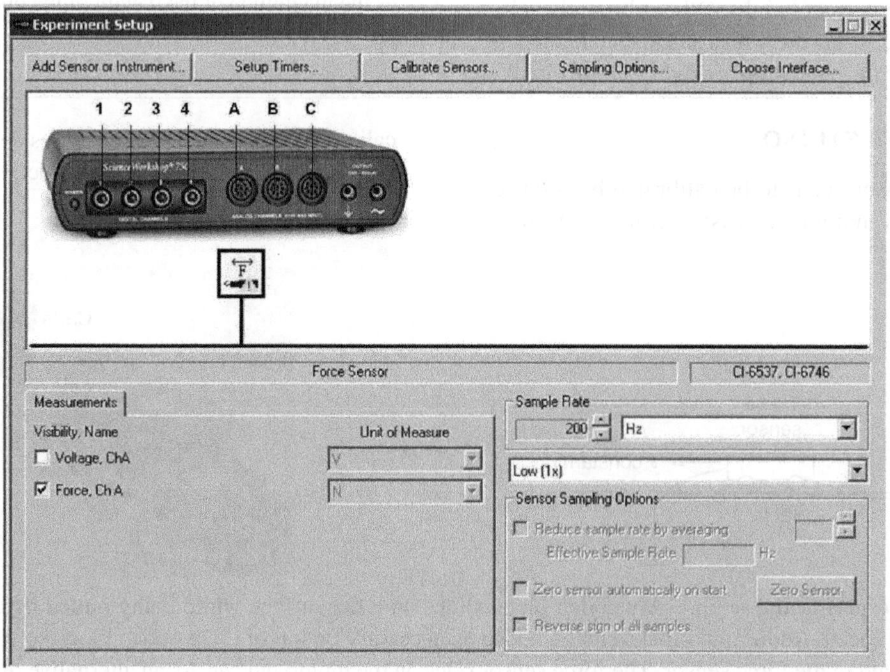

CI Figure 11.3 The Experiment Setup window. The force sensor is connected to analog channel A. The sample rate is set to 200 Hz.

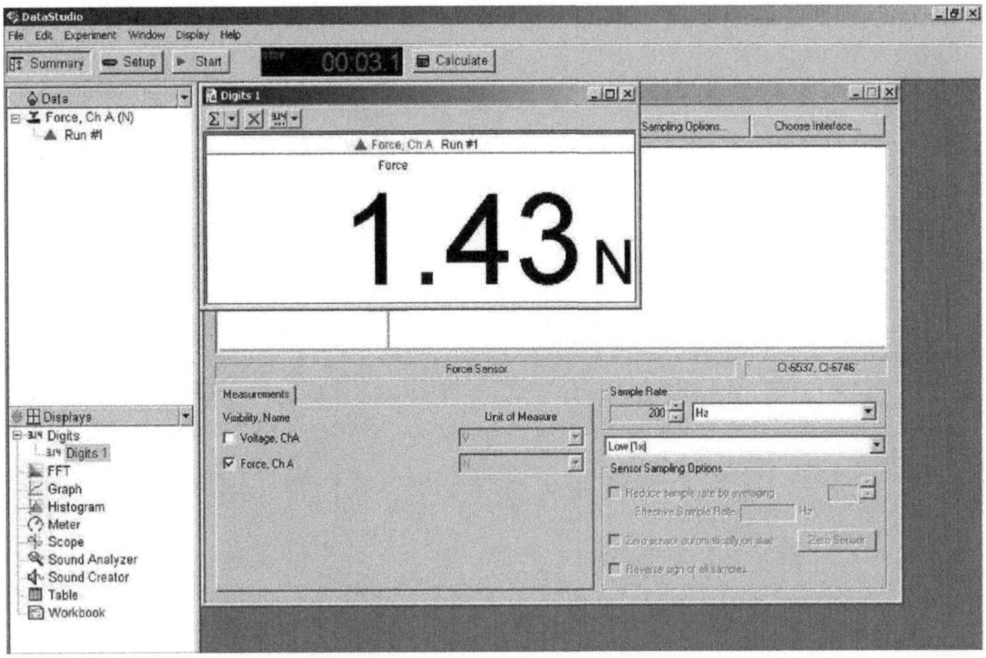

CI Figure 11.4 Data Studio Setup. A digits display will show the force reading of the sensor. Once data are collected, the size of the display window is adjusted to show two decimal places. (Data displayed using Data Studio Software. Reprinted courtesy of PASCO Scientific.)

2. Set up the equipment as shown in CI Fig. 11.1. It is important that the string connecting the force sensor to the pile of objects be *horizontal*. If using additional blocks instead of the PASCO cars, tape the blocks together so that they will not fall off.

3. Set the motorized car for a medium speed, and do not change it during the experiment.

4. ***Trial 1: The object with no extra load.***
   a. With the string slack, press the TARE button on the side of the force sensor to zero the sensor.
   b. Turn the motorized car on.
   c. Wait until the string tenses before pressing the START button to begin collecting data. Let the car move, pulling along the pile of blocks (the "object"), for about 20 cm, and then press the STOP button.
   d. Stop the car.
   e. Report the average fictional force reading in CI Data Table 1. Do not worry if the sensor reading is negative. That is a convention for direction (pull or push). In this experiment, we need only the magnitude.

5. ***Trials 2, 3, 4 and 5: The object with a load.***
   a. Place a load on top of the sliding object, and record the new mass of the sliding object in CI Data Table 1.
   b. Repeat the data collection process as described in steps (a) to (e) for Trial 1.
   c. Repeat by continuing to add mass on top of the object until the table is complete.

6. Calculate the normal force for each trial by determining the weight of the object plus load in each case. Record the results in CI Data Table 1.

7. Use a full page of graph paper to make a plot of friction versus normal force. Determine the slope of the best-fitting line for the plot, and enter the result in the table. Attach the graph to the Laboratory Report.

B. The Effect of the Speed

1. Set up the equipment as shown in CI Fig. 11.1. It is important that the string connecting the force sensor to the pile of objects be *horizontal*. If using additional blocks instead of the PASCO cars, tape the blocks together so that they will not fall off.
2. Set the motorized car for a slow speed.
3. Turn on the motorized car. Wait until the string tenses before pressing the START button to begin collecting data. Let the car move, pulling along the block, for about 20 cm, and then press the STOP button.
4. Stop the car.
5. Report the average frictional force reading in CI Data Table 2.
6. Increase the speed of the motorized car, and measure the average frictional force again. Repeat by increasing the speed for each trial until the table is complete.

Name ______________________ Section ______________ Date ______________

Lab Partner(s) ____

CI EXPERIMENT 11

Friction

CI Laboratory Report

A. The Effect of the Load

CI **DATA TABLE 1**

Purpose: To investigate the effect of changing the load on an object (and thus changing the normal force) on the magnitude of the frictional force.

	Trial	Total mass of sliding object	Frictional force (sensor reading)	Normal force $N = mg$
The object with no load	1			
The object with increasing load	2			
	3			
	4			
	5			

Slope of graph = ______________

Don't forget units

(continued)

B. The Effect of Speed on Friction

CI DATA TABLE 2

Purpose: To investigate the effect of speed on the frictional force.

Different speed trials (from low speed to high)	Average frictional force
1	
2	
3	
4	
5	
6	
7	
8	
9	
10	

Name ______________________ Section ______________ Date ______________

Lab Partner(s) ______________________________________

EXPERIMENT 11 **Friction** *Laboratory Report*

CI QUESTIONS

1. Is it true that the frictional force is directly proportional to the normal force? Discuss the experimental evidence.

2. What is the physical significance of the slope of the graph of friction versus normal force?

3. Is there a clear pattern for the frictional force as the speed of the object increases? (Compare to the pattern observed when increasing the load.) What can be concluded about the effect of the speed? Discuss.

4. Why was it so important that the string connecting the sensor and the object remain horizontal during the experiment? Discuss what would happen if it did not.

5. Refer to step 3 of the Experimental Procedure for Part A, which says, "Set the motorized car for a medium speed, and do not change it during the experiment." Given the results of Part B of the experiment, discuss whether changing the speed would have made a difference in the results of Experiment A. (See your textbook for modern theories of friction between two surfaces.)

Name ______________________ Section ______________ Date ______________

Lab Partner(s) ____

EXPERIMENT 12

Work and Energy

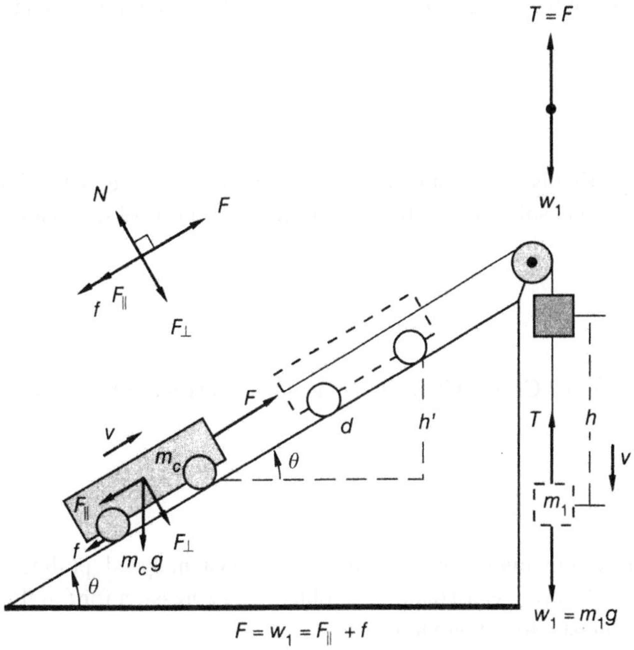

GL Figure 12.1 Going up. Work and energy considerations for Experimental Planning.

GL Experimental Planning

Work and energy are intimately related, as in heat and temperature. By doing work on an object, it can gain energy. Conversely, when energy is expended, work may be done. This experiment demonstrates the work–energy relationship in the context of work done by friction.

The work (W) done by a constant force ($\vec{\mathbf{F}}$) acting on an object and moving it through a parallel displacement ($\vec{\mathbf{d}}$) is given by the product of their magnitudes, $W = Fd$ (a scalar quantity). *Work then involves a force acting on an object and moving it through a distance.*

However, the constant force may not be acting parallel to the displacement. In this case, the magnitude of the component of the force parallel to the displacement is $F \cos\theta$, where θ is the angle between the force and displacement vectors. So in general,

$$W = F(\cos\theta)\, d$$

which is commonly written

$$W = Fd \cos\theta \qquad \textbf{(GL 12.1)}$$

In the case of friction, $W_f = -f\,d$, where f is the force of friction (assumed to be constant). Explain why there is a $(-)$ sign in this equation.

(continued)

A common experimental setup is shown in GL Fig. 12.1 for a car moving up an inclined plane at a *constant* speed pulled by a descending mass suspended over a pulley. Free-body diagrams for the forces acting on each object are also shown.

1. Write an equation for the sum of the forces acting on the car parallel to the plane and also for the sum of the forces acting on the descending mass. Note that the car and descending mass both move with constant velocities.

2. If the mass of the connecting string is small compared to the other masses, F and T will be approximately equal. Use this result to combine the equations and solve for the force of friction f in terms of the masses and angle.

   Did your result include a $\sin\theta$ term? Check with a classmate or the instructor to verify your result.

3. Now consider the case of the car moving down the plane with a constant speed, pulling a smaller mass upward. Draw the free-body diagrams and repeat the process used above to obtain an expression for f in this case. (Use m_2 for the ascending mass.) How does this result compare to the previous one?

   Note that $W_f = fd$ applies in both cases of the block moving up and down the plane, where d is the distance the block moves.

Name ______________________ Section ______________ Date ______________

Lab Partner(s) ____

4. Examine your equations for f, and determine what experimental quantities need to be measured to determine the work done by friction.

5. The previous strategy to calculate W_f was based on the definition of work (force-distance method). The work done by friction for this experimental setup can also be obtained by an energy method. Note in ● GL Fig. 12.1 that there is a *decrease* in potential energy of the descending mass (ΔU_w) and an *increase* in the potential energy of the cart (ΔU_c). Are these changes in potential energy equal in magnitude?

6. Since a nonconservative force is present (f), some energy is used in the work done to overcome friction (W_f), and this energy is no longer available as potential energy. Write the conservation of energy equation for this case in terms of the potential energies, and solve for W_f. Why have the kinetic energy terms been omitted in this analysis?

7. Check with a classmate or the instructor to verify your result. Then find a corresponding expression for W_f for the case of the car moving down the plane.

You now have two ways of determining W_f, a force-distance method and an energy method. Both of these methods will be used in the Experimental Procedure that follows.

Name ______________________ Section ______________ Date ______________

Lab Partner(s) ____

EXPERIMENT 12

Work and Energy

TI Advance Study Assignment

Read the experiment and answer the following questions.

1. Distinguish between the conservation of mechanical energy and the conservation of total energy.

2. Is mechanical energy conserved in real situations? Is the total energy conserved? Explain.

3. Discuss the relationship between work and energy for a car moving with a constant speed (a) up an incline and (b) down an incline.

4. Under what conditions would the frictional forces be expected to be equal in magnitude for a car moving up an incline and a car moving down an incline?

(continued)

5. Is the force of friction the same for different angles of incline if all other parameters are equal? Explain by specifically considering the angles used in the experiment.

6. What are possible sources of error in this experiment? Identify them as random or systematic errors. (See Experiment 2.)

PRE-LAB DEMONSTRATION | Two Ball Bounce (energy transfer)

Two balls of different sizes (basketball and tennis ball or racquetball) are dropped, and the energy loss on rebounding is estimated by the returning heights. Which ball loses more energy, and how much more? The smaller ball is then placed on top of the basketball and they are dropped simultaneously. What happens?

EXPERIMENT 12

Work and Energy

INTRODUCTION AND OBJECTIVES

Work and **energy** are intimately related, as emphasized in a common definition of energy as the ability to do work. That is, an object or system possessing energy has the *capability* of doing work. When work is done by a system, energy is expended—the system loses energy. Conversely, when there is work input to a system, the system gains energy.

In an ideal conservative system, mechanical energy is transferred back and forth between kinetic energy and potential energy. In such a system, the sum of the kinetic and potential energies is constant, as expressed by the *law of conservation of mechanical energy.* However, in actual systems, friction is always present, and these systems are nonconservative. That is, some energy is lost as a result of the work done against frictional forces. Even so, the *total* energy is conserved (*conservation of total energy*). The total energy is there in some form.

In this experiment, the conservation of energy will be used to study the relationship between work and energy in the cases of a car rolling up and down an inclined plane. The ever-present frictional forces and the work done against friction will be investigated and taken into account so as to provide a better understanding of the concept of work-energy. To simplify matters, experimental conditions with constant speeds will be used so that only the relationship between work and changes in gravitational potential energy will have to be considered.

After performing this experiment and analyzing the data, you should be able to:

1. Explain how work and energy are related.
2. Describe how frictional work can be determined experimentally using either force-distance or energy considerations.
3. Better appreciate the nonconservative aspects of real situations and the difference between the conservation of mechanical energy and the conservation of total energy.

EQUIPMENT NEEDED

- Inclined plane with a low-friction pulley and Hall's carriage (car)
- Weight hanger and slotted weights
- String
- Meterstick
- Protractor (if plane not so equipped)
- Laboratory balance

THEORY

A. Work of Friction: Force-Distance Method

CAR MOVING UP THE PLANE

The situation for a car moving up an inclined plane with a constant velocity is illustrated in GL Fig. 12.1. Since the car is not accelerating, the force up the plane (F) must be equal in magnitude to the sum of the forces down (parallel to) the plane, that is,

$$F = F_{\parallel} + f$$

where f is the force of friction, and $F_{\parallel} = m_c g \sin\theta$ is the component of the car's weight parallel to the plane. (See GL Fig. 12.1.)

Since the magnitude of F is equal to the weight w_1 of the suspended mass (m_1), then

$$F_{\parallel} + f = w_1$$

Solving for f and expressing the other forces in terms of the experimental parameters (GL Fig. 12.1),

$$f = w_1 - F_1$$

and

$$\boxed{f = m_1 g - m_c g \sin\theta} \qquad \textbf{(12.1)}$$

(car moving up)

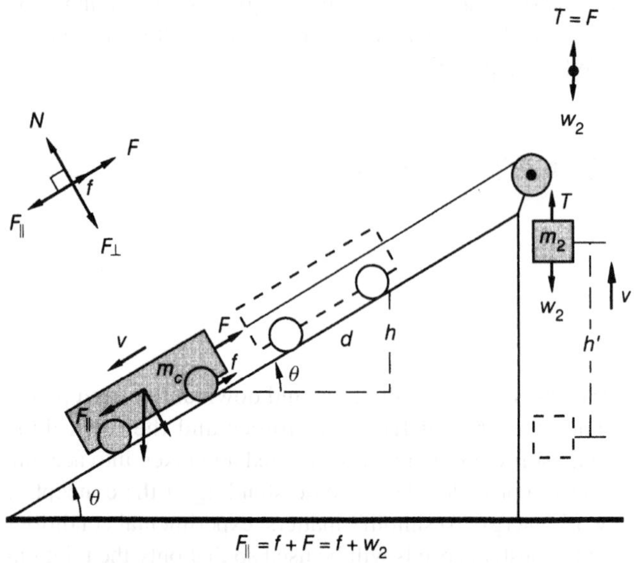

Figure 12.2 Car moving down the incline with the same constant speed as in GL Fig. 12.1. With no acceleration, the force on the car is zero, and $F_\parallel = F + f = w_2 + f$ (see free-body diagrams).

Car Moving Down the Plane

The situation for a car moving down an inclined plane with the same constant speed is illustrated in ● Fig. 12.2. Again, since the car is not accelerating, the sum of the forces up the plane must be equal in magnitude to the force down the plane (taken as positive), and

$$F = F_\parallel - f$$

where, in this case, the direction of f is up the plane. Since $F = w_2$,

$$F_\parallel = w_2 + f$$

and, expressing f as before,

$$\boxed{f = m_c g \sin\theta - m_2 g} \quad \textbf{(12.2)}$$

(car moving down)

Then, in either case, the frictional work is given by

$$\boxed{W_f = fd} \quad \textbf{(12.3)}$$

where d is the distance the car moves.

If the car moves approximately at the same constant speed in each case, it might be assumed that the magnitude of the frictional force f would be the same in each case (same angle of incline and load). This will be investigated experimentally.

B. Work of Friction: Energy Method

Another way of looking at the frictional work is in terms of energy.

Car Moving Up the Plane

For the case of the car moving up the plane, by the conservation of energy, the *decrease* in the potential energy of the descending weight on the weight hanger, $\Delta U_w = m_1 gh$, is equal to the *increase* in the potential energy of the car, $\Delta U_c = m_c gh'$, *plus* the energy lost to friction, which is equal to the work done against the force of friction, W_f (GL Fig. 12.1). That is,

$$\Delta U_w = \Delta U_c + W_f$$

or

$$W_f = \Delta U_w - \Delta U_c$$

and

$$\boxed{W_f = m_1 gh - m_c gh'} \quad \textbf{(12.4)}$$

(car moving up)

Car Moving Down the Plane

Similarly, for the case of the car moving down the plane, by the conservation of energy, the *decrease* in the potential energy of the descending car is equal to the *increase* in the potential energy of the ascending weight *plus* the work done against the force of friction (Fig. 12.2):

$$\Delta U_c = \Delta U_w + W_f$$

or

$$W_f = \Delta U_c - \Delta U_w$$

and

$$\boxed{W_f = m_c gh' - m_2 gh} \quad \textbf{(12.5)}$$

(car moving down)

In terms of the experimental parameters, the methods for determining W_f are equivalent.

EXPERIMENTAL PROCEDURE

Force-Distance Method

1. Using a laboratory balance, determine the mass of the car, m_c, and record it in the Laboratory Report.

2. Arrange the inclined plane and the car as shown in ● Fig. 12.3 with an angle of incline of $\theta = 30°$. Make certain that the pulley is adjusted so that the string attached to the car is parallel to the plane. (Should the car accelerate up the plane by the weight of the weight hanger alone, place some weights in the car so that the car is initially stationary. Then add the additional mass to that of the car in Data Table 1.)

3. Add enough weights to the weight hanger so that the car moves up the incline with a slow uniform speed

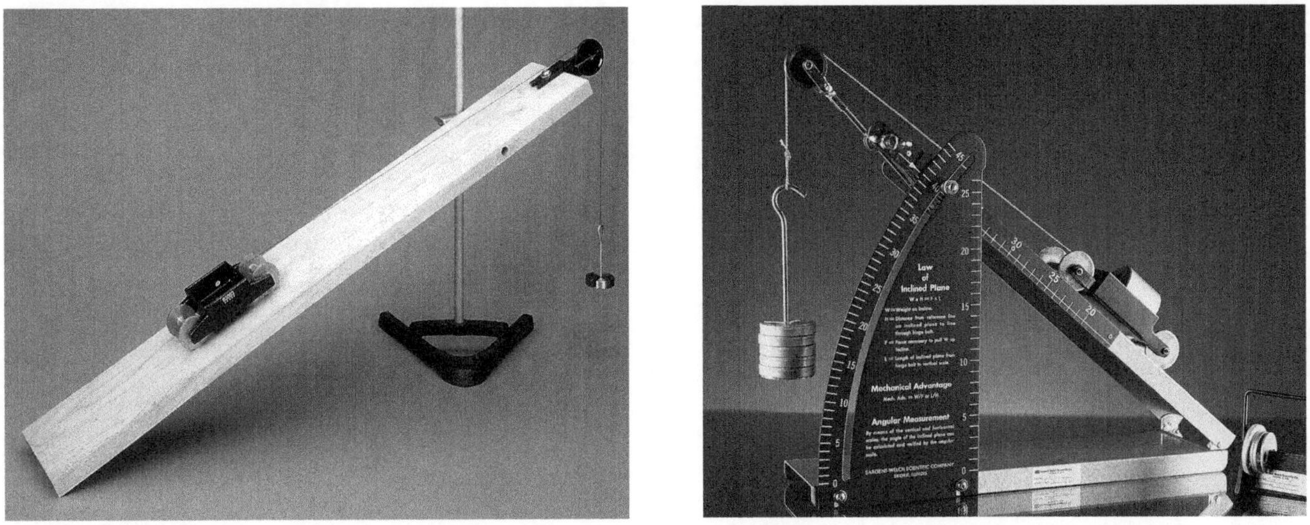

(a) (b)

Figure 12.3 Types of inclined planes. (a) Inclined plane with board and stand. (b) Calibrated inclined plane. (Photos Courtesy of Sargent-Welch.)

when the car is given a slight tap. Record the total suspended mass in Data Table 1.

4. With the car positioned near the bottom of the incline, mark the position of the car's front wheels and give the car a slight tap to set it into motion. Stop the car near the top of the plane after it moves up the plane (with a constant speed), and measure the distance d it moved up the plane as determined by the stopped position of the car's front wheels. Or measure the height h the weight hanger descends. This corresponds to the situation in GL Fig. 12.1. The lengths d and h are the same. Record this length in Data Table 1 as d.

5. With the car near the top of the plane, remove enough weights from the weight hanger so that the car rolls down the inclined plane with a slow uniform speed on being given a slight tap. Use as close to the same speed as for the upward case as is possible. This corresponds to the situation in Fig. 12.2. Record the total suspended mass in Data Table 1. For convenience, use the same d (or h) as in Procedure 4.

6. Compute the frictional force f [Eqs. (12.1) and (12.2)] and work done against friction W_f [Eq. (12.3)] for each case. Show your calculations, and record the results in Data Table 1.

7. Compare the frictional work for the two cases by computing the percent difference.

8. Adjust the angle of the inclined plane to $\theta = 45°$, and repeat Procedures 3 through 7, recording your measurements in Data Table 2.

Energy Method

9. Knowing that $d = h$, compute W_f for the previous cases using the energy method [Eqs. (12.4) and (12.5)] on the appropriate Laboratory Report pages.

10. Compare these values of W_f with those found using the force-distance method by computing the percent differences.

Name ________________ Section ________________ Date ________________

Lab Partner(s) ________________________________

EXPERIMENT 12

Work and Energy

TI *Laboratory Report*

Angle of incline ____________

DATA TABLE 1

Purpose: To determine work done against friction.

Mass of car m_c ____________

	Suspended mass ( )		d ( )	f ( )	W_f ( )
Car moving up incline	m_1				
Car moving down incline	m_2				

Calculation
(show work)

Percent difference in W_f ____________

Energy method calculations for W_f:

Don't forget units

(continued)

Angle of incline _______________

DATA TABLE 2

Mass of car m_c _______________

Purpose: To determine work done against friction.

	Suspended mass ( )		d ( )	f ( )	W_f ( )
Car moving up incline	m_1				
Car moving down incline	m_2				

Calculation
(show work)

Percent difference in W_f _______________

Energy method calculations for W_f:

Name ______________________ Section ______________ Date ______________

Lab Partner(s) ______________________________

EXPERIMENT 12 **Work and Energy** *Laboratory Report*

TI QUESTIONS

1. What was the work done by the suspended weight when the car (a) moved up the incline and (b) moved down the incline? (*Show your calculations.*)

	$\theta = 30°$	$\theta = 45°$
Car moving up incline	________	________
Car moving down incline	________	________

2. What was the work done by gravity acting on the car when it (a) moved up the incline and (b) moved down the incline? (*Show your calculations.*)

	$\theta = 30°$	$\theta = 45°$
Car moving up incline	________	________
Car moving down incline	________	________

3. (a) For the car going up the incline, what percentage of the work done by the suspended weight was lost to friction? (b) For the car moving down the incline, what percentage of the work done by gravity was lost to friction? (*Show your calculations.*)

	$\theta = 30°$	$\theta = 45°$
Car moving up incline	________	________
Car moving down incline	________	________

(continued)

4. Suppose the car accelerated up and down the incline. How would this affect the experimental determinations?

5. Is the assumption justified that f would be the same for both up and down cases for the same constant speed? If not, speculate as to why there is a difference.

6. Assuming that $f = \mu N$ (see Experiment 11), show that the coefficient of (rolling) friction for the car moving down the inclined plane with a constant speed is given by $\mu = \tan\theta - \dfrac{m_2}{m_c \cos\theta}$. (Use symbols, not numbers.)

Name ______________________ Section ______________ Date ______________

Lab Partner(s) ____

EXPERIMENT 13

Potential Energy of a Spring

TI Advance Study Assignment

Read the experiment and answer the following questions.

1. What does Hooke's law describe, and does it have limitations?

2. What do the spring constant and minus sign in Hooke's law indicate?

3. Why is the force of Hooke's law called a variable force?

4. For a mass vertically on a spring, what are the potential energy considerations?

5. How is the potential energy of the oscillating spring investigated when the total energy includes kinetic energy?

PRE-LAB DEMONSTRATION | The Slinky Slinky® (why doesn't it fall?)

A vertically extended Slinky, when dropped, seems to defy gravity—the bottom coils remain at rest—until what happens?

EXPERIMENT 13

Potential Energy of a Spring

INTRODUCTION AND OBJECTIVES

The spring force or restoring force in the elongation or compression of a spring is given by Hooke's law (after the English scientist Robert Hooke, 1635–1703, a contemporary of Newton).

Hooke's law:

$$F = -k\Delta x = -k(x - x_o) \text{ or } F = -kx \text{ (with } x_o = 0) \quad \mathbf{(13.1)}$$

The spring force varies with x, so it is a variable force and a function of position. The minus sign in the equation indicates that the spring force acts in a direction opposite to the displacement when a spring is either stretched or compressed. The constant k is called the spring constant or force constant. The greater the value of k, the stiffer or stronger the spring. The units of k, as can be seen from the equation $k = F/x$, are N/m.

This is an ideal relationship. However, real springs approximate this linear relationship within certain limits. (See ● Fig. 13.1.) Beyond the *elastic limit*, the spring is deformed and may eventually break.

As a mass oscillates vertically on a spring, work and energies are involved. The energies are kinetic energy and the potential energies of gravity and the spring. In this experiment, we will attempt to indirectly measure the maximum energy stored in the spring.

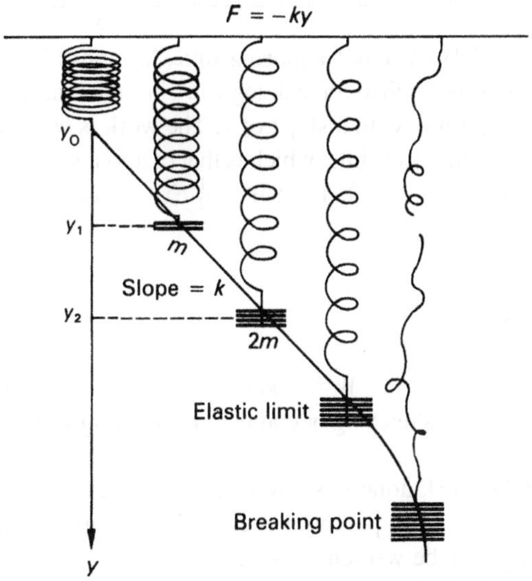

Figure 13.1 Hooke's law. An illustration in graphical form of spring elongation versus force. The greater the force, the greater the elongation, $F = -ky$. The Hooke's law relationship holds up to the elastic limit.

After performing this experiment and analyzing the data, you should be able to:

1. Describe how the spring constant of Hooke's law may be measured.
2. Understand there are two potential energies in an oscillating spring.
3. Explain how the spring potential energy may be measured at particular locations.

EQUIPMENT NEEDED

- Coil spring
- Slotted weights and hanger
- Meterstick
- Sheet of Cartesian graph paper

THEORY

Computing the work (and resulting energy) done by variable forces, such as in Hooke's law, generally requires calculus. But it is fortunate that the spring force is a special case that can be derived graphically. A plot of F versus x is shown in ● Fig. 13.2. The graph has a constant (straight-line) slope of k. The work (W) in the linear deformation of the spring is the area under the F-versus-x curve, and here it is in the form of a triangle. Then, computing this area,

$$\text{area} = W = \tfrac{1}{2}\,(\text{altitude} \times \text{base})$$

or

$$W = \tfrac{1}{2}Fx = \tfrac{1}{2}(kx)x = \tfrac{1}{2}kx^2$$

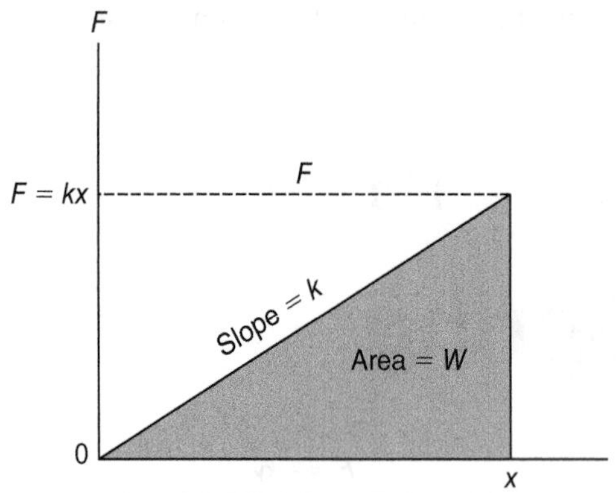

Figure 13.2 Work done by a spring force. A graph of F versus x, where F is the applied force doing work in stretching a spring, is a straight line with a slope of k. The work is equal to the area under the slope line, which is that of a triangle with area $= \frac{1}{2}$ (altitude × base). Then $W = \frac{1}{2} Fx = \frac{1}{2}(kx)x = \frac{1}{2} kx^2$. Then,

where $F = kx$. Or,

$$W = ½k(x - x_o)^2 \qquad \textbf{(13.2)}$$

(work done in stretching or compressing a spring with $x_o \neq 0$)

Since the work done in stretching or compressing a spring is stored as spring potential energy, the potential energy of a spring may be written $U_s = ½k(x - x_o)^2$

To check this equation, we will make use of an oscillating spring system, which will be considered conservative (● Fig. 13.3). In this case, the conservation of energy applies, $E_2 = E_1$, which are the sums of the kinetic (K) and potential (U) energies:

$$E_2 = E_1$$
$$K_2 + U_2 = K_1 + U_1$$

The kinetic energy term ($K = ½mv^2$) involves the velocity of the oscillating mass, which is difficult to measure. As is often done in physics, measurements are made at particular locations to simplify things. So, let's measure and compute the energies when $v = 0$ at the top and bottom of the oscillations.

Referring to the figure, there are two contributions of potential energy of the system at these points,

$$Us_1 + Ug_1 = Us_2 + Ug_2$$

where Us and Ug are the spring and gravitational potential energies, respectively. Then,

$$½k(y_1 - y_o)^2 + mgy_1 = ½k(y_2 - y_o)^2 + mgy_2$$

and with $y_2 - y_1 = \Delta y$ (change in height, change in potential energy), we have

$$½k(y_1 - y_o)^2 - ½k(y_2 - y_o)^2 = mg\Delta y \qquad \textbf{(13.3)}$$

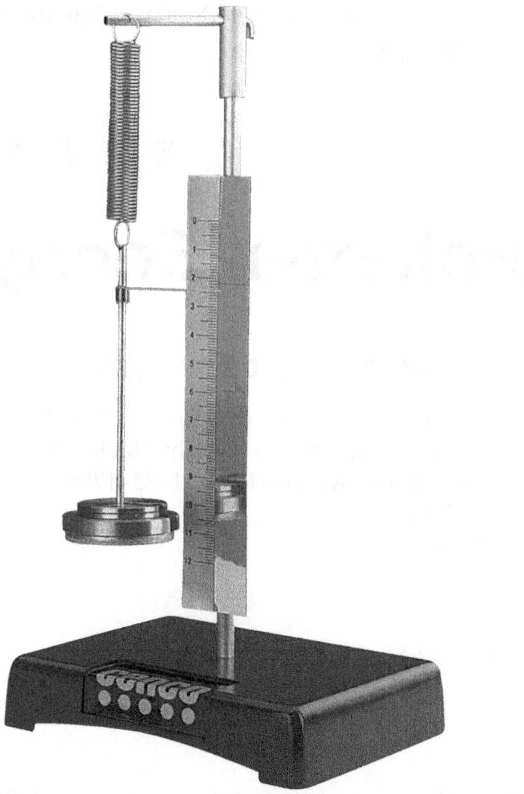

Figure 13.3 Hookes' law apparatus. The variables of Hooke's law ($F = mg$ and x) are measured using spring elongation. (Photo Courtesy of Sargent-Welch.)

EXPERIMENTAL PROCEDURE

A. Determining the spring constant

1. Hang the spring from a support, and suspend a weight hanger from the lower end. Add an appropriate weight to the hanger (for example, 100–200 g), and record the *total* suspended weight [m_1g in newtons (N)] in Data Table 1. Fix a meterstick vertically alongside the weight hanger, and note the position of the bottom of the weight hanger on the meterstick. Record this as y_1 in the data table.

2. Add appropriate weights (for example, 100 g) to the weight hanger one at a time for incremental elongations. Record the total suspended weight and the position (in meters) of the bottom of the weight hanger on the meterstick after each elongation (y_2, y_3, etc.). The added weights should be small enough so that seven or eight weights can be added without nearing the spring's elastic limit.

3. On a sheet of graph paper, plot the total suspended weight force versus elongation position (mg versus y), and draw the straight line that best fits the data points. Determine the slope of the line and record (k, the spring constant).

B. Energy Considerations

1. Resuspend the spring with no mass, and measure and record the unstretched distance y_o in Data Table 2.

2. Hang mass m_1 on the spring and pull down 5–10 cm. Record this position as y_2.

3. Release the mass, and estimate the highest position (y_1) the mass reaches before reversing direction.

4. Do this three times, and find an average y_1.

5. Repeat steps 2–4 for m_2.

6. Using the obtained data and Eq. 13.3, determine if the potential energy stored in the spring is correct in terms of the gravitational potential energy. (Does the equation balance?)

Name ____________ Section ____________ Date ____________

Lab Partner(s) ____________

EXPERIMENT 14

Torques, Equilibrium, and Center of Gravity

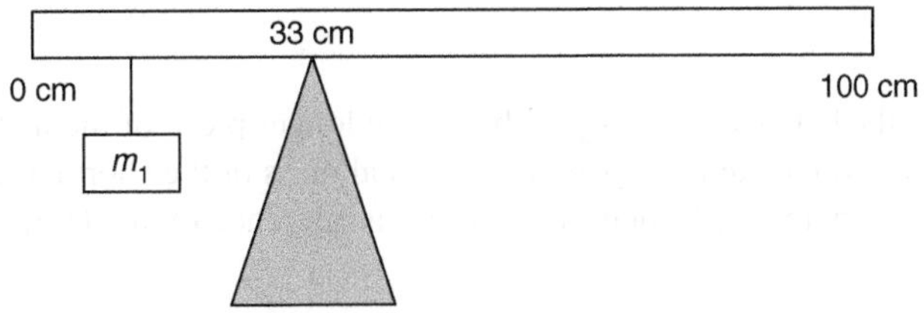

GL Figure 14.1 From broomstick to meterstick. See Experimental Planning text for description.

GL Experimental Planning

A torque gives rise to rotational motion of a rigid body through the application of a force at some distance from an axis of rotation. The magnitude of a torque (τ) may be found from the product of the force F and the perpendicular distance from the axis of rotation to the force's line of action, $r_\perp$ (called a lever arm): $\tau = r_\perp F$ (see Fig. 14.1 in the Theory section). When there is no net torque ($\Sigma\vec{\tau} = 0$) acting on a stationary rigid body, the body will be in *static rotational equilibrium* and there is no rotational motion.

As an example of the role of torque in static rotational equilibrium, consider a conventional kitchen broom. It is not very difficult to balance the broom lengthwise on one finger (and you *can* try this at home). Is the balance point in the middle or closer to one end?

Now suppose that the broom is cut into two pieces at the balance point. How would the masses of the two pieces compare? Are they the same or different, and if different, which piece would have a larger mass?

This situation can be modeled with the equipment for this experiment, and you will be able to verify your answer, or change your mind, as appropriate.

Given the following equipment:

- Meterstick and support stand
- String and one knife-edge clamp or two knife-edge clamps (one with wire loop)
- Laboratory balance
- Mass hanger and assorted masses (5 g, 10 g, 20 g, 50 g, 100 g)

(continued)

Set up an analogous situation to the balanced broomstick (see ● GL Fig. 14.1). Let's say the meterstick balanced at a point one-third of the distance from the zero end. You can adjust the mass (m_1 in GL Fig. 14.1) to get the balance point at the 33-cm mark on the meterstick. Now, instead of cutting the meterstick, we will do some physics and predict what the mass of each piece would be if the meterstick were cut at the balance point.

The shorter piece would have a mass of m_1 plus the mass of 33 cm of meterstick. If the meterstick is uniform, then 33 cm will have 33/100 (or 33%) of the total mass of the meterstick. If you measure the mass, take one-third and add m_1, you can determine the total mass of the short end of the meterstick. Do this now.

Now, the mass on the other side of the balance point is just that of the longer piece of the meterstick and is 67% of the total mass of the meterstick. Compute this value, and compare it to the total mass of the short end. Does your result match up with your answer to the broomstick question above? Explain your result in reference to the definition for torque.

Name ____________________ Section ____________ Date ____________

Lab Partner(s) ________________________________

EXPERIMENT 14

Torques, Equilibrium, and Center of Gravity

TI Advance Study Assignment

Read the experiment and answer the following questions.

1. What conditions must be present for (a) translational equilibrium and (b) rotational equilibrium of a rigid body?

2. If these conditions for equilibrium are satisfied, is the rigid body necessarily in static equilibrium? Explain.

3. What are the general definition and mathematical expression for torque?

Don't forget units

(continued)

4. If torque is a vector, with a specific direction in space, what is meant by clockwise and counterclockwise torques? If the sums of these torques on a rigid body are equal, what does this imply physically?

5. What defines the center of gravity of a rigid body, and how is it related to the center of mass?

6. Define the term *linear mass density.* Also, what is implied if it is assumed that the linear mass density of an object is uniform?

PRE-LAB DEMONSTRATION | Several good demonstrations—Instructor's Choice

Which Way Does the Yo-yo Roll? (torque—rotational motion) [● Fig. D6]
Candle Seesaw (torque—equilibrium)
Too Weak to Pick It Up? (center of gravity—equilibrium)
Stack Them Up (center of mass—equilibrium)

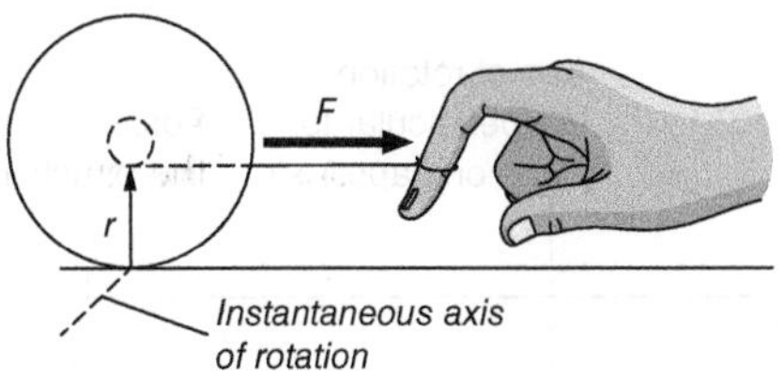

Figure D6 Which way does the yo-yo roll?

EXPERIMENT 14

Torques, Equilibrium, and Center of Gravity

INTRODUCTION AND OBJECTIVES

In introductory physics, forces act on particle "objects." That is, we consider an object to be a particle, which generally responds linearly to a force. In reality, an object is an extended collection of particles, and where a force is applied makes a difference. Rotational motion becomes relevant when the motion of a solid extended object or a rigid body is considered.

A **rigid body** is an object or system of particles in which the distances between particles are fixed and remain constant. A quantity of liquid water is *not* a rigid body, but the ice that would form if the water were frozen is.

Actually, the concept of a rigid body is an idealization. In reality, the particles (atoms and molecules) of a solid vibrate constantly. Also, solids can undergo deformations. Even so, most solids can be considered to be rigid bodies for the purposes of analyzing rotational motion.

An important condition of rigid bodies in many practical applications is **static equilibrium**. Examples include girders in bridges and the beam of a laboratory beam balance when taking a reading. They are at rest, or in static equilibrium. In particular, the balance beam is in *rotational static equilibrium* when "balanced" for a reading and not rotating about some point or axis of rotation.

The criterion for rotational static equilibrium is that the sum of the torques, or moments of force acting on a rigid body, be equal to zero. To study torques and rotational equilibrium, we will use a "beam" balance in the form of a meterstick and suspended weights. The torques of a setup will be determined experimentally by the "moment-of-force" method, and the values compared. Also, the concepts of center of gravity and center of mass will be investigated.

After performing this experiment and analyzing the data, you should be able to:

1. Explain mechanical equilibrium and how it is applied to rigid bodies.
2. Distinguish between center of mass and center of gravity.
3. Describe how a laboratory beam balance measures mass.

EQUIPMENT NEEDED

- Meterstick
- Support stand
- Laboratory balance
- String and one knife-edge clamp or four knife-edge clamps (three with wire hangers)
- Four hooked weights (50 g, two 100 g, and 200 g)
- Unknown mass with hook

THEORY

A. Equilibrium

The conditions for the mechanical equilibrium of a rigid body are

$$\Sigma \vec{\mathbf{F}} = 0 \quad \textbf{(14.1a)}$$

$$\Sigma \vec{\tau} = 0 \quad \textbf{(14.1b)}$$

That is, the (vector) sums of the forces $\vec{\mathbf{F}}$ and torques $\vec{\tau}$ acting on the body are zero.

The first condition, $\Sigma \vec{\mathbf{F}} = 0$, is concerned with **translational equilibrium** and ensures that the object is stationary (not moving linearly) or that it is moving with a uniform linear velocity (Newton's first law of motion). A stationary object is said to be in *translational static equilibrium.* In this experiment, the rigid body (a meterstick) is restricted from linear motion, so this is not a consideration.

To be in static equilibrium, a rigid body must also be in *rotational static equilibrium*. Although the sum of the forces on the object may be zero and it is not moving linearly, it is possible that it may be rotating about some fixed axis of rotation. However, if the sum of the torques

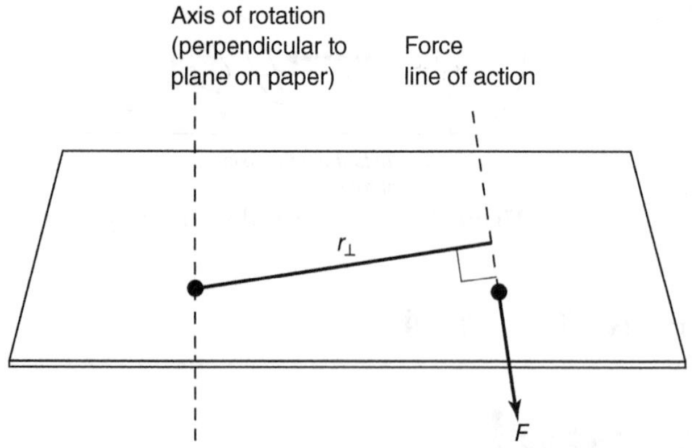

Figure 14.1 Torque. The magnitude of a torque is equal to the product of the magnitude of the force F and the perpendicular distance (lever arm), $r_\perp$, from the axis of rotation to the force's line of action; that is, $\tau = r_\perp F$.

is zero, $\Sigma\vec{\tau} = 0$, the object is in **rotational equilibrium**, and either it does not rotate (static case) or it rotates with a uniform angular velocity. (Forces produce linear motion, and torques produce rotational motion.)

Torque is a quantitative measure of the tendency of a force to cause or change the rotational motion of a rigid body. A *torque* (or *moment of force*) results from the application of a force acting at a distance from an axis of rotation (● Fig. 14.1). The magnitude of the torque is equal to the product of the force's magnitude F and the perpendicular distance $r_\perp$ from the axis of rotation to the force's line of action (a straight line through the force vector arrow). That is,

$$\boxed{\tau = r_\perp F} \qquad \textbf{(14.2)}$$

The perpendicular distance $r_\perp$ is called the **lever arm** or **moment arm**. The unit of torque can be seen to be the meter-newton (m-N). Notice that these units are the same as those of work ($W = Fd$), newton-meter (N-m) = joule (J). The unit of torque is commonly written as meter-newton (m-N) to emphasize the distinction. Keep in mind that although work and torque have the same units, they are not physically the same.

Torque is a vector quantity that points along the axis of rotation in one direction or the other. However, to distinguish torques and rotations, it is convenient to use a simple convention. If a torque tends to rotate the body in a counterclockwise direction (as viewed from above), then the torque is taken to be positive (+). If a torque tends to rotate the body in a clockwise direction, then the torque is taken to be negative (−). The plus and minus notation is helpful in torque calculations.

For example, in ● Fig. 14.2, taking the axis of rotation at the 50-cm position, F_1 and F_2 produce counterclockwise torques and F_3 and F_4 produce clockwise torques, but no rotation takes place if the torques are balanced and the system is in rotational static equilibrium.

It is convenient to sum the torques using magnitudes and directional signs, as determined by the counterclockwise (cc) and clockwise (cw) convention. In this case, the condition for rotational equilibrium [Eq. (14.1b)] becomes

$$\Sigma\tau_{\text{cw}} - \Sigma\tau_{\text{cc}} = 0$$

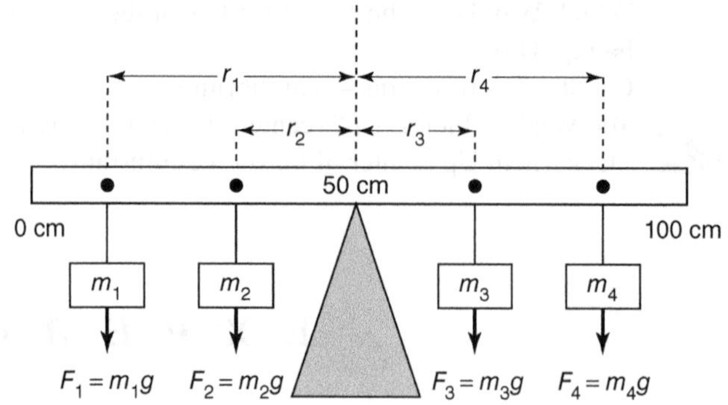

Figure 14.2 Torques in different directions. The forces F_1 and F_2 give rise to counterclockwise torques, and F_3 and F_4 clockwise torques, on the pivoted meterstick.

or

$$\boxed{\Sigma\tau_{\text{cc}} = \Sigma\tau_{\text{cw}}} \qquad \textbf{(14.3)}$$

(sum of counterclockwise torques = sum of clockwise torques)

Hence, we may simply equate the magnitudes of the cc and cw torques. For example, for the meterstick in Fig. 14.2 (writing the force first for convenience, that is, Fr),

Counterclockwise		Clockwise
$\tau_1 + \tau_2$	=	$\tau_3 + \tau_4$

or

$$F_1r_1 + F_2r_2 = F_3r_3 + F_4r_4$$

The forces are due to weights suspended from the rod, and with $F = mg$,

$$\boxed{m_1gr_1 + m_2gr_2 = m_3gr_3 + m_4gr_4} \qquad \textbf{(14.4)}$$

and, canceling g,

$$m_1r_1 + m_2r_2 = m_3r_3 + m_4r_4$$

Example 14.1 Let $m_1 = m_3 = 50$ g, $m_2 = m_4 = 100$ g in Fig. 14.2, where m_1, m_2, and m_3 are at the 10-, 40-, and 60-cm marks or positions, respectively, on the meterstick.* Where would m_4 have to be suspended for the stick to be in static equilibrium?

Solution In static equilibrium, the sum of the torques is zero, or the sum of the counterclockwise torques is equal to the sum of the clockwise torques [Eq. (14.3)],

$$\Sigma\tau_{\text{cc}} = \Sigma\tau_{\text{cw}}$$

In terms of forces and lever arms,

$$F_1r_1 + F_2r_2 = F_3r_3 + F_4r_4$$

* The official abbreviation for gram is g, and the commonly used symbol for acceleration due to gravity is *g*. The gravity *g* is written in italics, and the gram g is not. Look closely to avoid confusion.

where the forces are $F_i = m_i g$. The lever arms are measured from the 50-cm position of the meterstick, which is the pivot point or the location of the axis of rotation. In general, $r_i = (50\text{ cm} - x_i)$, where x_i is the centimeter location of a mass. Hence,

$$m_1g(50\text{ cm} - 10\text{ cm}) + m_2g(50\text{ cm} - 40\text{ cm}) = m_3g(60\text{ cm} - 50\text{ cm}) + m_4gr_4$$

and, canceling the g's,

$$m_1(40\text{ cm}) + m_2(10\text{ cm}) = m_3(10\text{ cm}) + m_4r_4$$

Then, putting in the mass values,

$$(50\text{ g})(40\text{ cm}) + (100\text{ g})(10\text{ cm}) = (50\text{ g})(10\text{ cm}) + (100\text{ g})r_4$$

and solving for r_4,

$$r_4 = \frac{2500\text{ g-cm}}{100\text{ g}} = 25\text{ cm}$$

Hence, for rotational equilibrium, m_4 is 25 cm from the support position (axis of rotation), or at the 75-cm position on the meterstick (measured from the zero end).

Here it is assumed that the meterstick is uniform (uniform mass distribution) so that the torques caused by the masses of the portions of the meterstick are the same on both sides of the support and therefore cancel.

B. Center of Gravity and Center of Mass

The gravitational torques due to "individual" mass particles of a rigid body define what is known as the body's center of gravity. The **center of gravity** is the "balance" point, the point of the body about which the sum of the gravitational torques about an axis through this point is zero. For example, consider the meterstick shown in ● Fig. 14.3. If the uniform meterstick is visualized as being made up of individual mass particles and the point of support is selected such that $\Sigma\vec{\tau} = 0$, then

$$\Sigma\tau_{cc} - \Sigma\tau_{cw}$$

or

$$\sum_{cc}(m_ig)r_i = \sum_{cw}(m_ig)r_i$$

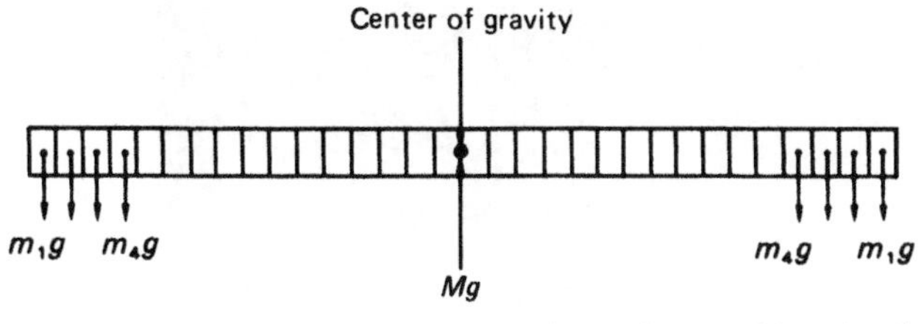

Figure 14.3 Center of gravity. A rod may be considered to be made up of individual masses in rotational equilibrium when the vertical support is directly through the center of gravity.

and

$$(m_1r_1 + m_2r_2 + m_3r_3 + \cdots)_{cc} = (m_1r_1 + m_2r_2 + m_3r_3 + \cdots)_{cw}$$

where g cancels. When the meterstick is in equilibrium, it is supported by a force equal to its weight, and the support force is directed through the center of gravity.

Hence, it is as though all of the object's weight (Mg) is concentrated at the center of gravity. That is, if you were blindfolded and supported an object at its center of gravity on your finger, you would not be able to tell, from its weight alone, whether it was a rod, a block, or an irregularly shaped object of equal mass. For a uniform meterstick, the center of gravity would be at the 50-cm position. (Why?)

If an object's weight is concentrated at its center of gravity, so should its mass be concentrated there. An object's center of gravity is often referred to as its **center of mass**. *These points are the same as long as the acceleration due to gravity g is constant (uniform gravitational field).* Notice how g can be factored and divided out of the previous *weight* equations, leaving *mass* equations.

Also, it should be evident that for a symmetric object with a uniform mass distribution, the center of gravity and center of mass are located at the center of symmetry. For example, if a rod has a uniform mass distribution, its centers of mass and gravity are located at the center of the rod's length. For a uniform sphere, the centers are at the center of the sphere.

Linear Mass Density

In part of the experiment, the masses of certain lengths of the meterstick will need to be known. These may be obtained from the **linear mass density (μ)** of the stick—that is, the mass (m) per unit length (L),

$$\boxed{\mu = \frac{m}{L}} \qquad \textbf{(14.5)}$$

with units of grams/centimeter (g/cm) or kilograms/meter (kg/m). For example, suppose a meterstick is measured to have a mass of 50 g on a balance. Then, since the stick is 100 cm long ($L = 100$ cm), the linear mass density of the stick is $\mu = m/L = 50\text{ g}/100\text{ cm} = 0.50$ g/cm. If the mass distribution of the stick were uniform, then every centimeter would have a mass of 0.50 g. However, metersticks are not uniform, so this is an average value.

Example 14.2 If a meterstick has a linear mass density of 0.50 g/cm, what is the mass of a 16-cm length of the stick?

Solution With $\mu = m/L$, then $m = \mu L$, and for $\mu = 0.50$ g/cm and $L = 16$ cm,

$$m = \mu L = (0.50\text{ g/cm})(16\text{ cm}) = 8.0\text{ g}$$

EXPERIMENTAL PROCEDURE

(Here the equilibrium conditions will be determined by the summing of torques or moments of force, hence the term "moments-of-force" method.)

A. Apparatus with Support Point at Center of Gravity

1. A general experimental setup is illustrated in ● Fig. 14.4, where the masses or weights are suspended by knife-edge clamp weight hangers. The hooked masses may also be suspended from small loops of string, which can be slid easily along the meterstick. The string allows the position of a mass to be read easily and may be held in place by a small piece of masking tape.
   (a) Determine the mass of the meterstick (without any clamps), and record it in the Laboratory Report.
   (b) Weights may be suspended by loops of string or clamps with weight hangers. The string method is simpler; however, if you choose or are instructed to use weight hangers, weigh the three clamps together on a laboratory balance and compute the average mass of a clamp. Record in the Laboratory Report.

2. With a knife-edge clamp on the meterstick near its center, place the meterstick (without any suspended weights) on the support stand. Make certain that the knife edges are on the support stand. (The tightening screw head on the clamp will be down.)

   Adjust the meterstick through the clamp until the stick is balanced on the stand. Tighten the clamp screw, and record in Data Table 1 the meterstick reading or the distance of the balancing point x_o from the zero end of the meterstick.

3. *Case 1: Two known masses*
   (a) With the meterstick on the support stand at x_o, suspend a mass $m_1 = 100$ g at the 15-cm position on the meterstick—that is, 15 cm from the zero end of the meterstick.
   (b) Set up the conditions for static equilibrium by adjusting the moment arm of a mass $m_2 = 200$ g suspended on the side of the meterstick opposite m_1. Record the masses and moment arms in Data Table 1. If clamps are used instead of string, do not forget to add the masses of the clamps. Remember the moment arms are the distances from the pivot point to the masses (that is, $r_i = |x_i - x_o|$).
   (c) Compute the torques, and find the percent difference in the computed values (that is, compare the clockwise torque with the counterclockwise torque).

4. *Case 2: Three known masses*
   Case (a)
   (i) With the meterstick on the support stand at x_o, suspend $m_1 = 100$ g at the 30-cm position and $m_2 = 200$ g at the 70-cm position. Suspend $m_3 = 50$ g, and adjust the moment arm of this mass so that the meterstick is in static equilibrium. Record the data in Data Table 1.
   (ii) Compute the torques and compare as in Procedure 3.
   Case (b)
   (i) Calculate theoretically the lever arm (r_3) for the mass $m_3 = 50$ g for the system to be in equilibrium if $m_1 = 100$ g is at the 20-cm position and $m_2 = 200$ g is at the 60-cm position. (Remember to add the masses of the hanger clamps if used.) Record this value in the data table.
   (ii) Check your results experimentally, and compute the percent error of the experimental value of r_3, taking the previously calculated value as the accepted value.

5. *Case 3: Unknown mass—The balance principle.* A balance (scale) essentially uses the method of moments to compare an unknown mass with a known mass. Some balances have constant and equal lever arms, and others do not (see Experiment 3, Fig. 3.1). This procedure will illustrate the balance principle.

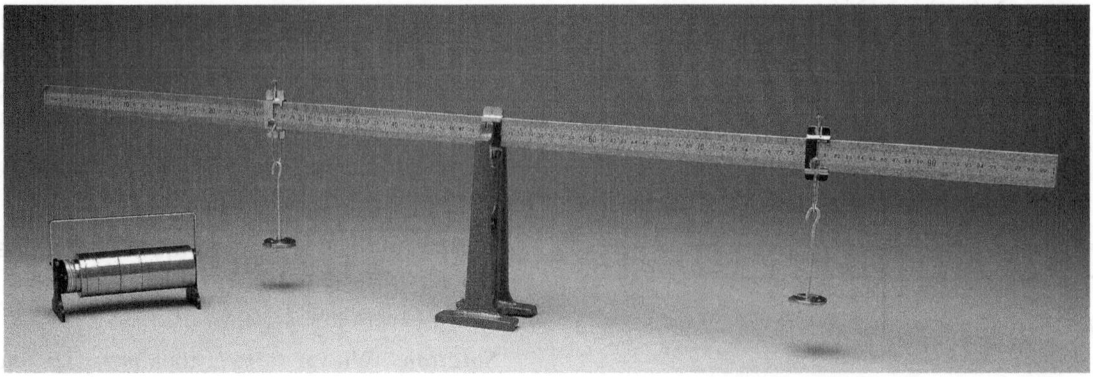

Figure 14.4 Torque apparatus. Example of experimental setup and equilibrium conditions. (Photo Courtesy of Sargent-Welch.)

(a) With the meterstick on the support stand at x_o, suspend the unknown mass (m_1) near one end of the meterstick (for example, at the 10-cm position). Suspend from the other side of the meterstick an appropriate known countermass m_2 (for example, 200 g) and adjust its position until the meterstick is "in balance" or equilibrium. Record the value of the known mass and the moment arms in Data Table 1.

(b) Remove the unknown mass, and determine its mass on a laboratory balance.

(c) Compute the value of the unknown mass by the method of moments, and compare it with the measured value by calculating the percent error.

6. *Case 4: Instructor's choice (optional).* Your instructor may have a particular case he or she would like you to investigate. If so, the conditions will be given. Space has been provided in the data table for reporting your findings.

B. Apparatus Supported at Different Pivot Points

In the previous cases, the mass of the meterstick was not explicitly taken into account since the fulcrum or the position of the support was at the meterstick's center of gravity or center of mass. In effect, the torques due to the mass of the meterstick on either side of the support position canceled each other. The centers of gravity of the lengths of the stick on either side of the support are equidistant from the support (for example, at the 25-cm and 75-cm positions for a uniform stick) and have equal masses and moment arms.

For the following cases, the meterstick will not be supported at its center-of-gravity position (x_o) but at some other pivot points (designated in general by x_o'; for example, see ● Fig. 14.5). In these cases, the mass of the meterstick needs to be taken into account. To illustrate this very vividly, let's start off with a case with only one suspended mass.

7. *Case 5: Meterstick with one mass.* Suspend a mass $m_1 = 100$ g at or near the zero end of the meterstick (Fig. 14.5). Record the mass position x_1 in Data Table 2. If a string loop is used, a piece of tape to hold the string in position helps. Move the meterstick in the support clamp until the system is in equilibrium. (This case is analogous to the solitary seesaw—sitting on one side of a balanced seesaw with no one on the other side.) Record the support position x_o' in Data Table 2.

Since the meterstick is in balance (static equilibrium), the point of support must be at the center of gravity of the system; that is, the torques (clockwise and counterclockwise) on either side of the meterstick must be equal. But where is the mass or force on the side of the meterstick opposite the suspended mass? The balancing torque must be due to the mass of length L_2 of the meterstick (Fig. 14.5). To investigate this:

(a) Using the total mass m of the meterstick (measured previously) as m_2, with a moment arm r_2 (see the diagram in Data Table 2), compute the counterclockwise and clockwise torques, and compare them by computing the percent difference. Record it in Data Table 2.

(b) Now the masses of the lengths of meterstick will be taken into account. Compute the average linear mass density of the meterstick (see Theory, Section B), and record it in the data table.

If we assume that the mass of the meterstick is uniformly distributed, the center of mass (or center of gravity) of the length of meterstick L_2 on the *opposite* side of the support from m_1 is at its center position (see Fig. 14.5). Compute the mass m_2 of this length of stick (see Example 14.2) and record. Also, record the center position of L_2, where this mass is considered concentrated (x_2), and find the length of the lever arm r_2. It should be evident that $r_2 = L_2/2$.

Compute the torque due to m_2, and record it as τ_{cw}. From the linear mass density, compute the m_3 of the portion of the meterstick remaining to the left of the pivot. Calculate the torque due to this portion of the meterstick, add it to the torque due to mass m_1 to find the total counterclockwise torque, and record it as τ_{cc}. Compare the torque *differences* with those found in Case 5(a).

8. *Case 6: Center of gravity.*

(a) With a mass $m_1 = 100$ g positioned at or near one end of the meterstick as in Case 5, suspend a mass $m_2 = 100$ g on the opposite side of the support stand at the 60-cm position. Adjust the meterstick in the support-stand clamp until the stick is in balance. This locates the center of gravity x_o' of the system. Record in Data Table 2, and find r_1 and r_2.

(b) Repeat the procedure with m_2 positioned at 70 cm.

(c) Repeat the procedure with m_2 positioned at 80 cm. Notice how the position of the center of gravity moves as the mass distribution is varied.

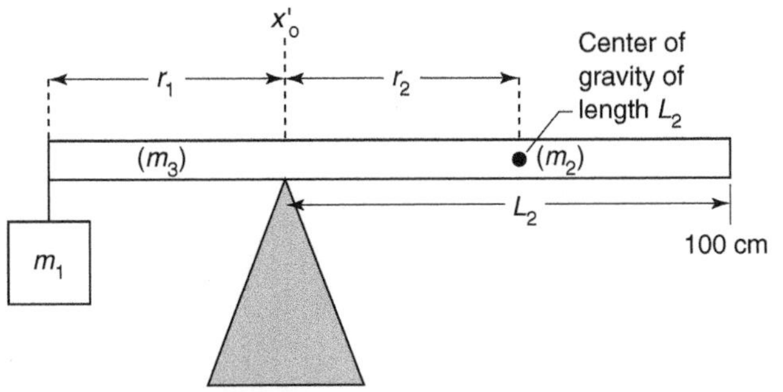

Figure 14.5 Equilibrium. A meterstick in equilibrium with one suspended mass. See text for description.

(d) Based on the experimental data, what would you predict the position of the center of gravity x_o' of the system would be if m_2 were moved to the 90-cm position? Record your prediction in the data table.

Using your prediction, compute the counterclockwise and clockwise torques, taking into account the mass of the meterstick as in Procedure 7(c). Compare the torques by computing the percent difference.

Experimentally determine the position of the center of gravity of the system, and compute the percent difference between the experimental and predicted values.

Name ______________________ Section ______________ Date ______________

Lab Partner(s) ____

EXPERIMENT 14 **Torques, Equilibrium, and Center of Gravity** *Laboratory Report*

Diagram*	Values (add m_c if applicable)	Moment (lever) arms	Results†
Case 6(a)	m_1 ______ $x_1 = 0$ cm m_2 ______ $x_2 = 60$ cm x'_o ______	r_1 ______ r_2 ______	
Case 6(b)	same except $x_2 = 70$ cm x'_o ______	r_1 ______ r_2 ______	
Case 6(c)	same except $x_2 = 80$ cm x'_o ______	r_1 ______ r_2 ______	
Case 6(d)	same except $x_2 = 90$ cm x'_o ______ (predicted)	τ_{cc} ______ τ_{cw} ______ Percent diff. ______	x'_o ______ (measured) Percent diff. ______

*Draw a diagram to illustrate each case, using the Case 5(a) diagram as an example. Put the mass of a length of stick in parentheses as in that diagram.
†Attach a sheet to the Laboratory Report showing calculations for each case.

TI QUESTIONS

1. Was the center of gravity of the meterstick exactly at the 50-cm mark? If not, why? Explain how the condition $\Sigma\vec{\mathbf{F}}_i = 0$ is satisfied for the meterstick in part A of the experiment.

(continued)

2. Why are clockwise and counterclockwise referred to as "senses," rather than directions?

3. Suppose in a situation like Case 2(a) in the experiment, $m_1 = 200$ g were at the 20-cm position and $m_2 = 100$ g at the 65-cm position. Would there be a problem in experimentally balancing the system with $m_3 = 50$ g? Explain. If so, how might the problem be resolved?

4. Describe the effects of taking the mass of the meterstick into account when the balancing position is not near the 50-cm position.

5. (*Optional*) A uniform meterstick is in static rotational equilibrium when a mass of 220 g is suspended from the 5.0-cm mark, a mass of 120 g is suspended from the 90-cm mark, and the support stand is placed at the 40-cm mark. What is the mass of the meterstick?

Name ______________________ Section ______________ Date ______________

Lab Partner(s) ____

EXPERIMENT 15

Simple Machines: Mechanical Advantage

GL Experimental Planning

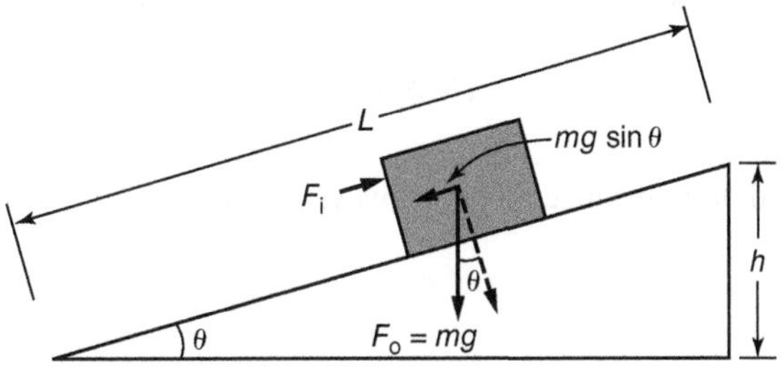

GL Figure 15.1 A simple machine—the inclined plane. Less input force, F_i, is required to move a load a vertical distance h, but the load must be moved through a greater distance L. See Experimental Planning text for description.

Machines are used every day. Although most common machines, such as can openers, lawn mowers, or automobile engines, are thought of as complex mechanical devices, they all utilize the basic principles and components of simple machines.

Machines make it easier to do work. But how is this done? You should know from conservation principles that you don't get work done without using energy. So how do machines make work easier? The idea of mechanical advantage will be explored in this section.

A **simple machine** is a device that can change the magnitude (or direction) of an applied force. The work done by the force depends on the force component (F) parallel to the displacement d, that is, $W = Fd$. (See Experiment 12.) The conservation of energy principle does not allow the energy or work output of a machine to exceed the energy or work input. In the *ideal* case of being equal,

$$\text{Work input} = \text{work output}$$

$$W_i = W_o$$

$$F_i d_i = F_o d_o \qquad \textbf{(GL 15.1)}$$

1. If a simple machine produces an output force that is larger than the input force ($F_o > F_i$), what does GL Eq. 15.1 tell you about the input and output distances?

2. A ramp or inclined plane is an example of a simple machine that makes it easier to raise objects to a higher elevation (● GL Fig. 15.1). What is the applied force needed to push a block up a frictionless inclined plane at a constant speed? (Express your result in terms of the weight of the block and the angle of incline.)

(continued)

3. How does this force (F_i) compare to the weight of the block (F_o)? (The output force in this case is the force needed to lift the object directly a vertical distance h (without the ramp.)

4. Now consider d_i and d_o. If the output distance d_o is the vertical distance h, what is d_i (the distance through which the applied force F_i acts)? Express your result in terms of the height h and the angle of incline.

5. The **theoretical mechanical advantage** (TMA) of a simple machine for the ideal (frictionless) case is defined as:

$$\text{TMA} = \frac{F_o}{F_i} \qquad \textbf{(GL 15.2)}$$

(but F_o and F_i cannot be measured directly since there is always friction).*

6. Note that GL Eqs. 15.1 and 15.2 can be combined to show that

$$\text{TMA} = \frac{d_i}{d_o} \qquad \textbf{(GL 15.3)}$$

Use your results from Question 4 above to express the TMA with this equation.

Thus, the TMA can be determined in terms of either the forces or the distances. In the case of the inclined plane, the TMA may be obtained directly from the angle of incline. Show that, for an inclined plane, the TMA $= 1/\sin\theta$.

In the not-so-ideal case where friction is present, the **actual mechanical advantage (AMA)** is determined the same way:

$$\text{AMA} = \frac{F_o}{F_i} \qquad \textbf{(GL 15.4)}$$

However, in this case, the conservation of energy principle includes the work associated with friction (W_f):

$$\text{Total work input} = \text{total work output}$$

$$W_i = W_o + W_f$$

or

$$F_i d_i = F_o d_o + W_f \qquad \textbf{(GL 15.5)}$$

where W_f is the magnitude of the energy used to overcome friction. From GL Eq. 15.5 it can be seen that $W_o < W_i$ and $F_o d_o < F_i d_i$. How will the AMA compare to the TMA for any simple machine that is not frictionless?

*The theoretical mechanical advantage (TMA) is sometimes called the **ideal mechanical advantage (IMA)**.

7. Machines are often rated in terms of their **efficiency** (ε), which is defined as the ratio of the work output to the work input:

$$\varepsilon = \frac{\text{work output } (W_o)}{\text{work input } (W_i)} \qquad \textbf{(GL 15.6)}$$

As a ratio of work, efficiency is unitless and is often expressed as a percentage. The efficiency is always less than 1 or 100%. Explain why this is the case, and show that

$$\varepsilon = \frac{\text{AMA}}{\text{TMA}} \qquad \textbf{(GL 15.7)}$$

(*Hint*: Use the distance form of the TMA.)

8. Note that efficiency tells how much of the work input ends up as useful work output:

$$\text{Useful work output } (W_o) = \varepsilon \text{ (work input, } W_i)$$

If the efficiency of a machine is 0.75, how many joules of energy are required to do 1000 J of work?

Name ______________________ Section ______________ Date ______________
Lab Partner(s) ____

EXPERIMENT 15

Simple Machines: Mechanical Advantage

TI Advance Study Assignment

Read the experiment and answer the following questions.

1. Why is the actual mechanical advantage (AMA) a force multiplication factor?

2. Can work be multiplied by a machine? Explain.

3. Why is the theoretical mechanical advantage (TMA) theoretical or ideal?

4. Which is larger, the AMA or the TMA of a machine, and why?

(continued)

5. What factors comprise the ratio of the efficiency of a machine, and how is efficiency related to the AMA and TMA?

6. Why is efficiency always less that 100%?

EXPERIMENT 15

Simple Machines: Mechanical Advantage

INTRODUCTION AND OBJECTIVES

Machines are used daily to "do work." Upon analysis, all mechanical machines, however complex, are combinations of simple machines of which there are six mechanical classes: (1) inclined planes, (2) levers, (3) pulleys, (4) wheel and axles, (5) wedges, and (6) screws.

Although used to perform work, a simple machine is basically a device that is used to change the magnitude (or direction) of a force. Essentially, a machine is a *force multiplier*.

The magnitude of this multiplication is given by a machine's **mechanical advantage**, that is, the **actual mechanical advantage (AMA)**, which takes into account frictional losses, and the **theoretical mechanical advantage (TMA)**, which expresses the ideal, nonfictional case. The relative amount of useful work done by a machine is expressed by the ratio of the useful work output and the work input, which is called the **efficiency** (ε). Basically, efficiency tells "what you get out for what you put in." The rest of the input is lost mainly to work done against friction.

In this experiment, the AMAs, TMAs, and the efficiencies of some simple machines will be experimentally determined to illustrate these concepts and to show the parameters on which the force multiplications of machines depend.

After performing the experiment and analyzing the data, you should be able to do the following:

1. Describe how machines "do work" for us.
2. Distinguish between TMA and AMA.
3. Explain how the TMAs can be measured for:
   (a) an inclined plane.
   (b) a lever, and why the TMA gives a good approximation of the AMA,
   (c) pulley(s).

EQUIPMENT NEEDED

- 2 single pulleys and 2 double- or triple-sheave pulleys*
- Wheel and axle (Fig. 15.5)
- 2 weight hangers, slotted weights, and single weight
- Spring scale (calibrated in newtons)
- Meterstick
- Vernier caliper
- String
- Tape

* The single pulleys are not really necessary, as one sheave of the multiple pulleys can be used as a single pulley. The single pulleys are convenient for instruction.

THEORY

The **actual mechanical advantage (AMA)** of a machine is defined as

$$\text{AMA} = \frac{F_o}{F_i} \tag{15.1}$$

where F_o and F_i are the output and input forces, respectively. The AMA is the force multiplication factor of the machine. For example, if AMA $= 2 = F_o/F_i$, then $F_o = 2F_i$, or the output force is twice the input force.

But in no case is work multiplied by a machine. If this were the case, with more work output than work input, energy would have to be created. However, the total work (energy) is conserved:

Total work in = total work out
= useful work + work done against friction

or

$$W_i = W_o + W_f$$

$$F_i d_i = F_o d_o + W_f \tag{15.2}$$

where d_i and d_o are parallel distances through which the respective forces, or component of forces, act (work = force × parallel distance, $W = Fd$) and W_f is the work done against friction.

In actual situations, there is always some work (energy) input lost to friction. If a machine were frictionless, $W_f = 0$, then $F_i d_i = F_o d_o$. For this theoretical situation, a **theoretical mechanical advantage (TMA)** can be expressed as

$$\text{TMA} = \frac{F_o}{F_i} = \frac{d_i}{d_o} \tag{15.3}$$

This is an ideal situation, and the theoretical mechanical advantage is sometimes called the *ideal mechanical advantage* (IMA).

Note that the TMA is the ratio of the distances through which the forces act and thus depends on the geometrical configuration of the machine. That is, the distances can be determined by measurement from the machine and, hence, the TMA.

The **efficiency** (ε) of a machine is defined as the ratio of its work output and its work input, and

$$\varepsilon = \frac{\text{work output } (W_o)}{\text{work input } (W_i)} = \frac{F_o d_o}{F_i d_i} = \frac{F_o/F_i}{d_i/d_o} = \frac{\text{AMA}}{\text{TMA}} \quad \textbf{(15.4)}$$

The efficiency is often expressed as a percentage. Because of friction, AMA < TMA, and the efficiency is always less than 1, or 100%. The efficiency tells what part of the work input goes into useful work output:

$$\text{Useful work output } (W_o) = \varepsilon \text{ (work input, } W_i)$$

for example, if $\varepsilon = 0.7$, or 70%, then 70% of the work input is used by the machine to do useful work. The rest of the work input, 0.3, or 30%, is lost to friction.

A. Inclined Plane

The theory of the **inclined plane** (● GL Fig. 15.1) was presented in the TI Experimental Planning with the result

$$\text{TMA} = \frac{F_o}{F_i} = \frac{d_i}{d_o} = \frac{1}{\sin\theta} \quad \textbf{(15.5)}$$

(inclined plane)

B. Lever

The **lever** is a very efficient simple machine. It consists of a rigid bar that is pivoted to rotate about a point or line called the *fulcrum* (● Fig. 15.2). The input force F_i, commonly called the *effort*, is applied to the end of the lever to maintain or lift a *load* (w). The input force, F_i, must be equal to (static case, Fig. 15.2a) or greater than the weight of the load when lifted (Fig. 15.2b). The input and output lever arms, L_i and L_o, are the distances from the fulcrum to the effort (F_i) and from the fulcrum to the load (F_o), respectively.

The theoretical mechanical advantage (TMA) of a lever can be calculated from work considerations (friction neglected):

$$W_{in} = W_{out}$$

and

$$F_i s_i = F_o s_o \quad \text{or} \quad \frac{F_o}{F_i} = \frac{s_i}{s_o}$$

where s represents arc length distance.

The two arcs subtend the same angle θ, so $s_i = L_i\theta$ and $s_o = L_o\theta$ and $s_i/s_o = L_i/L_o$. Then by Eq. 15.3,

$$\text{TMA} = \frac{F_o}{F_i} = \frac{d_i}{d_o} = \frac{s_i}{s_o} = \frac{L_i}{L_o}$$

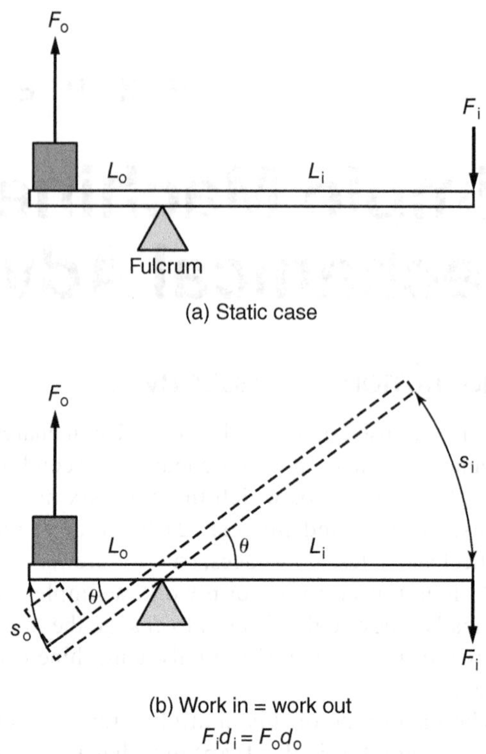

Figure 15.2 The lever. (a) A static case of maintaining a load. (b) In lifting a load, the work in equals the work out (neglecting friction). See text for description.

and

$$\text{TMA} = \frac{L_i}{L_o} \quad \textbf{(15.6)}$$

(lever)

Note that the TMA is given by the geometry of the system. For example, if $L_i = 75$ cm and $L_o = 25$ cm, then TMA $= L_i/L_o = 75 \text{ cm}/25 \text{ cm} = 3$, and the lever ideally multiplies the force by a factor of 3.

Frictional losses are normally quite small in the lever action. So, for most practical purposes, the actual mechanical advantage (AMA) can be taken to be approximately equal to the TMA. However, the AMA can be determined experimentally by

$$\text{AMA} = \frac{F_o}{F_i} \quad \textbf{(15.7)}$$

C. Pulleys

A **pulley** is actually a continuous lever with equal lever arms (● Fig. 15.3a). When a pulley or system of pulleys is used to lift a load of weight w by an applied force F, the AMA is

$$\text{AMA} = \frac{F_o}{F_i} = \frac{w}{F} \quad \textbf{(15.8)}$$

[pulley(s)]

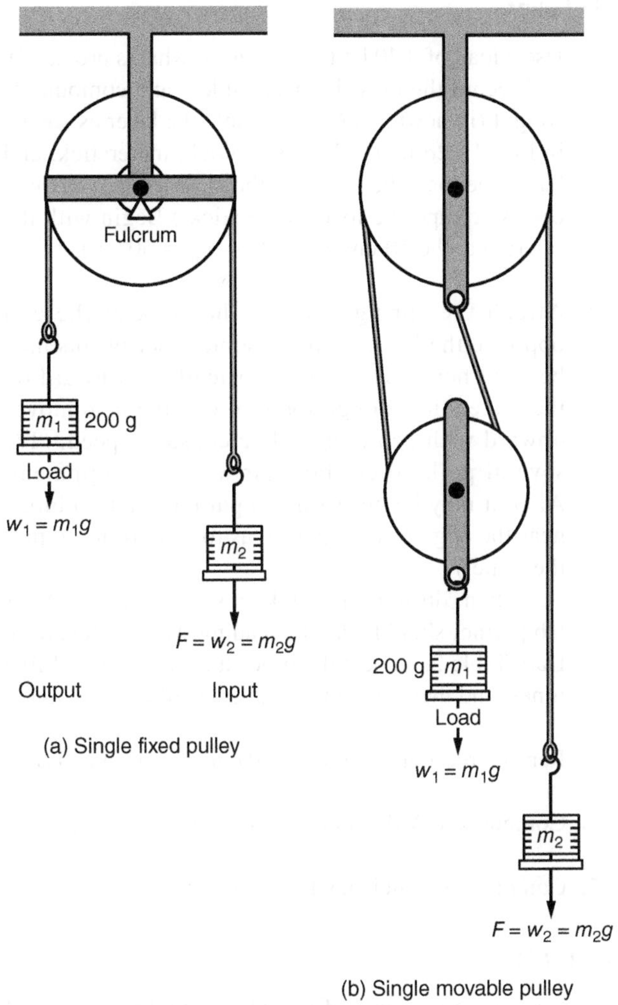

Figure 15.3 Pulley arrangements. (a) A single fixed pulley. (b) A single movable pulley. Notice that the weight of the movable pulley is part of the load.

The TMA is the ratio of the distances through which the forces act, and since ideally $W_o = W_i$ as in the previous case,

$$\text{TMA} = \frac{d_i}{d_o} = \frac{h_i}{h_o} \qquad \textbf{(15.9)}$$

[*pulley(s)*]

where h_i and h_o are the vertical heights through which the input and output forces act, respectively.

For example, for a single movable pulley (Fig. 15.3b), suppose the load is moved a distance h upward. To move the load an up distance h, each suspending string must be "shortened" a distance h, and therefore the applied force must move downward a distance $2h$. Note that the suspended movable pulley adds to the weight of the load, since the weight of the pulley is also lifted by the applied force.

The AMA is measured when the load is lifted with a uniform speed so that acceleration is not a consideration. The mechanical advantage is based on the minimum input force, which may be obtained when the system is in static equilibrium, in which case the net vertical force is zero. (To lift the load, a slight tap or force would have to be given to put the system in motion.)

A set of fixed and movable pulleys is called a **block and tackle**. The pulleys, called *sheaves* (pronounced "shevs"), may be arranged in tandem, which is the configuration commonly used in the laboratory to make it easier to thread the pulleys (● Fig. 15.4a). Or, the pulleys may have a common axis of rotation for compactness in many practical applications (Fig. 15.4b).

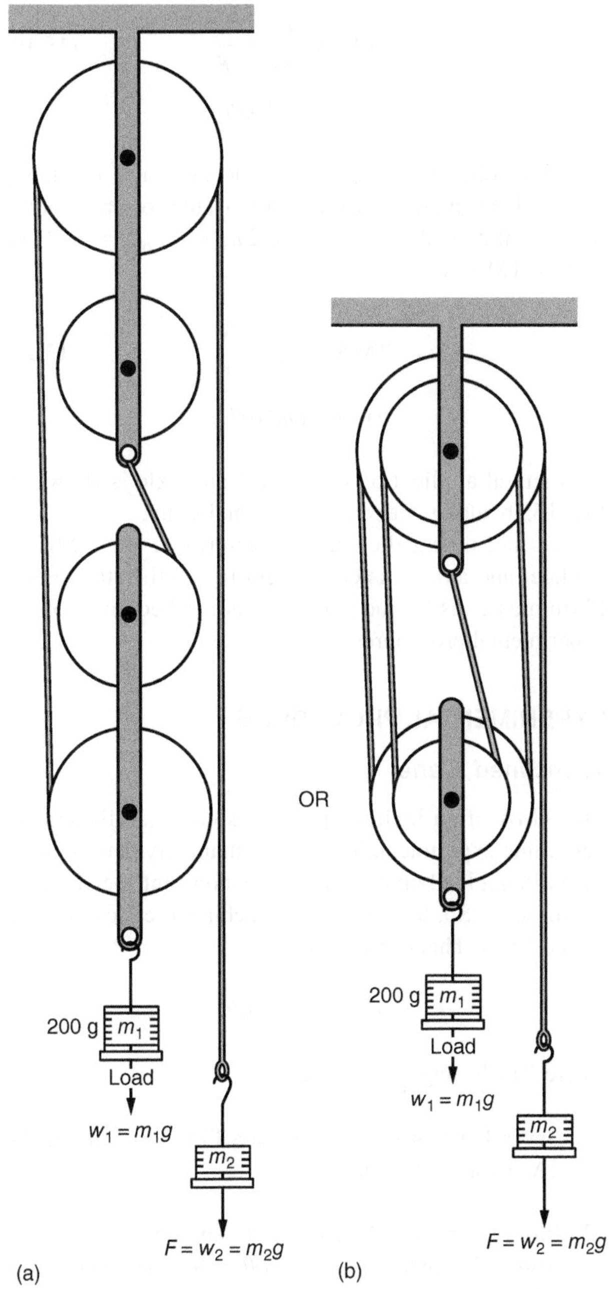

Figure 15.4 Pulley arrangement with double movable pulleys. (a) The pulleys may be arranged in tandem or (b) have a common axis for compactness.

D. Wheel and Axle

The combination of a **wheel and axle** has many practical applications. For example, when you open a door by turning a doorknob, you are using a wheel and axle. This simple machine consists of a wheel fixed to a shaft or axle with the same axis of rotation (● Fig. 15.5a). Essentially, it is equivalent to a lever with unequal lever arms.

A force F_i applied tangentially to the wheel with a radius R can lift a load w (F_o) by means of a string or rope wrapped around the axle (radius r). The AMA of the wheel and axle is

$$\text{AMA} = \frac{F_o}{F_i} = \frac{w}{F} \qquad (15.10)$$

(wheel and axle)

In one revolution, the input force acts through a distance $2\pi R$ and the output force through a distance of $2\pi r$. For the ideal, nonfictional case, $F_i d_i = F(2\pi R) = F_o d_o = w(2\pi r)$, and the TMA is

$$\text{TMA} = \frac{d_i}{d_o} = \frac{R}{r} \qquad (15.11)$$

(wheel and axle)

A practical application of a wheel and axle is shown in Fig. 15.5b, along with an experimental setup.

In measuring the force to determine the AMA of a wheel and axle, it is convenient to use the static equilibrium case, as for the pulley system in Section C of the experimental procedures.

EXPERIMENTAL PROCEDURE

A. Inclined Plane

The AMA of an inclined plane is somewhat difficult to determine experimentally. That is, manually determine F_i by pulling a load up the incline at a constant speed with a spring scale. So, for this simple machine, the focus will be on the TMA. This is given by

$$\text{TMA} = 1/\sin\theta$$

where θ is the angle of incline.

1. Several angles are listed in Data Table 1. Compute the TMA for each angle.

2. Plot a graph of TMA versus θ, and comment on the result. Compute the TMA for $\theta = 90°$. What does this tell you?

B. Lever

3. Use a load of 0.40 kg to 0.50 kg (or what is provided), and record the mass in Data Table 2 and compute its weight (in newtons, N). Assemble the lever as shown in Fig. 15.2b using the pivot block, meterstick, and load. The load should be on the table. (It may be necessary to tape the load to the stick.) Begin with the fulcrum at the 50-cm mark ($L_i = L_o = 50$ cm).

4. Attach the spring scale to the stick at the end opposite the load. (Tape or some other means may be used here.) Then, pull vertically downward on the scale with enough force so that the load moves upward with approximately a constant speed. (Try several pulls to get familiar with the equipment. Also, it may be necessary to put the fulcrum block near the edge of the table so as to have room to pull the scale.)

   Then, during a pull take a scale reading. (Your lab partner should take the reading. Why?) Record in Data Table 2. Repeat the procedure for a total of five times, and then compute the average of the readings.

5. Repeat Procedure 4 for $L_i = 60$ cm, 70 cm, and 80 cm.

6. Compute the AMA and the TMA for each case.

7. Compute the efficiency for each case.

C. Pulleys

8. Determine the mass of one of the single pulleys and one of the multiple-sheave pulleys on a laboratory balance, and record in Data Table 3. These pulleys will be used as the movable pulleys in the following situations, and their weights must be included in the loads.

9. Assemble a single fixed pulley as illustrated in Fig. 15.3a, with enough weights on the force input hanger so that it moves downward with a slow uniform speed when given a slight tap. (A single pulley or one pulley of multiple sheaves may be used.) Record the forces due to these masses in Data Table 3.

10. The next step is to measure the distances d_i (or h_i) and d_o (or h_o) with a meterstick. Pull down the weight hanger supplying the input force F a distance (d_i) of 20 cm or more, and note the distance the load moves upward (d_o). Record these distances in Data Table 3.

11. Calculate the AMA, TMA, and efficiency (ε) for this case.

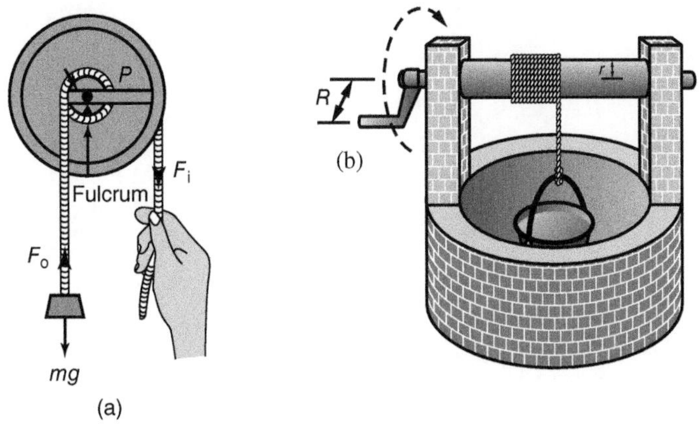

Figure 15.5 Wheel and axle. (a) The wheel and axle consists of a larger wheel fixed to a smaller shaft or axle with the same axis of rotation. It is equivalent to a lever with unequal lever arms. (b) A practical example of a wheel and axle.

12. Assemble a pulley system as illustrated in Fig. 15.3b, and repeat Procedures 9 through 11 for this case. (Don't forget to include the mass of the movable pulley as part of the load, since it, too, is being raised.)

13. Assemble a pulley system as illustrated in Fig. 15.4a (or b), and repeat Procedures 9 through 11.

D. Wheel and Axle

14. Using the vernier caliper, determine the radii of the wheel (largest diameter) and of the larger and smaller axles. The wheel and axle apparatus commonly used has three sizes.*

15. Fixing and wrapping strings around the wheel and axle, set up the apparatus with weight hangers and enough weights on the input force weight hanger so that it descends with a slow uniform speed. Start with the larger-diameter axle if your wheel and axle has multiple axles. Record the masses of the applied force and the load in Data Table 4.

16. Calculate the AMA, TMA, and efficiency (ε) for this case.

17. Repeat Procedures 15 and 16 with the load suspended from the smaller axle (if your wheel and axle has multiple axles).

*See Experiment 3 for the use of vernier caliper.

Name ______________________ Section ______________ Date ______________

Lab Partner(s) ____

EXPERIMENT 15

Simple Machines: Mechanical Advantage

TI Laboratory Report

A. Inclined Planes

DATA TABLE 1

θ	TMA
5°	
10°	
15°	
20°	
25°	
30°	
35°	
40°	
45°	

Comment on graph:

Comment on $\theta = 90°$

B. Levers

DATA TABLE 2

Load mass ______________

Load weight (F_o) ______________

L_i length	50 cm	60 cm	70 cm	80 cm
	F_i	F_i	F_i	F_i
Trial 1				
2				
3				
4				
5				
Average F_i				

	AMA	TMA	Efficiency
50 cm			
60 cm			
70 cm			
80 cm			

Calculations (show work)

Don't forget units

(continued)

C. Pulleys

DATA TABLE 3

Pulleys	Output force or load $w_1 = m_1g$ ( )	Input force, F, $w_2 = m_2g$ ( )	Output distance, d_o or h_o ( )	Input distance, d_i or h_i ( )	AMA	TMA	Eff
Single fixed							
Single movable							
Double movable							

Calculations
(show work)

D. Wheel and Axle

DATA TABLE 4

Axle radius ( )	Wheel radius ( )	Output force or load $w_1 = m_1g$ ( )	Input force, F, $w_2 = m_2g$ ( )	AMA	TMA	Eff

Calculations
(show work)

Name ______________________ Section ______________ Date ______________

Lab Partner(s) ____

TI QUESTIONS

1. Simple machines are sometimes divided into two basic classes, inclined planes and levers, where the wedge and screw are included in one class, and the pulley and the wheel and axle are included in another. Explain why these four simple machines can be included in the basic classes of inclined planes and levers.

2. A machine multiplies force, but what is reduced or "sacrificed" for this force multiplication? Give a specific example.

3. (a) State how the AMA and TMA of an inclined plane vary with the inclination of the plane.

   (b) State how the efficiency of an inclined plane varies with the inclination of the plane, and explain the reason for this variation.

4. In going up stairs, the climb seems easier when going up in a zig-zag fashion, rather than straight up. Why is this?

5. Show that the TMA of a lever can be derived from torque considerations. (See Fig. 15.2a.)

6. A single fixed pull is often called a "direction changer" rather than a force multiplier. Explain why this is an appropriate name.

(continued)

7. The TMA of a pulley system with movable pulley(s), or a block and tackle, is equal to the number of supporting strands of the movable pulley or block.
   (a) Do your experimental results support this statement?

   (b) Explain the physical basis of this statement.

8. Give three common applications of the wheel and axle. (*Hint*: Is a screwdriver a wheel and axle?)

9. Estimating the radii of a common doorknob and its shaft, how much force is applied to the shaft mechanism when the knob is turned with an applied force of 2.0 N?

Name ______________________ Section ______________ Date ______________

Lab Partner(s) ____

EXPERIMENT 16

Simple Harmonic Motion

TI *Advance Study Assignment*

Read the experiment and answer the following questions.

1. What are Hooke's law and simple harmonic motion, and how are they related?

2. What is the physical significance of the spring constant of a spring? What does it tell you?

3. How is the spring constant determined in this experiment?

4. In the equation of motion for simple harmonic motion (Eq. TI 16.2), what physically determines A and T?

(continued)

5. How is the period of a mass oscillating on a spring related to the spring constant? (Express your answer mathematically and verbally.)

CI Advance Study Assignment

Read the experiment and answer the following questions.

1. What are the requirements for an object to move with simple harmonic motion?

2. Why is simple harmonic motion an idealization?

3. What is a simple pendulum?

4. Under what conditions can a pendulum be considered a simple harmonic oscillator?

5. Why is it important to start taking data when the pendulum is still at rest in its equilibrium position?

PRE-LAB DEMONSTRATION | **How about a demonstration on rotational motion?–Instructor's choice**

Cylinder Derby (rotational motion moment of inertia)
Egg Spin (rigid-body rotation)

EXPERIMENT 16

Simple Harmonic Motion

OVERVIEW

Experiment 16 considers simple harmonic motion (SHM) with TI and/or CI procedures. The TI procedure examines Hooke's law using rubber-band and spring elongations. Simple harmonic motion is investigated through the period of oscillation of a mass on a spring.

The CI procedure investigates the SHM of a simple pendulum and the resulting conversion of energy (kinetic and potential) that occurs during the motion. An electronic sensor measures the angular speed, $\omega = \Delta\theta/\Delta t$, of the pendulum, from which the tangential speed is computed and the energies calculated.

INTRODUCTION AND OBJECTIVES

Elasticity implies a restoring force that can give rise to vibrations or oscillations. For many elastic materials, the restoring force is proportional to the amount of deformation, if the deformation is not too great.

This is best demonstrated for a coil spring. The restoring force F exerted by a stretched (or compressed) spring is proportional to the stretching (compressing) distance x, or $F \propto x$. In equation form, this is known as **Hooke's law**,

$$F = -kx$$

where x is the distance of one end of the spring from its unstretched ($x = 0$) position, k is a positive constant of proportionality, and the minus sign indicates that the displacement and force are in opposite directions. The constant k is called the **spring** (or **force**) **constant** and is a relative indication of the "stiffness" of the spring.

A particle or object in motion under the influence of a linear restoring force, such as that described by Hooke's law, undergoes what is known as **simple harmonic motion (SHM)**. This periodic oscillatory motion is one of the common types found in nature.

In this experiment, Hooke's law will be investigated, along with the parameters and description of simple harmonic motion.

After performing this experiment and analyzing the data, you should be able to:

TI OBJECTIVES

1. Tell how Hooke's law is represented graphically and cite an example of an elastic object that does *not* follow Hooke's law.
2. Explain why simple harmonic motion (SHM) is simple and harmonic.
3. Better understand how the period of a mass oscillating on a spring varies with the mass and the spring constant.

CI OBJECTIVES

1. Explain the energy conversion that happens during the simple harmonic motion of a pendulum.
2. Experimentally verify the law of conservation of mechanical energy.

TI EXPERIMENT 16

Simple Harmonic Motion

TI EQUIPMENT NEEDED

- Coil spring
- Wide rubber band
- Slotted weights and weight hanger
- Laboratory timer or stopwatch
- Meterstick
- Laboratory balance
- 2 sheets of Cartesian graph paper

TI THEORY

A. Hooke's Law

The fact that for many elastic substances the restoring force that resists the deformation is directly proportional to the deformation was first demonstrated by Robert Hooke (1635–1703), an English physicist and contemporary of Isaac Newton. For one dimension, this relationship, known as **Hooke's law**, is expressed mathematically as

$$F = -k\Delta x = -k(x - x_o) \qquad \textbf{(TI 16.1)}$$

or

$$F = -kx \qquad (\text{with } x_o = 0)$$

where Δx is the linear deformation or displacement of the spring and x_o is its initial position. The minus sign indicates that the force and displacement are in opposite directions.

For coil springs, the constant k is called the **spring** or **force constant**. The spring constant is sometimes called the "stiffness constant," because it gives an indication of the relative stiffness of a spring—the greater k, the greater the stiffness. As can be seen from TI Eq. 16.1, k has units of N/m or lb/in.

According to Hooke's law, the elongation of a spring is directly proportional to the magnitude of the stretching force.* For example, as illustrated in ● TI Fig. 16.1, if a spring has an initial length y_o, and a suspended weight of mass m stretches the spring to a length y_1, then in equilibrium the weight is balanced by the spring force and

$$F_1 = mg = k(y_1 - y_o)$$

(with directional minus sign omitted).

Here y is used to indicate the vertical direction, instead of x as in TI Eq. 16.1, which is usually used to mean the horizontal direction. Similarly, if another mass m is added and the spring is stretched to a length y_2, then

$$F_2 = 2mg = k(y_2 - y_o)$$

and so on for more added weights. The linear relationship of Hooke's law holds, provided that the deformation or elongation is not too great. Beyond the *elastic limit*, a spring is permanently deformed and eventually breaks with increasing force.

Notice that Hooke's law has the form of an equation for a straight line:

$$F = k(y - y_o)$$

or

$$F = ky - ky_o$$

which is of the general form $y = x + b$

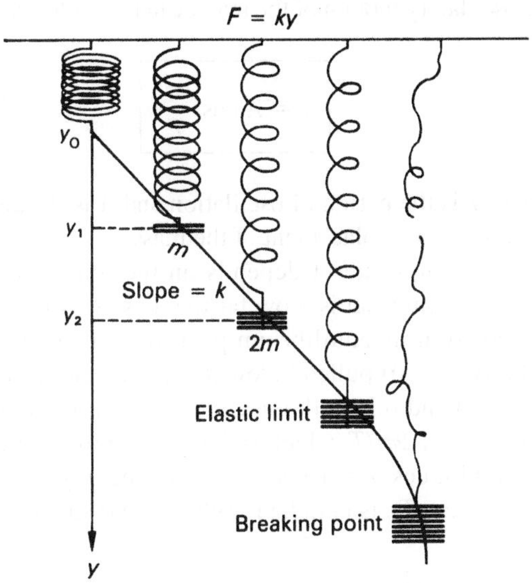

TI Figure 16.1 Hooke's law. An illustration in graphical form of spring elongation versus force. The greater the force, the greater the elongation, $F = -ky$. This Hooke's law relationship holds up to the elastic limit.

*The restoring spring force and the stretching force are equal in magnitude and opposite in direction (Newton's third law).

B. Simple Harmonic Motion

When the motion of an object is repeated in regular time intervals or periods, it is called **periodic motion**. Examples include the oscillations of a pendulum with a path back and forth along a circular arc and a mass *oscillating* linearly up and down on a spring. The latter is under the influence of the type of force described by Hooke's law, and its motion is called **simple harmonic motion (SHM)**—*simple* because the restoring force has the simplest form and *harmonic* because the motion can be described by harmonic functions (sines and cosines).

As illustrated in ● TI Fig. 16.2, a mass oscillating on a spring would trace out a wavy, time-varying curve on a moving roll of paper. The equation for this curve, which describes the oscillatory motion of the mass, can be written as

$$y = A\cos\frac{2\pi t}{T} \qquad \textbf{(TI 16.2)}$$

where T is the period of oscillation and A is the amplitude or maximum displacement of the mass.

The amplitude A depends on the initial conditions of the system (that is, how far the mass was initially displaced from its equilibrium position). If the mass were initially ($t = 0$) pulled below its equilibrium position (to $y = -A$) and released, the equation of motion would be $y = -A \cos 2\pi t/T$, which satisfies the initial condition at $t = 0$ with $\cos 0 = 1$ and $y = -A$. The argument of the cosine, $(2\pi t/T)$, is in radians rather than degrees.

In actual practice, the amplitude decreases slowly as energy is lost to friction, and the oscillatory motion is slowly "damped." In some applications, the simple harmonic motion of an object is intentionally damped, for example the spring-loaded needle indicator of an electrical measurement instrument or the dial on a common bathroom scale. Otherwise, the needle or dial would oscillate about the equilibrium position for some time, making it difficult to obtain a quick and accurate reading. (See Experiment 4 for the damping of a simple pendulum.)

The period of oscillation depends on the parameters of the system and, for a mass on a spring, is given by

$$T = 2\pi\sqrt{\frac{m}{k}} \qquad \textbf{(TI 16.3)}$$

(period of mass oscillating on a spring)

TI EXPERIMENTAL PROCEDURE

A. Rubber-Band Elongation

1. Hang a rubber band on a support, and suspend a weight hanger from the rubber band. Add an appropriate weight to the weight hanger (for example, 100 g–300 g), and record the total suspended weight (m_1g) in TI Data Table 1. Fix a meterstick vertically alongside the weight hanger, and note the position of the bottom of the weight hanger on the meterstick. Record this as y_1 in the data table.

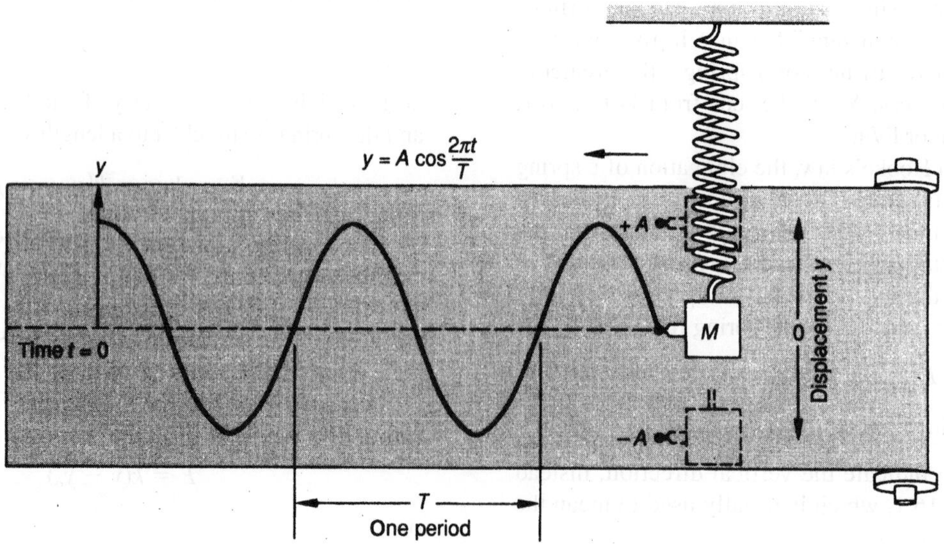

TI Figure 16.2 Simple harmonic motion. A marker on a mass oscillating on a spring traces out a curve, as illustrated, on the moving paper. The curve may be represented as a function of displacement (magnitude y) versus time, such as $y = A \cos 2\pi t/T$, where $y = A$ at $t = 0$.

2. Add appropriate masses weights (for example, 100 g) to the weight hanger one at a time, and record the total suspended weight and the position of the bottom of the weight hanger on the meterstick after each elongation (y_2, y_3, etc.). The masses should be small enough so that seven or eight masses can be added without overstretching the rubber band.

3. Plot the total suspended weight force versus elongation position (mg in newtons versus y in meters), and draw a smooth curve that best fits the data points.

B. Spring Elongation

4. Repeat Procedures 1 and 2 for a coil spring, and record the results in TI Data Table 2. Choose appropriate mass increments for the spring stiffness. (A commercially available Hooke's law apparatus is shown in ● TI Fig. 16.3.)

5. Plot mg versus y on the same sheet of graph paper used in Procedure 3 (double-label axes, if necessary), and draw a straight line that best fits the data. Determine the slope of the line (the spring constant k), and record it in the data table. Answer TI Questions 1 through 3 following the data tables.

C. Period of Oscillation

6. **(a)** On the weight hanger suspended from the spring, place a mass just great enough to prevent the spring from oscillating too fast and to prevent the hanger from moving relative to the end of the spring during oscillations when it is pulled down (for example, 5 to 10 cm) and released. Record the total mass in TI Data Table 3.

   (b) Using a laboratory timer or stopwatch, release the spring weight hanger from the predetermined initial displacement and determine the time it takes for the mass to make a number (5 to 10) of complete oscillations or cycles.

   The number of cycles timed will depend on how quickly the system loses energy or is damped. Make an effort to time enough cycles to get a good average period of oscillation. Record in the data table the total time and the number of oscillations.

   Divide the total time by the number of oscillations to determine the average period.

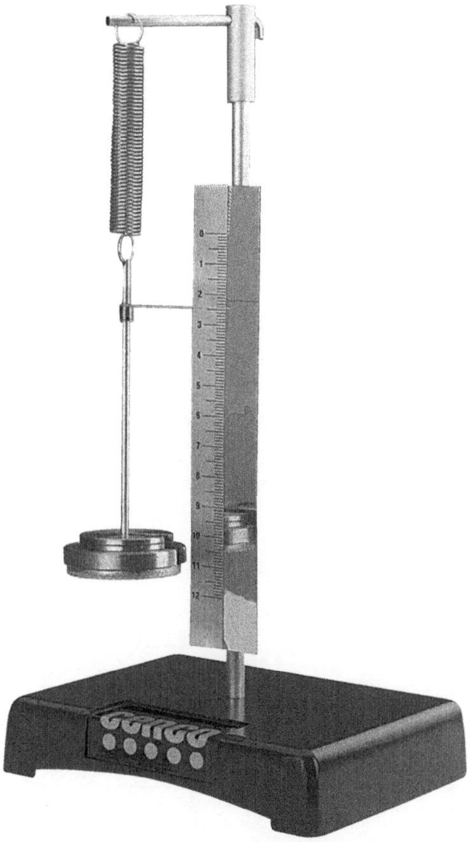

TI Figure 16.3 Hooke's law apparatus. The variables of Hooke's law ($F = mg$ and y) are measured using spring elongation. (Photo Courtesy of Sargent-Welch.)

7. Repeat Procedure 6 for four more mass values, each of which is several times larger than the smallest mass, and record the results in TI Data Table 3. The initial displacement may be varied, if necessary. (This should have no effect on the period. Why?)

8. Plot a graph of the average period squared (T^2) versus the mass (m), and draw a straight line that best fits the data points. Determine the slope of the line and compute the spring constant k. [Note from TI Eq. 16.3 that k is not simply equal to the slope; rather, $k = (2\pi)^2/\text{slope}$.]

   Compare this value of k with that determined from the slope of the spring elongation graph in Part B by computing the percent difference, and finish answering the TI Questions.

Name ______________________ Section ______________ Date ______________
Lab Partner(s) ______________________

TI EXPERIMENT 16

Simple Harmonic Motion

TI Laboratory Report

A. Rubber-Band Elongation

TI DATA TABLE 1

Total suspended weight* ( )		Scale reading ( )	
m_1g		y_1	
m_2g		y_2	
m_3g		y_3	
m_4g		y_4	
m_5g		y_5	
m_6g		y_6	
m_7g		y_7	
m_8g		y_8	

*It is convenient to leave g, the acceleration due to gravity, in symbolic form; that is, if $m_1 = 100$ g or 0.100 kg, then weight $= m_1g = (0.100 \text{ kg})g$ N, but your instructor may prefer otherwise. Be careful not to confuse the symbol for acceleration due to gravity, g (italic), with the abbreviation for gram g (roman).

Calculations
(show work)

B. Spring Elongation

TI DATA TABLE 2

Total suspended weight* ( )		Scale reading ( )	
m_1g		y_1	
m_2g		y_2	
m_3g		y_3	
m_4g		y_4	
m_5g		y_5	
m_6g		y_6	
m_7g		y_7	
m_8g		y_8	

* It is convenient to leave g in symbol form, even when graphing.

k (slope of graph) ______________
(units)

Don't forget units

(continued)

C. Period of Oscillation

TI DATA TABLE 3

	Total suspended mass ( )	Total time ( )	Number of oscillations ( )	Average period T	T^2 ( )
m_1					
m_2					
m_3					
m_4					
m_5					

Calculations
(show work)

Slope of graph ____________

Computed spring constant k ____________

Percent difference (of k's in B and C) ____________

TI QUESTIONS

1. Interpret the intercepts of the straight line for the spring elongation in the mg-versus-y graph of Part B.

Name ______________________ Section ______________ Date ______________

Lab Partner(s) ______________________

EXPERIMENT 16 **Simple Harmonic Motion** ***Laboratory Report***

2. Is the elastic property of the rubber band a good example of Hooke's law? Explain.

3. Draw a horizontal line through the y-intercept of the straight-line graph of Part B, and form a triangle by drawing a vertical line through the last data point.
   (a) Prove that the area of the triangle is the work done in stretching the spring. (*Hint*: $W = \frac{1}{2}kx^2$, and area of triangle $A = \frac{1}{2}ab$, that is, one-half the altitude (a) times the base (b).)

   (b) From the graph, compute the work done in stretching the spring.

4. Interpret the x-intercept of the straight line of the T^2-versus-m graph of Part C.

(continued)

5. For a mass oscillating on a spring, at what positions do the (a) velocity and (b) acceleration of the mass have maximum values?

6. What is the form of the equation of motion for the SHM of a mass suspended on a spring when the mass is initially (a) released 10 cm above the equilibrium position; (b) given an upward push from the equilibrium position so that it undergoes a maximum displacement of 8 cm; and (c) given a downward push from the equilibrium position so that it undergoes a maximum displacement of 12 cm? (*Hint*: Sketch the curve for the motion as in TI Fig. 16.2 and fit the appropriate trigonometric function to the curve.)

7. For case (a) in Question 6 only, what is the displacement y of the mass at times (a) $t = T/2$; (b) $t = 3T/2$; (c) $t = 3T$?

CI EXPERIMENT 16

Simple Harmonic Motion

EQUIPMENT NEEDED

- Rotary Motion Sensor (PASCO CI-6538)
- Mini-rotational accessory (PASCO CI-6691. This set includes a brass mass and a light rod to make the pendulum.)
- Support rods and clamps

THEORY

In this experiment, the simple harmonic motion of a pendulum will be investigated by examining the energy conversions that occur during the motion. Simple harmonic motion is the motion executed by an object of mass m subject to two conditions:

- The object is subject to a restoring force that is proportional to the displacement from equilibrium of the object.
- No dissipative forces act during the motion, so there is no energy loss.

Notice that as it is described in the theory, simple harmonic motion is an idealization because of the assumption of no frictional forces acting on the particle.

In this experiment, the simple harmonic motion of a pendulum will be investigated. A simple pendulum consists of a mass (called a bob) suspended by a "massless" string from a point of support. The pendulum swings in a plane.

The restoring force on a simple pendulum is the component of its weight that tends to move the pendulum back to its equilibrium position. As can be seen from ● CI Fig. 16.1, the magnitude of the force is

$$\boxed{F = mg\sin\theta} \qquad \textbf{(CI 16.1)}$$

Note, however, that this force is not proportional to the angular displacement θ of the pendulum, as required for SHM, but is proportional to the $\sin\theta$ instead. A pendulum can be approximated to be in SHM motion only if the angle θ is small, in which case $\sin\theta \approx \theta$ (where θ is in radians). Thus,

$$\boxed{F = mg\sin\theta \approx mg\theta} \qquad \textbf{(CI 16.2)}$$

Notice that in this approximation, the force is directly proportional to the displacement θ.

As the pendulum swings, kinetic energy is converted into potential energy as the pendulum rises. This potential energy is converted back to kinetic energy as the pendulum swings downward. The kinetic and potential energies of the pendulum at any moment during its motion can easily be determined. The kinetic energy of a pendulum of mass m moving with a linear speed v is given by

$$\boxed{K = \tfrac{1}{2}mv^2} \qquad \textbf{(CI 16.3)}$$

The potential energy, measured with respect to the equilibrium position, depends on the height above the equilibrium at a particular time. That is,

$$\boxed{U = mgh = mg(L - L\cos\theta)} \qquad \textbf{(CI 16.4)}$$

(See ● CI Fig. 16.2.)

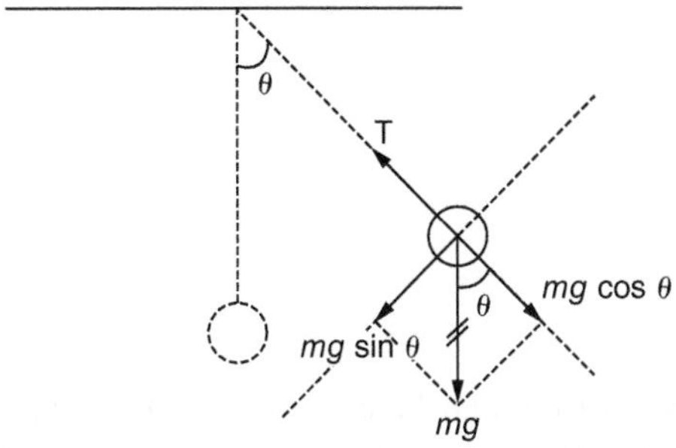

CI Figure 16.1 Forces acting on a swinging pendulum. The restoring force acting on a pendulum is the component $mg\sin\theta$ of gravity, which attempts to bring the pendulum back to the equilibrium position.

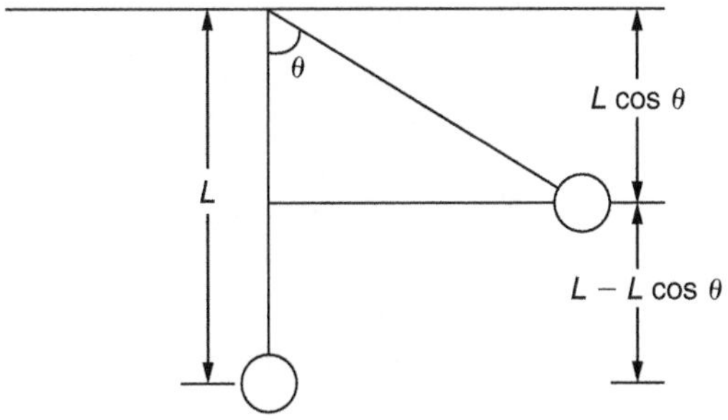

CI Figure 16.2 The elevation of a pendulum with respect to the equilibrium position. The elevation of a pendulum with respect to the equilibrium (lowest) position can be expressed in terms of L, the length of the pendulum, and of θ, as $L - L\cos\theta$. (The angular displacement has been exaggerated in the illustration. For simple harmonic motion, θ must be small.)

In this experiment, a sensor will keep track of the angular position, θ, of the pendulum as it swings. The sensor will also keep track of the angular speed, $\omega = \Delta\theta/\Delta t$, of the pendulum. The linear speed (v) can then be determined as $v = \omega L$, where L is the length of the pendulum and also the radius of the circular arc described by its motion. The kinetic and potential energies of the pendulum at any time can then be calculated.

BEFORE YOU BEGIN

1. Measure the mass of the pendulum bob (M), and record it in the Laboratory Report, in kilograms.
2. Measure the length of the pendulum (L), in meters, from the center of rotation to the center of the bob. Record it in the report.

This information will be needed during the setup of Data Studio.

SETTING UP DATA STUDIO

1. Open Data Studio, and choose "Create Experiment."
2. The Experiment Setup window will open and you will see a picture of the Science Workshop interface. There are seven channels to choose from. (Digital Channels 1, 2, 3, and 4 are the small buttons on the left; analog Channels A, B, and C are the larger buttons on the right, as shown in ● CI Fig. 16.3.)
3. Click on the Channel 1 button in the picture. A window with a list of sensors will open.
4. Choose the Rotary Motion Sensor from the list, and press OK.
5. The diagram now shows you the properties of the RMS sensor directly under the picture of the interface. (See CI Fig. 16.3.)
6. Connect the sensor to Channels 1 and 2 of the interface, as shown on the computer screen.
7. Adjust the properties of the RMS as follows:
   First Measurements tab: select Angular Position, Chapters 1 and 2, and select the unit of measure to be degrees. Also select Angular Velocity, Channels 1 and 2, in rad/s.
   Rotary Motion Sensor tab: set the Resolution to high (1440 divisions/rotations); and set the Linear Scale to Large Pulley (Groove).

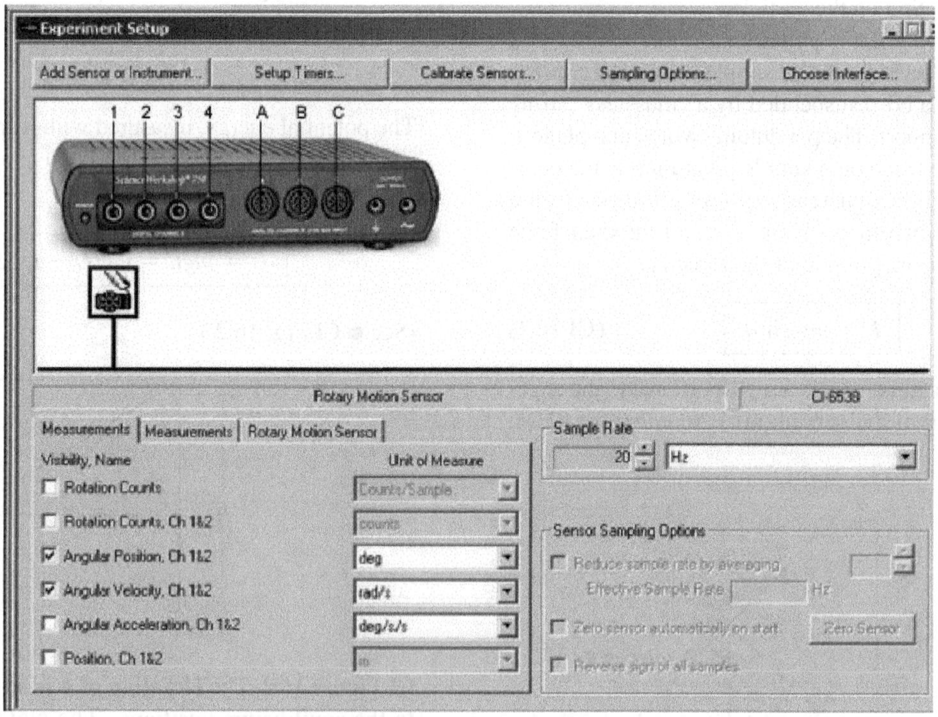

CI Figure 16.3 Experimental setup. The seven available channels are numbered 1 through 4 and A, B, or C. The rotary motion sensor, connected to Channels 1 and 2, will measure the angular position and the angular velocity of the pendulum. Make sure that the angular position is being measured in degrees, but the angular velocity in rad/s. (Reprinted courtesy of PASCO Scientific.)

Set the Sample Rate to 20 Hz.
The Data list on the left of the screen should now have two icons: one for the angular position data, the other for the angular velocity data.

8. Open the program's calculator by clicking on the Calculate button, on the top main menu. Usually, a small version of the calculator opens, as shown in ● CI Fig. 16.4. Expand the calculator window by clicking on the button marked Experiment Constants.

9. The expanded window (shown in ● CI Fig. 16.5) is used to establish values of parameters that will remain constant throughout the experiment. In this case, these are the length of the pendulum (L) and the mass of the pendulum (M), which have already been measured. This is how to do it:

(a) Click on the lower New button (within the "Experiment Constants" section of the calculator window) and enter the name of the constant as L, the value as the length of the pendulum measured before, and the units as meters (m).

(b) Click the lower Accept button.

(c) Click on the New button again, and enter the name of the constant as M, the value as the mass of the pendulum measured before, and the units as kilograms (kg).

(d) Click the lower Accept button.

(e) Close the experiment constants portion of the calculator window by pressing the button marked Experiment Constants again.

10. Calculation of the linear speed:

(a) In the same calculator window, clear the definition box and enter the following equation:

V = L * smooth (6, w)

This is the calculation of the linear speed $v = \omega L$, which will be called V. Note that the length L of the pendulum is multiplied by the angular speed, which is called w here. The smooth function is to produce a sharper graph.

(b) Press the Accept button after entering the formula. The variables L and w will appear in a list. L will have the value defined before, but w will be waiting to be defined.

(c) To define the variable w, click on the drop menu button on the left side of the variable. A list of options will show, asking what type of variable this is.

- Define w as a Data Measurement and, when prompted, choose Angular Velocity (rad/s).

11. Calculation of the kinetic energy:

(a) Still in the same calculator window, press the New button again to enter a new equation.

(b) Clear the definition box and enter the following equation: KE = 0.5 * M * v^2. This is the calculation of the kinetic energy $K = \frac{1}{2}mv^2$, that will be called KE.

(c) Press the Accept button after entering the formula. The variables M and v will appear in a list;

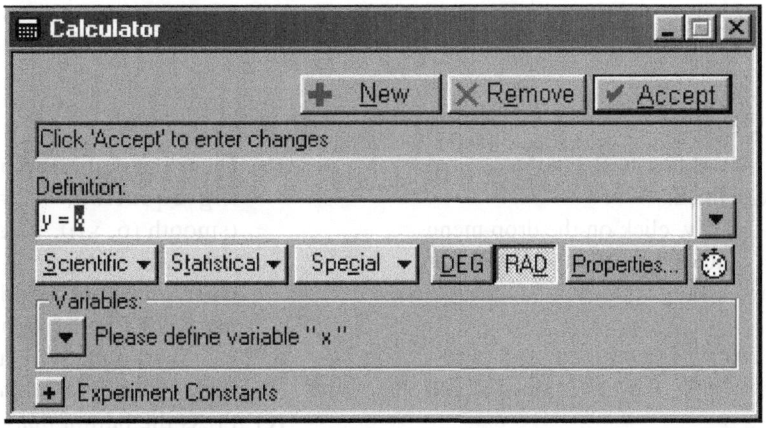

CI Figure 16.4 The calculator window. This small version of the calculator window opens when the Calculate button is pressed. The calculator will be used to enter equations that handle the values measured by the sensor. The computer will perform the calculations automatically as the sensor takes data. (Reprinted courtesy of PASCO Scientific.)

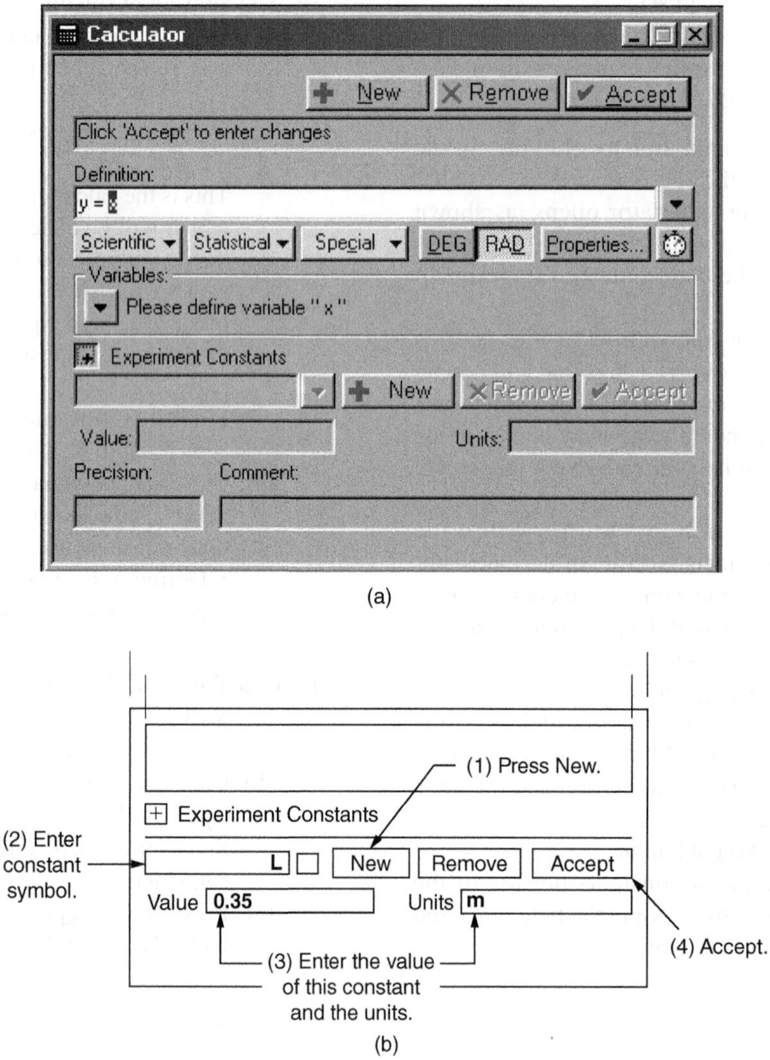

CI Figure 16.5 The expanded calculator window. (a) After the button marked Experiment Constants is pressed, the calculator window expands to full size. (b) The "Experiment Constants" section is the lower part of the expanded calculator window. This section is used to define parameters that are to remain constant during the experiment. The diagram shows the steps needed to enter experimental constants into the calculator. (Reprinted courtesy of PASCO Scientific.)

M is the value entered before for the mass, and v is waiting to be defined.

(d) To define the variable v, click on the drop menu button on the left side of the variable. The list of options will show, asking what type of variable this is.

- Define v as a Data Measurement and, when prompted, choose V, the equation defined previously.

12. Calculation of the potential energy:

(a) Press the <u>New</u> button once again to enter a new equation.

(b) Clear the definition box, and enter the following equation: PE = M * 9.81 * (L − L * cos (smooth (6, x))). This is the calculation of the potential energy $U = mgh = mg(L - L\cos\theta)$, which will be called PE. Note that M is the mass, 9.81 is the value of g, and the variable x in this formula will stand for the angular position θ of the pendulum, in degrees.

(c) Press the button marked DEG that is under the definition box. This will make sure the calculation of the cosine is done in degrees, not in radians.

(d) Press the Accept button after entering the equation. The variables M, L, and x will appear in a list, with x waiting to be defined.
(e) Define x as a Data Measurement and, when prompted, choose Angular Position (deg).
(f) Press the Accept button.

13. Close the calculator window.

14. The data list on the upper left of the screen should now include icons for the three quantities that are calculated: V, KE, and PE. A small calculator icon will show on the left of the calculated data quantities.

15. Create a graph by dragging the Angular Position (deg) data icon and dropping it on top of the "Graph" icon on the displays list. A graph of angular position (deg) versus time will open. The window will be called Graph 1.

16. Drag the KE equation icon and drop it somewhere on top of the graph created in step 15. The graph will then split in two, with the graph of angular position versus time on top and the graph of KE versus time on the bottom. The graphs will have matching time axes.

17. Repeat step 15 to create a second graph window. Graph 2 will also be a graph of angular position (deg) versus time.

18. Drag the PE equation icon and drop it on Graph 2. Graph 2 will then split in two, showing both the position and the PE of the pendulum at any time *t*, with matching time axes.

19. It is not necessary to be able to see both graph windows at the same time, but they can be moved around the screen so that both are visible. Their sizes may also be adjusted so that when they are active, they occupy the full screen individually. It is easy to change from viewing one to viewing the other by clicking on the particular graph to bring it to the front. ● CI Fig. 16.6 shows what the screen will look like after all the setup is finished.

CI EXPERIMENTAL PROCEDURE

1. Put the rotary motion sensor on a support rod. Install the mass on the light rod, and then install the pendulum on the front screw of the rotary motion sensor. A diagram of the equipment setup is shown in ● CI Fig. 16.7.

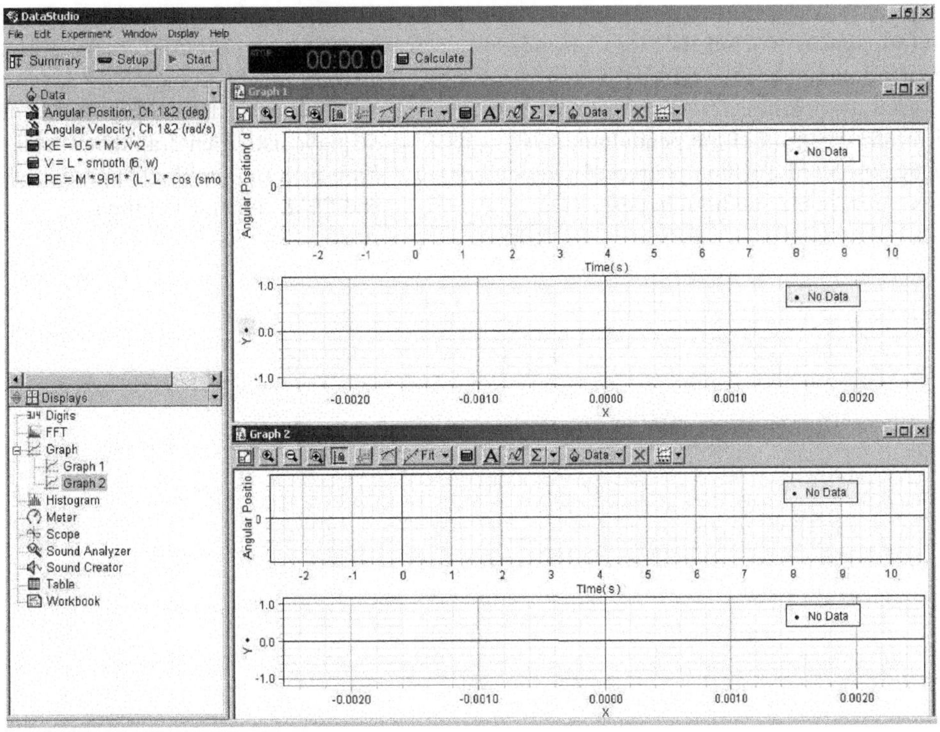

CI Figure 16.6 Data Studio setup. Graph displays are generated for angular position, kinetic energy, and potential energy. The individual graph windows can be viewed together (as in this picture) or independently, if resized to fit the full screen. (Reprinted courtesy of PASCO Scientific.)

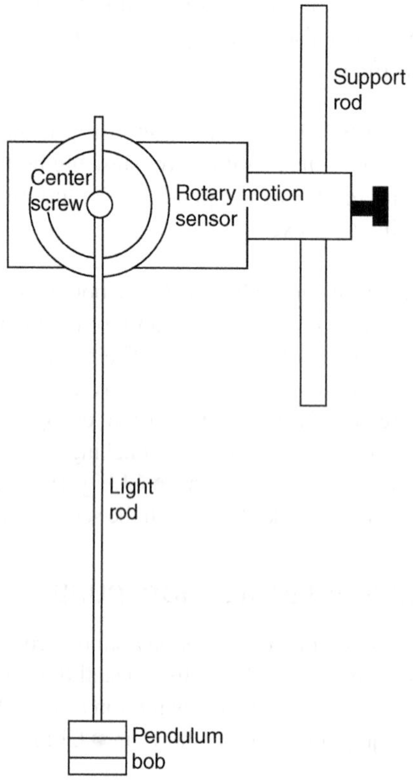

CI Figure 16.7 The experimental setup. The light rod with the bob at the end is attached to the front screw of the rotary motion sensor.

2. The rotary motion sensor will set its "zero" at the location of the pendulum when the START button is pressed. If we want the position $\theta = 0$ to correspond with the equilibrium position of the pendulum, it is *very important* that the START button be pressed while the pendulum is at rest in the equilibrium position.

3. *After* pressing the start button, displace the pendulum a small angle ($\leq 10°$) to the side and let it go.

4. Collect data for about 5 or 6 seconds, and then press the STOP button.

5. Print the graphs and paste them to the Laboratory Report.

6. Read from any of the position graphs what was the maximum amplitude of the pendulum, and record it in CI Data Table 1.

7. Determine from the graph the period of oscillation of the pendulum, and record it in the table.

8. From the kinetic energy graph, look at the first clear complete cycle of the motion, and find the maximum kinetic energy during that cycle. Record it in the table. Record also the position of the pendulum when the maximum kinetic energy was reached.

9. From the potential energy graph, look at the first clear complete cycle of the motion, and find the maximum potential energy during that cycle. Record it in the table. Record also the position of the pendulum when the maximum potential energy was reached.

10. Repeat for the minimum values of kinetic and potential energies.

11. To further reinforce the idea of conversions between kinetic and potential energy, create a new graph ("Graph 3") by dragging the kinetic energy data icon and dropping it on top of the "Graph" icon on the displays list. Then drag the potential energy icon and drop it in the graph. This graph will show both KE and PE as functions of time.

Name ______________________ Section ______________ Date ______________

Lab Partner(s) ____

CI EXPERIMENT 16

Simple Harmonic Motion

CI Laboratory Report

CI DATA TABLE 1

Purpose: To examine the variations of kinetic and potential energy as a pendulum swings.

Mass of pendulum, *M* ______________ kg

Max. amplitude ______________ °

Length, *L* ______________ m

Period ______________ s

	Value	Position of pendulum (deg)
KE max		
PE max		

KE min		
PE min		

Don't forget to attach the graphs to the Laboratory Report.

Don't forget units

(continued)

CI QUESTIONS

1. Compare the values of the maximum kinetic energy and the maximum potential energy. Discuss them in terms of the conservation of energy.

2. The following diagram illustrates three different positions of the pendulum as it moves in simple harmonic motion. (The angular displacement has been exaggerated for illustration purposes.) Label in the diagram which position corresponds to maximum KE, which to maximum PE, which to minimum KE, and which to minimum PE.

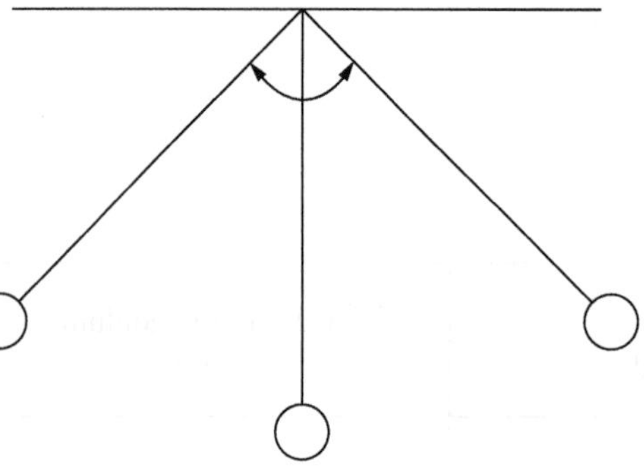

3. Was the amplitude of the pendulum constant? Explain.

4. The period of a simple pendulum in SHM is given by $T = 2\pi\sqrt{\frac{L}{g}}$. Use the measured length of the pendulum to calculate its period using this formula. Then compare to the period you determined from the graph. Discuss what causes the percent error.

5. *Optional Exercise*: Create a new calculation (in the calculator window) that will determine the total energy of the pendulum. That is, calculate KE + PE. Then plot the total energy as a function of time. Was the total energy constant? Explain.

Name ______________________ Section ______________ Date ______________
Lab Partner(s) ____

EXPERIMENT 17

Standing Waves in a String

TI Advance Study Assignment

Read the experiment and answer the following questions.

1. How is wave speed related to frequency and wavelength? How is the period of oscillation related to wave speed?

2. What is a standing wave, and what are nodes and antinodes?

3. What are normal modes?

(continued)

4. How does the wavelength of a standing wave in a vibrating string vary with the tension force in the string and/or the linear mass density of the string?

5. Standing waves in a string can be produced by oscillating the string at the various natural frequencies. However, in this experiment, the string vibrator has only one frequency. How, then, are standing waves with different wavelengths produced?

PRE-LAB DEMONSTRATION | Several good wave demonstrations—Instructor's choice

Transverse and Longitudinal Waves (vibrating and singing rod)
Musical Straws (make your own kind of music)
Singing Wine Glass (resonance)
Whirly Tube (Bloogle resonator)

EXPERIMENT 17

Standing Waves in a String

INTRODUCTION AND OBJECTIVES

A **wave** is the propagation of a disturbance or energy. When a stretched cord or string is given a shake, the wave travels along the string with a speed that depends on the tension in the string and its linear mass density. Upon reaching a fixed end, the wave is reflected back along the string.

For a continuous disturbance, the propagating waves interfere with the oppositely moving reflected waves, and a standing- (or stationary-) wave pattern is formed under certain conditions. These standing-wave patterns can be visually observed. The visual observation and measurement of standing waves serve to provide a better understanding of wave properties and characteristics.

In this experiment, the relationship between the tension force and the wavelength in a vibrating string will be studied, as applied to the *natural frequencies* or *normal modes* of oscillation of the string.

After performing this experiment and analyzing the data, you should be able to:

1. Explain how standing waves are formed.
2. Distinguish between nodes and antinodes.
3. Tell what determines the natural frequencies of a vibrating string system.

EQUIPMENT NEEDED

- Electric string vibrator
- Clamps and support rod
- Pulley with rod support
- String
- Weight hanger and slotted weights
- Meterstick
- Laboratory balance
- 1 sheet of Cartesian graph paper

THEORY

A wave is characterized by its **wavelength λ** (in meters), the **frequency** (of oscillation) f, in Hz or $1/\text{s} = \text{s}^{-1}$, and **wave speed** v (m/s). (See ● Fig. 17.1.) These quantities are related by the expression

$$\boxed{\lambda f = v} \qquad (17.1)$$

(Check to see whether the equation is dimensionally correct.)

Waves in a stretched string are transverse waves; that is, the "particle" displacement is perpendicular to the direction of propagation. In longitudinal waves, the particle displacement is in the direction of wave propagation, for example in sound waves. The maximum displacements of the particle oscillation are $+A$ and $-A$. The magnitude of the maximum displacement, called the **amplitude** A, is related to the energy of the wave. The **period** (of oscillation) T is related to the frequency of oscillation, $T = 1/f$ (Fig. 17.1).

When two waves meet, they interfere and the combined wave form is a superposition of the two interfering waves. The superposition of two waves of equal amplitude and frequency traveling in opposite directions gives rise to what is known as a **standing** or **stationary wave**.

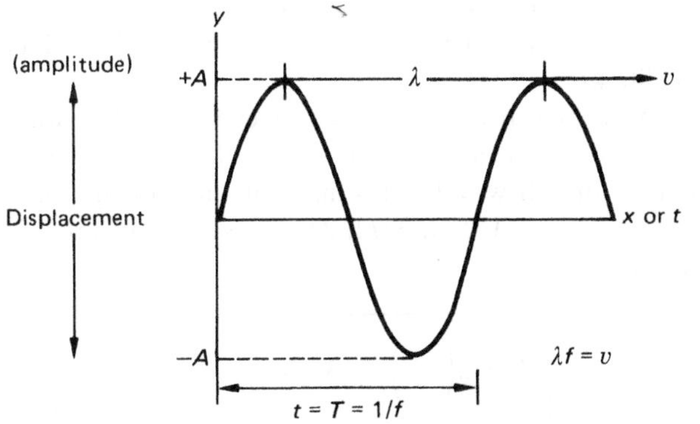

Figure 17.1 Wave description. The parameters involved in describing a wave. See text for description.

The periodic constructive and destructive interference causes the formation of a standing-wave pattern as illustrated in ● Fig. 17.2. Notice that some of the "particles"

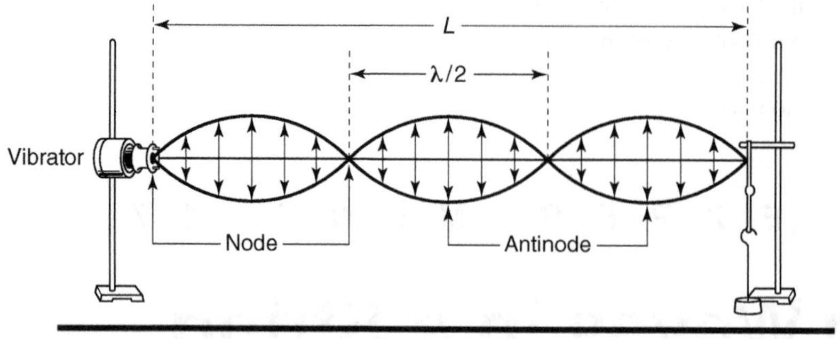

Figure 17.2 Standing wave. Periodic constructive and destructive interferences give rise to a standing wave form, as illustrated here. The length of one loop of the standing wave is equal to one-half the standing wave's wavelength. Note the positions of the nodes and antinodes.

on the axis are stationary. These positions are called *nodal points* or **nodes**, and the points of maximum displacement are called **antinodes**. The energy of the particles in a standing-wave envelope alternates between the kinetic and potential energies.

In a stretched string being oscillated or shaken at one end, waves traveling outward from the oscillator interfere with waves that have been reflected at the other fixed end. However, standing waves in a given length of string occur only for *certain* wave frequencies. That is, for a given stretching tension or force, the string must be driven or oscillated with certain vibrational frequencies to produce standing waves.

The frequencies at which large-amplitude standing waves are produced are called **natural frequencies** or **resonant frequencies**. The resulting standing-wave patterns are called **normal** or **resonant modes of vibration**. In general, all systems that oscillate have one or more natural frequencies, which depend on such factors as mass, elasticity or restoring force, and geometry (boundary conditions).

Since the string is fixed at each end, a standing wave must have a node at each end. As a result, only an integral number of half wavelengths may "fit" into a length L of the string, $L = \lambda/2, \lambda, 3\lambda/2, 2\lambda$, and so on, such that in general

$$L = n\left(\frac{\lambda_n}{2}\right) \text{ or } \lambda_n = \frac{2L}{n} \qquad n = 1, 2, 3, 4, \ldots \tag{17.2}$$

Fig. 17.2 illustrates the case for $L = 3\lambda/2$.

The wave speed in a stretched string is given by

$$v = \sqrt{\frac{F}{\mu}} \tag{17.3}$$

(wave speed in a stretched string)

where F is the magnitude of the tension force in the string and μ is the linear mass density (mass per unit length, $\mu = m/L$) of the string.* Using Eqs. 17.2 and 17.3 in $\lambda f = v$ [Eq. (17.1)] yields

$$f_n = \frac{v}{\lambda_n} = \frac{n}{2L}\sqrt{\frac{F}{\mu}} \qquad n = 1, 2, 3, \ldots \tag{17.4}$$

(resonant frequencies)

where f_n and λ_n are the frequency and wavelength, respectively, for a given integer n.

Setting $n = 1$ in Eq. (17.5) gives the lowest possible frequency, which is known as the **fundamental frequency**:

$$f_1 = \frac{1}{2L}\sqrt{\frac{F}{\mu}} \tag{17.5}$$

(fundamental frequency)

so Eq. (17.4) may be written in terms of the fundamental frequency as

$$f_n = \frac{v}{\lambda_n} = \frac{n}{2L}\sqrt{\frac{F}{\mu}} = nf_1 \qquad n = 1, 2, 3, \ldots \tag{17.6}$$

Moreover, only certain frequencies produce standing waves for a given string tension, linear density, and length.

As noted above, the lowest natural frequency f_1 [Eq. (17.5)] is called the *fundamental frequency.* All other natural frequencies are integral multiples of the fundamental frequency: $f_n = nf_1$ (for $n = 1, 2, 3, \ldots$). The set of frequencies f_1, $f_2 = 2f_1$, $f_3 = 3f_1$, ... is called a **harmonic series**: f_1 (the fundamental frequency) is the *first harmonic,* f_2 the *second harmonic,* and so on.

*See Experiment 14, Theory B, for a discussion of linear mass density.

In this experiment, the electrically driven string vibrator has a fixed frequency, so the driving frequency cannot be varied to produce different normal-mode standing-wave patterns. However, varying the string tension can vary the wave speed so as to produce different standing-wave patterns. Since $v = \sqrt{F/\mu}$ [Eq. (17.3)],

$$\lambda = \frac{v}{f} = \frac{1}{f}\sqrt{\frac{F}{\mu}} \tag{17.7}$$

where f and μ are constant. Hence, by varying F, one can select the appropriate wavelengths that will "fit" into a given string length L to produce standing waves.

EXPERIMENTAL PROCEDURE

1. If one has not been provided, cut a piece of string long enough to be used in the experimental setup—long enough to be looped at each end so as to be attached to the vibrator and a weight hanger suspended from the end running over the pulley (● Fig. 17.3). The vibrator and pulley should be clamped to support posts at the opposite ends of the laboratory table to give an active string length of about 150 cm. (This length may vary for a given setup.)

   Measure the *total* length of the string, and determine its mass on a laboratory balance. Record these values in the data table, and compute the linear mass density $\mu = m/L_o$. (*Note*: L_o is the total length of the string.)

2. Attach the string to the vibrator and suspend a weight hanger from the other end as shown in Fig. 17.3. Make certain that the string is aligned properly and that it is parallel to the table surface. Measure the distance between the vibrator arm and the point of contact of the string on the pulley. Record this length L in the data table.

   Turn on the vibrator. Try to produce different standing-wave patterns in the string by alternately lifting and carefully pulling down on the weight hanger. It is helpful to fold a thin strip of paper in half and hang it on the string to observe vibrating action. The number of loops should increase with less tension. (Why?) Also, try grasping the string at a node and antinode of a given pattern to see what happens.

3. When you are familiar with the operation of the apparatus, add enough weights to the weight hanger so that a standing-wave pattern of two loops is formed in the string (nodal point at the center). Adjust the tension by adding or removing some small weights until the loops are of maximum amplitude.

   If sufficiently small weights are not available, a fine adjustment can be made by loosening the clamp holding the vibrator rod and sliding it slightly back and forth so as to find the optimum string length between the ends that gives the maximum loop width or amplitude for a given tension.

   When this is accomplished, measure with a meterstick the distance from the point where the string contacts the pulley to the center nodal point. The meterstick can be held alongside the vibrating string, or you may find it more convenient to grasp the string at the nodal point with your fingers, shut off the vibrator, and measure the distance from the pulley contact to the nodal point along the nonvibrating string.

   Make certain not to pull the string toward the vibrator, for that would increase the length by raising the weight hanger. Apply a slight tension in the string *away* from the vibrator if necessary.

   Record this length L_1 and the total suspended mass in the data table. Since the length of one loop is one-half of a wavelength, $L_1 = \lambda/2$.

4. Remove enough weights from the weight hanger and adjust so that a standing-wave pattern of maximum amplitude with three loops (two nodal points in the string) is formed. Measure the distance from the pulley contact to the nodal point nearest the vibrator. (The fixed-end nodal point *at* the vibrator is not used because in vibrating up and down, it is not a "true" nodal point.)

   Record this length L_2 and the total suspended mass in the data table. Since the length of two loops is equal to one wavelength, $L_2 = \lambda$.

5. Repeat Procedure 4 for consecutive standing-wave patterns up to eight measured loops, if possible. [The weight hanger by itself may supply too much tension for higher-order patterns, so it may have to be removed and smaller weight(s) suspended.] Compute the wavelength for each case.

   It should become evident that in general, $\lambda = 2L_N/N$, or $L_N = N\lambda/2$, where N is the number of loops in a given L_N. Notice the similarity of the latter form of this equation to Eq. (17.2), wherein the length L is the total vibrating length of the string.

6. Note that Eq. (17.7) can be rewritten as

$$\lambda = \frac{1}{f}\sqrt{\frac{F}{\mu}} = \left(\frac{1}{f\sqrt{\mu}}\right)\sqrt{F} \tag{17.7a}$$

   where f and μ are constants. It has the form of an equation of a straight line, $y = mx + b$, with $x = \sqrt{F}$ and $b = 0$.

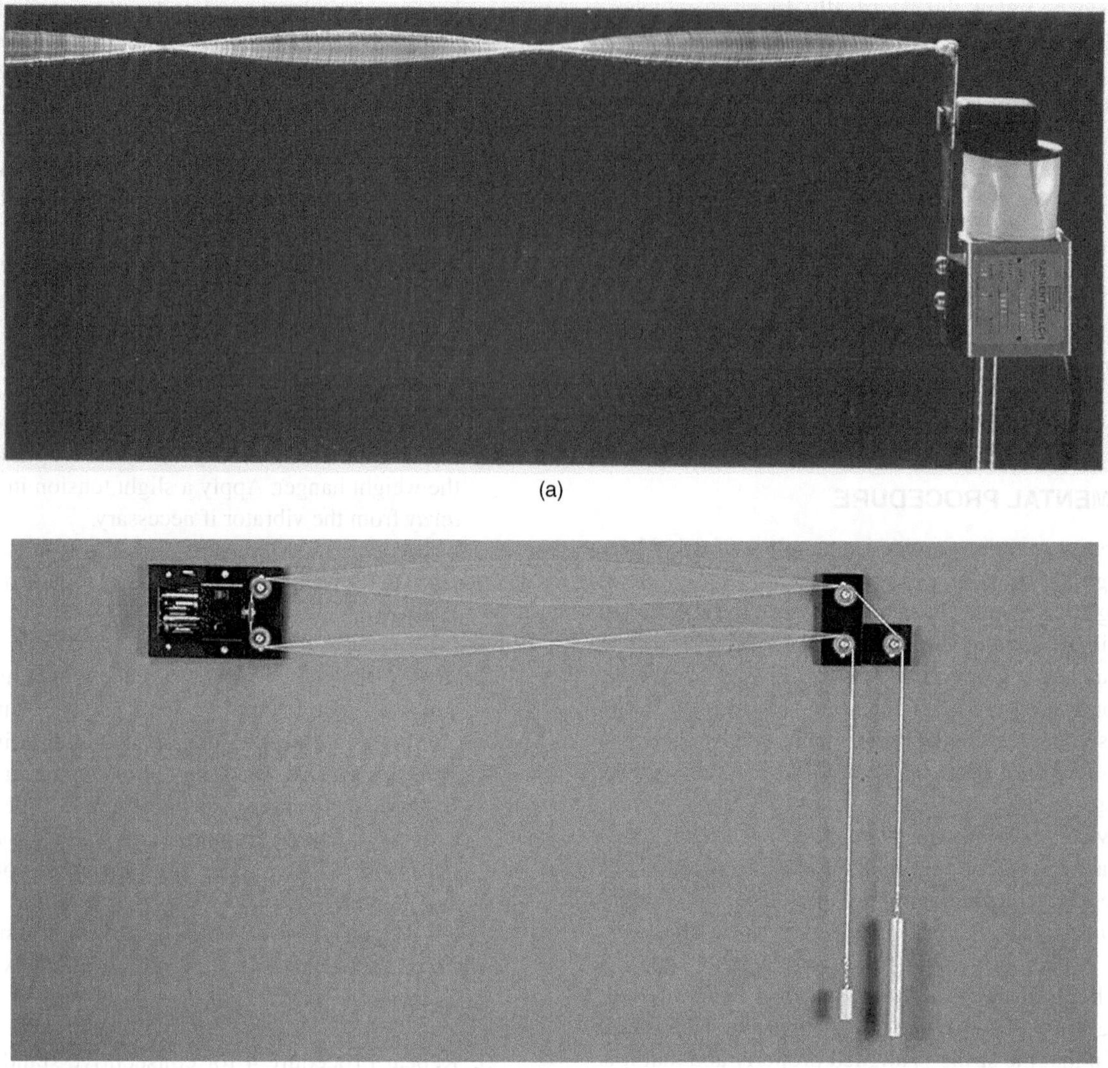

(a)

(b)

Figure 17.3 Standing wave apparatus. (a) A string vibrator oscillates the string. Different standing waves are produced by varying the tension in the string. (b) A dual string vibrator. Different tensions produce different normal modes. (Photos Courtesy of Sargent-Welch.)

Plot the experimental data on a graph of λ versus $\sqrt{F}$. Draw the straight line that best fits the data, and determine the slope of the line. From this value and the previously determined value of μ, compute the average frequency f of the oscillations.*

* If you have some scattered data points far from the straight line, see Question 2 in the Laboratory Report.

The string vibrator operates on 60-cycle ac current. The vibrating action is accomplished by means of an electromagnet operated by the input current. The vibrator arm is attracted toward an electromagnet during each half-cycle, or twice each cycle, so the vibrating frequency is $2 \times 60 = 120$ Hz (cycles per second). Using this as the accepted value of the vibrational frequency, compute the percent error of the experimentally determined value.

Name ______________________ Section ______________ Date ______________

Lab Partner(s) ____

EXPERIMENT 17

Standing Waves in a String

TI *Laboratory Report*

Mass of string ______________

Total length of string L_o ______________

Linear mass density μ ______________

Length of string between vibrator and pulley L ______________

DATA TABLE

Purpose: To determine the frequency of oscillation from normal modes.

Number of loops measured N	Suspended mass ( )	Tension force F* ( )		Measured length L_N for N loops ( )		Wavelength λ ( )	$\sqrt{F}$ ( )
1		F_1		L_1			
2		F_2		L_2			
3		F_3		L_3			
4		F_4		L_4			
5		F_5		L_5			
6		F_6		L_6			
7		F_7		L_7			
8		F_8		L_8			

*For convenience, the tension weight force may be expressed in terms of g (that is, if $m = 0.10$ kg, then $F = mg = 0.10\,g$ N).

Calculations
(show work)

Slope of graph ______________

Computed frequency f ______________

Accepted frequency ______________

Percent error ______________

Don't forget units

(continued)

Calculations
(show work)

TI QUESTIONS

1. The length, L_1, is not the wavelength of the fundamental frequency of the string.
   (a) With the tension equal to F_1, to which natural frequency does the wavelength equal to L_1 correspond?

   (b) What tension in the string would be required to produce a standing wave with a wavelength equal to L_1? (*Hint*: Use Eq. 17.7.)

2. Theoretically, the vibrator frequency is 120 Hz. However, sometimes the vibrator resonates with the string at a "subharmonic" of 60 Hz.
   (a) If this were the case in all instances, how would it affect the slope of the graph?

(b) If you have some scattered data points far from the straight line on your graph, analyze the data for these points using Eq. 17.7 to determine the frequency.

3. How many normal modes of oscillation or natural frequencies does each of the following have: (a) a simple pendulum, (b) a clothes line, and (c) a mass oscillating on a spring?

4. Stringed musical instruments, such as violins and guitars, use stretched strings. Explain (a) how tightening and loosening the strings tunes them to their designated tone pitch or frequency; (b) why the strings of lower tones are thicker or heavier; and (c) why notes of higher pitch or frequency are produced when the fingers are placed on the strings.

5. (*Optional*) Consider a long whip antenna of the type used on some automobiles and trucks. Show that the natural frequencies of oscillation for the antenna are $f_m = mv/4L$, where $m = 1, 3, 5, \ldots$, v is the wave speed, and L is the length of the antenna. (*Hint*: The boundary conditions are a node and an antinode.)

Name ______________________ Section ______________ Date ______________

Lab Partner(s) ____

EXPERIMENT 18

Temperature and Thermometer Calibration

TI Advance Study Assignment

Read the experiment and answer the following questions.

1. What is meant by saying that temperature is a relative measure?

2. Is our sense of touch a reliable measure of temperature? Explain.

3. How does a liquid-in-glass thermometer operate or indicate temperature?

4. What is needed to calibrate a thermometer?

PRE-LAB DEMONSTRATION | Burning Dollar Bill (money to burn)

A dollar bill is immersed in a flammable liquid, then held up and set on fire—money to burn?

EXPERIMENT 18

Temperature and Thermometer Calibration

INTRODUCTION AND OBJECTIVES

Temperature tells us whether something is hot or cold. That is, temperature is a *relative* measure, or indication, of hotness or coldness. For example, we know that water coming out of the hot water tap is hotter than that coming from the cold water tap. Humans have temperature perception in the sense of touch. However, this perception is unreliable and may vary a great deal among people. Our sense of touch doesn't enable us to measure temperature accurately or quantitatively. The quantitative measurement of temperature is generally accomplished through the use of a thermometer.

After performing the experiment and analyzing the results, you should be able to do the following:

1. Explain what is needed to calibrate a thermometer.
2. Explain why thermometer scales have different degree number readings.
3. Convert from one temperature scale to another.

THEORY

A thermometer is an instrument that utilizes the physical properties of materials to determine temperature values. The temperature-dependent property most commonly used is *thermal expansion*. Nearly all substances expand with increasing temperature and contract with decreasing temperature. The change in length or volume of a substance is quite small, but the effects of thermal expansion can be made evident by using special arrangements. The most common type of thermometer is the *liquid-in-glass thermometer*, with which you are no doubt familiar. It consists of a glass bulb on a glass stem with a capillary bore and a sealed upper end. A liquid in the bulb (usually mercury, or alcohol colored with a red dye to make it visible) expands on heating, and a column of liquid is forced up the capillary tube. The glass expands somewhat, but the liquid expands much more.

Thermometers are calibrated with scales having numerical values that can be assigned to different temperatures. Most common are the Fahrenheit and Celsius temperature scales. In this experiment, you will calibrate a thermometer with your own scale and compare the temperature on your scale to that on one of the common scales.

EQUIPMENT NEEDED

- Two thermometers, one with Celsius calibration and one uncalibrated
- Small glass beaker
- Crushed ice
- Hot plate or Bunsen burner and striker (ring stand and pad if burner is used)
- Ruler
- Pencil and sheet of paper
- Fine-tip magic marker or grease pencil that will mark on glass

THEORY

To calibrate a thermometer, two reference points are needed. These are commonly the ice and steam points of water. The *ice point* is the temperature of a mixture of ice and water at normal atmospheric pressure. The *steam point* is the temperature at which water boils at normal atmospheric pressure. The ice point and steam point are commonly referred to as the *freezing point* and *boiling point*, respectively.

The familiar Fahrenheit scale has an ice point of 32° (32 degrees) and a steam point of 212°. The interval between the ice point and steam point is evenly divided into 180 equal units or degrees. The Celsius scale is based on an ice point of 0° and a steam point of 100°, and there are 100 equal units or degrees between these points.*

Another temperature scale is the Kelvin or absolute scale. The Kelvin temperature (T_K) is related to the Celsius temperature by $T_K = T_C + 273$. (The Kelvin temperature is expressed in kelvins, K not °K.)

*A temperature can be easily converted from one scale to the other:

$$T_F = \frac{9}{5}T_C + 32 \text{ (Celsius to Fahrenheit) and}$$

$$T_C = \frac{5}{9}(T_F - 32) \text{ (Fahrenheit to Celsius)}$$

EXPERIMENTAL PROCEDURE

A. Ice and Steam Points

1. Lay the uncalibrated thermometer on a sheet of paper, and trace a pencil outline.
2. Fill the beaker about two-thirds full of crushed ice; add enough water just to cover the ice and stir. Let the ice-water mixture stand for a few minutes so as to come to an even temperature.
3. Insert the uncalibrated thermometer into the mixture and hold it so that the thermometer bulb does not touch the beaker. When the thermometer liquid stabilizes, mark the level on the thermometer with a glass marker. Remove the thermometer, and dry it with a paper towel.
4. Place the thermometer on the outline paper, and with a pencil make a mark on the paper that corresponds to the marked liquid level (your ice point).
5. Empty the beaker; half-fill it with water and heat to boiling. Insert the uncalibrated thermometer, again not touching the beaker. When the reading stabilizes, mark the liquid level. Remove and dry the thermometer. (Be careful of the hot end.) Place it on the outline paper again and mark the height of the marked level (your steam point). Lay the thermometer aside and let it cool to room temperature.

B. Your Own Temperature Scale

6. With a ruler and a pencil, divide the length between the marks on the paper into arbitrary numbered equal intervals, for example, 5, 10, 15, 20, or as great a number as is convenient, starting with the ice point as zero, but *not* using 100 for the steam point. (This is your scale.)

THE TEST

7. After the thermometer has come to room temperature equilibrium, lay it on the paper and record your scale reading in the Laboratory Report. You can give the scale a name if you like, maybe your name—Name scale (°N), for example Smith (°S).
8. Read the room temperature on the calibrated Celsius thermometer and record it.
9. Compare the readings as described in the Laboratory Report.

Name ______________________ Section ______________ Date ______________

Lab Partner(s) ____

EXPERIMENT 18

Temperature and Thermometer Calibration

TI Laboratory Report

Temperature reading on your thermometer scale ______________

Temperature reading on the Celsius thermometer ______________

1. How do the readings compare? That is, convert your thermometer reading to Celsius degrees. [*Hint*: In the conversion equation for Celsius (T_C) to Fahrenheit (T_F), the 9/5 ratio arises due to the different number of degrees on the scales. (See previous footnote.) That is, between the ice point and steam point there are 180° on the Fahrenheit scale and 100° on the Celsius scale, so there is a ratio of $T_F/T_C = 180/100 = 9/5$. (The 32 comes in the conversion equation because the ice point on the Fahrenheit scale is not zero, and an adjustment must be made.) Find the ratio for your scale to make the conversion.]

2. Compute the percent difference of the readings.

3. What is the equivalent Fahrenheit temperature of the temperature reading on your scale?

4. What is the equivalent Kelvin temperature of the temperature reading on your scale?

5. What are the Kelvin temperatures for the ice and steam point of water? What does the difference tell you about the size of kelvins?

Don't forget units

(continued)

TI QUESTIONS

1. When a thermometer is placed in a hot substance, on close inspection you can note that there is a slight delay before the liquid starts up the capillary bore. Why is this?

2. Why is water not used for the liquid in a liquid-in-glass thermometer?

3. The two common liquids used in liquid-in-glass thermometers are alcohol (ethanol) and mercury, which have freezing points and boiling points of –114 °C and 79 °C, and –39 °C and 357 °C, respectively. Would either of these thermometers be better for low temperature or high temperature measurement?

4. What would be the equation for the conversion of the temperature on your scale to Fahrenheit?

Name ______________________ Section ______________ Date ______________

Lab Partner(s) ____

EXPERIMENT 19

The Thermal Coefficient of Linear Expansion

TI Advance Study Assignment

Read the experiment and answer the following questions.

1. What is the cause of thermal expansion on the molecular level?

2. Distinguish between linear expansion and isotropic expansion.

3. How is the thermal coefficient of linear expansion determined experimentally?

4. What are the units of the thermal coefficient of linear expansion?

(continued)

5. What is meant by the fractional change in length?

PRE-LAB DEMONSTRATION | Instructor's Choice

Drinking Bird Engine (perpetual motion?) (● Fig. D7)
Salt and Ice (melting and freezing)

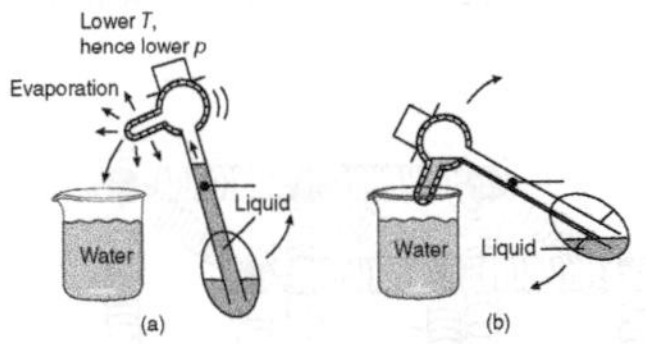

Figure D7 Drinking Bird – Perpetual Motion?

EXPERIMENT 19

The Thermal Coefficient of Linear Expansion

INTRODUCTION AND OBJECTIVES

With few exceptions, solids increase in size or dimensions as the temperature increases. Although this effect is relatively small, it is very important in applications involving materials that undergo heating and cooling. Unless these changes are taken into account, material and structural damage can result; for example, a piston may become too tight in its cylinder, a rivet could loosen, or a bridge girder could produce damaging stress.

The expansion properties of a material depend on its internal makeup and structure. Macroscopically, the thermal expansion is expressed in terms of temperature coefficients of expansion, which are experimental quantities that represent the change in the dimensions of a material per degree of temperature change. In this experiment, the thermal expansion of some metals will be investigated and their temperature coefficients of linear expansion determined.

After performing this experiment and analyzing the data, you should be able to:

1. Tell how the thermal coefficient of linear expansion describes such expansion.
2. Explain how the thermal coefficient of linear expansion is measured, and give an order of magnitude of its values for metals.
3. Describe and give examples of how thermal expansion considerations are important in applications of materials.

EQUIPMENT NEEDED

- Linear expansion apparatus and accessories
- Steam generator and stand
- Bunsen burner and striker or electric hot plate
- Rubber tubing
- Beaker
- Meterstick
- Thermometer (0 °C to 110 °C)
- Two or three kinds of metal rods (for example, iron and aluminum)

THEORY

Changes in the dimensions and volumes of materials are common effects. The thermal expansion of gases is very obvious and is generally described by gas laws. But the thermal expansion of liquids and solids is no less important. For example, such expansions are used to measure temperature in liquid-in-glass thermometers and bimetallic (oven) thermometers.

In general for solids, a temperature increase leads to the thermal expansion of an object as a whole. This expansion results from a change in the average distance separating the atoms (or molecules) of a substance. The atoms are held together by bonding forces, which can be represented simplistically as springs in a simple model of a solid (● Fig. 19.1).

The atoms vibrate back and forth; and with increased temperature (more internal energy), they become increasingly active and vibrate over greater distances. With wider vibrations in all dimensions, the solid expands as a whole. This may be different in different directions; however, if the expansion is the same in all directions, it is referred to as *isotopic expansion*.

The change in one dimension (length, width, or thickness) of a solid is called **linear expansion**. For small temperature changes, linear expansion is approximately proportional to ΔT, or the change in temperature, $\Delta T = T - T_o$ (● Fig. 19.2). The *fractional* change in length is $(L - L_o)/L_o$ or $\Delta L/L_o$, where L_o is the original length of the solid at the initial temperature. This ratio is related to the change in temperature by

$$\frac{\Delta L}{L_o} = \alpha \Delta T \quad \text{or} \quad \Delta L = \alpha L_o \Delta T \qquad \textbf{(19.1)}$$

where $\Delta L = L - L_o$ and $\Delta T = T - T_o$ and α is the **thermal coefficient of linear expansion**, with units of inverse

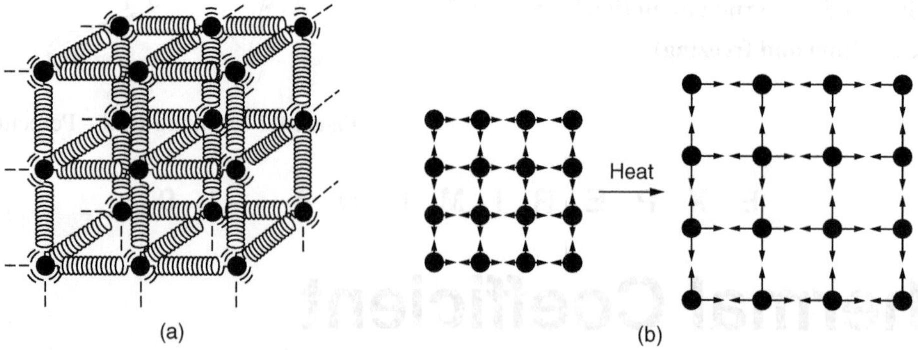

Figure 19.1 A springy solid. (a) The elastic nature of interatomic forces is indicated by simplistically representing them as springs, which, like the forces, resist deformation. (b) Heat causes the molecules to vibrate with greater amplitudes in the lattice, thereby increasing the volume of the solid (right). The arrows represent the molecular bonds, and the drawing is obviously not to scale. (Shipman, Wilson, and Higgins, *An Introduction to Physical Science,* Thirteenth Edition. Copyright © 2008 by Houghton Mifflin Company. Reprinted with permission. From Wilson/Buffa/Lou, *College Physics,* Seventh Edition. Copyright © 2007. Reprinted by permission of Pearson Education.)

temperature—that is, 1/°C. Note that with a temperature decrease and a contraction, ΔL would be negative, or a negative expansion.

As Eq. 19.1 shows, α is the *fractional change* in length per degree temperature change, $\Delta L/L_o$.* This thermal coefficient of expansion may vary slightly for different temperature ranges, but this variation is usually negligible for common applications, and α is considered to be constant.

By Eq. 19.1, α is defined in terms of experimentally measurable quantities:

$$\alpha = \frac{\Delta L}{L_o \Delta T} \tag{19.2}$$

Hence, by measuring the initial length L_o of an object (for example, a metal rod) at an initial temperature T_o and the change in its length ΔL for a corresponding temperature change ΔT, α can be computed.

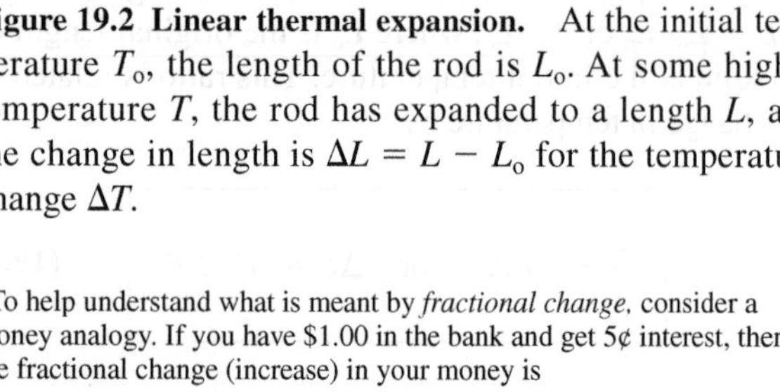

Figure 19.2 Linear thermal expansion. At the initial temperature T_o, the length of the rod is L_o. At some higher temperature T, the rod has expanded to a length L, and the change in length is $\Delta L = L - L_o$ for the temperature change ΔT.

*To help understand what is meant by *fractional change*, consider a money analogy. If you have \$1.00 in the bank and get 5¢ interest, then the fractional change (increase) in your money is

$$\Delta\$/\$_o = 5 \text{ cents}/100 \text{ cents} = 1/20 = 0.050 \text{ (or } 5.0\%\text{)}.$$

This development may be extended to two dimensions. The linear expansion expression [Eq. 19.1] may be written

$$L = L_o(1 + \alpha \Delta T) \tag{19.3}$$

and for an isotropic material, its area is $A = L \times L$, or

$$\begin{aligned} A &= L^2 \\ &= L_o^2(1 + \alpha\Delta T)^2 \\ &= A_o(1 + 2\alpha\Delta T + \alpha^2\Delta T^2) \end{aligned}$$

where $A_o = L_o^2$. Since typical α's are of the order of $10^{-6}/°C$, the α^2 term may be dropped with negligible error, and to a good approximation,

$$A = A_o(1 + 2\alpha\Delta T) \tag{19.4}$$

Comparing this expression with Eq. 19.3, the thermal coefficient of area expansion is seen to be approximately twice the coefficient of linear expansion (that is, 2α).

A similar development can be carried out for the coefficient of volume expansion, which is approximately equal to 3α.

EXPERIMENTAL PROCEDURE

1. A typical arrangement for determining thermal coefficients of linear expansion is shown in ● Fig. 19.3. The apparatus consists of a steam jacket with a micrometer attachment for measuring ΔL of a metal rod. A thermometer in the steam jacket measures the temperature of the rod. Steam is supplied to the jacket by a steam generator, and a beaker is used to catch the condensate.

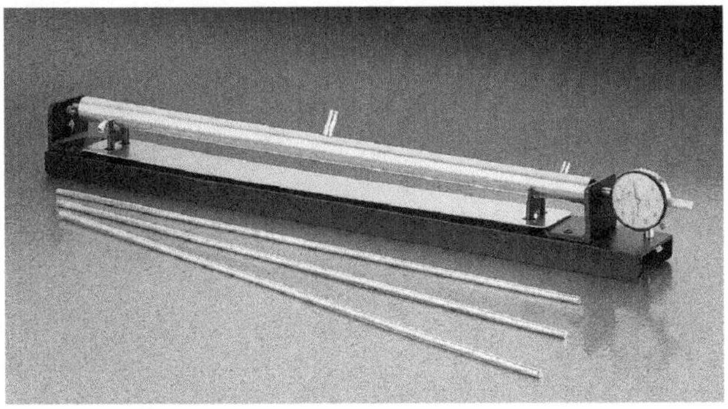

(a)

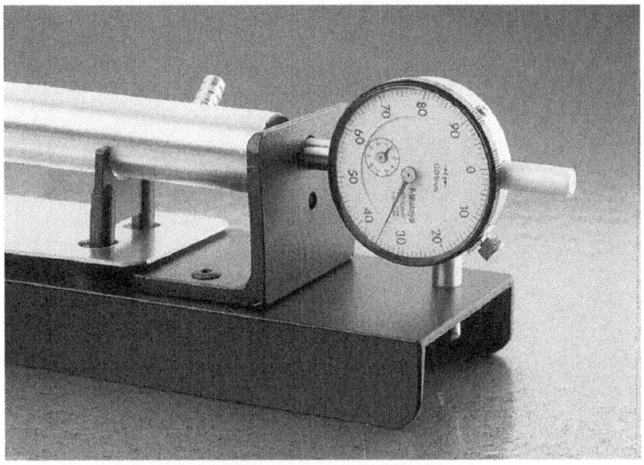

(b)

Figure 19.3 Linear thermal expansion apparatus. (a) The heat of steam admitted to the steam jacket causes a metal rod to expand. Rods of different metals may be used. (b) The expansion is measured with a dial indicator. (Photos Courtesy of Sargent-Welch.)

2. Before assembling the apparatus, measure the lengths L_0 of the metal rods with a meterstick to the nearest 0.1 mm, and record these lengths in the Data Table. Avoid handling the rods with your bare hands in order not to raise their temperature. Use a paper towel or cloth.

3. Assemble the apparatus, placing one of the rods in the steam jacket. Initially, have one end of the rod placed *firmly* against the fixed end screw and the other end not touching the micrometer screw.

   Carefully turn the micrometer screw until it just makes contact with the rod. Avoid mechanical backlash (and electrical spark-gap ionization, see below) by always turning the screw *toward* the rod just before reading. Do not force the screw. Record the micrometer setting. Do this three times and take the average as the initial setting. As soon as the initial micrometer reading is taken, read and record the initial temperature T_0.

   (The linear expansion apparatus may be equipped with an electrical circuit that uses a bell, light, or voltmeter to indicate when contact is made. The averaging process is unnecessary in this case.)

4. Turn the micrometer screw back from the end of the rod several millimeters to allow for the thermal expansion of the rod with increasing temperature. With the steam generator about one-half full, turn on the hot plate (or light the Bunsen burner) and boil the water so that steam passes through the jacket. The thermometer in the steam jacket should just touch the metal rod.

   Allow steam to pass through the jacket until the thermometer reading stabilizes (several minutes). When equilibrium has been reached, record the thermometer reading. Then carefully advance the micrometer screw until it touches the end of the rod, and record the micrometer setting. Do this three times, and take the average of the micrometer readings unless contact is indicated by electrical circuit. Turn off the heat source.

5. Repeat Procedures 3 and 4 for the other metal rods.

   Caution: *Be careful not to burn yourself with the condensed hot water in the steam jacket or the hot rod when you remove it. Take proper precautions.*

6. Compute ΔL and ΔT, and find the coefficient of linear expansion for each metal. Compare these α's with the accepted values given in Appendix A, Table A3, by computing the percent errors.

Name ______________________ Section ______________ Date ______________

Lab Partner(s) ______________________________________

EXPERIMENT 19

The Thermal Coefficient of Linear Expansion

TI Laboratory Report

DATA TABLE

Purpose: To determine the thermal coefficients of expansion of metal samples.

	Initial length L_o ( )	Initial micrometer setting	Final micrometer setting	ΔL ( )	Initial temp. T_o ( )	Final temp. T ( )	ΔT ( )	α meas. ( )	α accepted ( )
1. Type of rod ________									
2. Type of rod ________									
3. Type of rod ________									

Calculations
(show work)

Metal	Percent error
________	________
________	________
________	________

Don't forget units

(continued)

TI QUESTIONS

1. What are the probable sources of error in this experiment? Which would cause the largest error?

2. Would the numerical values of the thermal coefficients of linear expansion have been the same if the temperatures had been measured in degrees Fahrenheit? Explain, and give an example.

3. For a contraction with a negative fractional change, would the coefficient of thermal expansion be negative? Explain.

4. When a mercury-in-glass thermometer is placed in hot water, the thermometer reading first drops slightly and then rises. Explain why.

5. If flat strips of iron and brass were bonded together and this bimetallic strip were heated, what would be observed? Justify your answer, and draw a sketch of the situation. (*Hint*: See Appendix A, Table A3, for α's.)

6. A Pyrex graduated cylinder has a volume of exactly 200 mL at 0 °C. If its temperature is increased to 100 °C, will its volume increase or decrease? Compute the change in volume.

7. Assume a metal rod with an initial length L_o is heated through a temperature increase of ΔT to a length L_1 and then cooled to its initial temperature—that is, through a temperature decrease of $-\Delta T$ (same ΔT increase and decrease). Call the final length of the rod L_2 after this thermal cycle.
   (a) Show that Eq. 19.3 implies that $L_2 = L_o[1 - (\alpha\Delta T)^2]$, that is, $L_2 \neq L_o$.

   (b) What is the implication if the rod were taken through a number of such thermal cycles?

(continued)

(c) Obviously, something is wrong. Can you explain what it is? (*Hint*: Think of *basis,* or reference. For example, if you had an investment that appreciated 100% in value one day, and you lost 100% of your investment the next, would you still have any money left?)

Name ______________ Section ______________ Date ______________

Lab Partner(s) ______________

EXPERIMENT 20

Specific Heats of Metals

TI Advance Study Assignment

Read the experiment and answer the following questions.

1. Distinguish between heat capacity and specific heat.

2. Why is the specific heat of water equal to unity, that is, 1.0 cal/g-°C or 1.0 kcal/kg-°C?

3. Given that the specific heat of one material is twice that of another, compare the relative amounts of heat required to raise the temperature of equal masses of each material by 1 °C.

4. Say the same amount of heat was added to samples of the materials in Question 3, and each sample had the same increase in temperature. Compare the relative masses of the samples.

(continued)

5. What is the method of mixtures, and how is it used to determine specific heat?

6. On what does the accuracy of the method of mixtures depend? That is, what are possible sources of error? Would these be random or systematic errors? (See Experiment 2.)

PRE-LAB DEMONSTRATION | Take the Heat (specific heat)

The different specific heats of two liquids are demonstrated by holding a flame to the bottoms of styrofoam cups.

EXPERIMENT 20

Specific Heats of Metals

INTRODUCTION AND OBJECTIVES

Different substances require different amounts of heat to produce a given temperature change. For example, about three and one-half times as much heat is needed to raise the temperature of 1 kg of iron through a given temperature interval, ΔT, as is needed to raise the temperature of 1 kg of lead by the same amount.

This material behavior is characterized quantitatively by **specific heat**, *which is the amount of heat necessary to raise the temperature of a unit mass of a substance by one unit temperature interval, that is, to raise 1 g or 1 kg of a substance 1 °C*. Thus, in the previous example, iron has a greater specific heat than does lead.

The specific heat of a material is *specific,* or characteristic, for that material. As can be seen from the definition, the specific heat of a given material can be determined by adding a known amount of heat to a known mass of material and noting the corresponding temperature change. It is the purpose of this experiment to determine the specific heats of some common metals by calorimetry methods.

After performing this experiment and analyzing the data, you should be able to:

1. Tell what is meant by the specific heat of a substance, and compare the effects of different specific heats.
2. Calculate the heat necessary to raise the temperature of a given mass of a substance a particular number of degrees.
3. Describe and explain calorimetry and the method of mixtures.

EQUIPMENT NEEDED

- Calorimeter
- Boiler and stand
- Hot plate or Bunsen burner and striker
- 2 thermometers (0 °C to 110 °C)
- Two kinds of metal (shot form or slugs with attached strings)
- Laboratory balance
- Ice
- Safety glasses
- Strainer

THEORY

The change in temperature, ΔT, of a substance is proportional to the amount of heat, ΔQ, added (or removed) from it:

$$\boxed{\Delta Q \propto \Delta T}$$

In equation form, we may write

$$\boxed{\Delta Q = C\Delta T} \tag{20.1}$$

where the constant of proportionality C is called the **heat capacity** of the substance.

However, the amount of heat required to change the temperature of an object is also proportional to the mass of the object. Hence, it is convenient to define a *specific heat capacity* (or simply **specific heat**) c:

$$\boxed{c = \frac{C}{m}} \tag{20.2}$$

which is the heat capacity per unit mass of a substance. Thus, Eq. 20.1 becomes $\Delta Q = mc\Delta T$, and

$$\boxed{c = \frac{\Delta Q}{m\Delta T}} \tag{20.3}$$

(specific heat)

The specific heat is then the amount of heat required to change the temperature of 1 g of a substance by 1 °C.

The calorie (cal) unit of heat is defined as the amount of heat required to raise the temperature of 1 g of water by 1 °C. By definition, then, water has a specific heat of 1 cal/g-°C.

$$c = \frac{\Delta Q}{m\Delta T} = \frac{1 \text{ cal}}{(1 \text{ g})(1\ °\text{C})} = 1 \text{ cal/g-°C}.$$

[A kilocalorie (kcal) is the unit of heat defined as the amount of heat required to raise the temperature of 1 kg of water by 1 °C. In these units, water has a specific heat of 1 kcal/kg-°C, or, in SI units, 4.18×10^3 J/kg-°C. Your instructor may recommend that you use one of these units.]

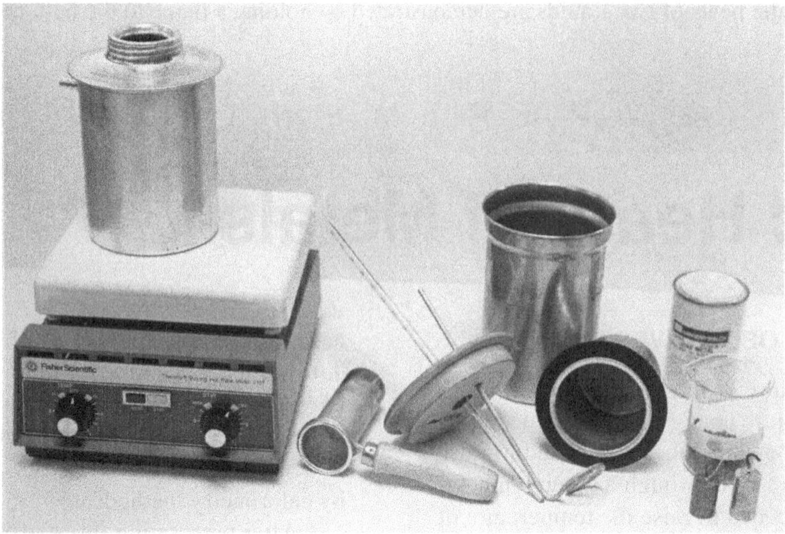

Figure 20.1 Apparatus for measurement of specific heats. Metal shot or a piece of metal (right) is heated with boiling water in the container on the hot plate. The metal is then placed in a known amount of water in the calorimeter, which insulates the system from losing heat. The inner calorimeter cup is shown with its dark, insulating ring lying in front of the outer cup. A thermometer and stirrer extend through the calorimeter cover. (© Cengage Learning.)

The specific heat of a material can be determined experimentally by measuring the temperature change of a given mass of material produced by a quantity of heat. This is done indirectly by a calorimetry procedure known as the **method of mixtures**.

If several substances at various temperatures are brought together, the hotter substances lose heat and the colder substances gain heat until all the substances reach a common equilibrium temperature. If the system is insulated so that no heat is lost to or gained from the surroundings, then, by the conservation of energy, the heat lost is equal to the heat gained.

In this experiment, hot metal is added to water in a calorimeter cup, and the mixture is stirred until the system is in thermal equilibrium. The calorimeter insulates the system from losing heat (● Fig. 20.1). By the conservation of energy, the heat lost by the metal is equal to the heat gained by the water and cup and stirrer. In equation form,

$$\text{heat lost} = \text{heat gained}$$

or

$$\Delta Q_{\text{metal}} = \Delta Q_{\text{water}} + \Delta Q_{\text{cup and stirrer}}$$

and

$$\begin{aligned} m_m c_m (T_m - T_f) &= m_w c_w (T_f - T_w) + m_{cs} c_{cs} (T_f - T_w) \\ &= (m_w c_w + m_{cs} c_{cs})(T_f - T_w) \end{aligned}$$

Solving for c_m,

$$c_m = \frac{(m_w c_w + m_{cs} c_{cs})(T_f - T_w)}{m_m (T_m - T_f)} \qquad \textbf{(20.4)}$$

where T_f is the final intermediate equilibrium temperature of the system. The other subscripts indicate the masses, specific heats, and initial temperatures of the respective components. Hence, Eq. 20.4 may be used to determine the specific heat, c_m, of the metal if all the other quantities are known.

EXPERIMENTAL PROCEDURE

1. Weigh out 400 g to 500 g (0.4 kg to 0.5 kg) of one kind of dry metal shot. [Do this by first determining the mass of the empty boiler cup (in which the metal shot is heated), and then adding an appropriate amount of metal shot to the cup and reweighing.]

   Record the mass of the metal, m_m, and the room temperature, T_r, in the Data Table. Your instructor may prefer to use a solid piece of metal with a string attached instead of metal shot. In this case, it is necessary to weigh only the piece of metal.

2. Insert a thermometer well into the metal shot (or into the cup with a piece of metal, if used), place the cup and shot in the boiler, and start heating the boiler water.

   Caution: *If a mercury thermometer is used, special care must be taken. If the thermometer should break and mercury spill into the hot metal, immediately notify your instructor. The cup should be removed* *from the room (to an exhaust hood or outdoors). Mercury fumes are* ***highly*** *toxic. If not using mercury, still be careful of broken glass.*

The boiler should be about half full of water. Keep steam or water from dampening the dry metal by shielding the cup with a cardboard lid (with a hole for the thermometer).

3. While the boiler is heating, determine and record the mass of the inner calorimeter cup and the stirrer (without the ring). Record the total mass, $m_{cs.}$ Also, note and record the type of metal and specific heat of the cup and stirrer, which is usually stamped on the cup.* (The specific heat may be found in Appendix A, Table A4, if it is not stamped on the cup.)

4. Fill the calorimeter cup about one-half to two-thirds full of cold tap water, and weigh the cup, stirrer, and water to determine the mass of the water, m_w.

(If a solid piece of metal is used, which usually has less mass than the recommended amount of shot, less water should be used so as to obtain an appreciable ΔT temperature change. This may also be the case at high elevations, where the temperature of boiling water is substantially less than 100 °C).

Place the calorimeter cup with the water and stirrer in the calorimeter jacket, and put on the lid, with a thermometer extending into the water.

5. After the water in the boiler boils and the thermometer in the metal has stabilized (allow several minutes), read and record the temperature of the metal, T_m.

Start with the water and stirrer in the cup at a temperature T_w several degrees below room temperature T_r. Adjust the temperature of the inner calorimeter cup and its contents by placing it in a beaker of ice water. Measure and record the temperature T_w.

6. Remove the thermometer from the metal. Then remove the lid from the calorimeter and quickly, but carefully, lift the cup with the hot metal from the boiler and pour the metal shot into the calorimeter cup with as little splashing as possible so as not to splash out and lose any water. (If a solid piece of metal is used, carefully lower the metal piece into the calorimeter cup by means of the attached string.)

Replace the lid with the thermometer, and stir the mixture gently. The thermometer should not touch the metal. While stirring, watch the thermometer and record the temperature when a maximum equilibrium temperature is reached (T_f).

For best results, the final temperature T_f should be above room temperature T_r by about as many degrees as T_w was below it. If this is not approximately the case, repeat Procedures 4 through 6, adjusting T_w until the relationship $T_f - T_r \approx T_r - T_w$ is satisfied.

7. Repeat Procedures 1 through 6 for another kind of metal sample. Make certain that you use fresh water in the calorimeter cup. (Dump the previous metal shot and water into a strainer in a sink so that it may be dried and used by others doing the experiment later.)

8. Compute the specific heat of each metal, using Eq. (20.4). Look up the accepted values in Appendix A, Table A4, and compute the percent errors.

*If the cup and stirrer are not of the same material, they must be treated separately, and the denominator term in Eq. (20.4) becomes $(m_w c_w + m_c c_c + m_s c_s)(T_f - T_w)$.

Name ______________ Section ______________ Date ______________

Lab Partner(s) ______________

EXPERIMENT 20

Specific Heats of Metals

TI Laboratory Report

DATA TABLE

Purpose: To determine the specific heats of metal samples.

Room temperature T_r ______________

Type of metal	Mass of metal m_m ( )	Mass of calorimeter and stirrer m_{cs} ( )	Specific heat of calorimeter and stirrer c_{cs} ( )	Mass of water m_w ( )	T_m ( )	T_w ( )	T_f ( )

Calculations (show work)

Type of metal	c_m (experimental)	c_m (accepted)	Percent error

Don't forget units

(continued)

TI QUESTIONS

1. (a) The percent errors of your experimental values of the specific heats may be quite large. Identify several sources of experimental error.

   (b) Why does it improve the accuracy of the experiment if $T_f - T_r \approx T_r - T_w$?

2. The specific heat of aluminum is 0.22 cal/g-°C. What is the value of the specific heat in (a) kcal/kg-°C, (b) J/kg-°C? (Show your calculations.)

3. (a) If wet shot had been poured into the calorimeter cup, how would the experimental value of the specific heat have been affected?

(b) If some water had splashed out as you were pouring dry shot into the cup, how would the experimental value of the specific heat have been affected?

4. In solar heating applications, heat energy is stored in some medium until it is needed (for example, to heat a home at night). Should this medium have a high or a low specific heat? Suggest a substance that would be appropriate for use as a heat storage medium, and explain its advantages.

5. Explain why specific heat is *specific,* and how it gives a relative indication of molecular configuration and bonding.

Name ______________________ Section ______________ Date ______________

Lab Partner(s) ____

EXPERIMENT 21

Archimedes' Principle: Buoyancy and Density

TI Advance Study Assignment

Read the experiment and answer the following questions.

1. Describe the physical reason for the buoyant force in terms of pressure.

2. Show that the buoyant force is given by $F_b = \rho_f g V_f$ using the development in the Theory section.

3. Give the conditions on densities that determine whether an object will sink or float in a fluid.

4. Distinguish between density and specific gravity, and explain why it is convenient to express these quantities in cgs units.

(continued)

5. Describe how the density of an object less dense than water can be determined using Archimedes' principle. How about the density of a liquid?

6. Why is it important to make certain that no air bubbles adhere to objects during the submerged weighing procedures? How would the experimental results be affected if bubbles were present? Is this a random or systematic error? (See Experiment 2.)

PRE-LAB DEMONSTRATION | Poke a Hole (and no leaks)

Wooden pencils are pushed completely through both sides of a plastic bag filled with water—and none leaks.

E X P E R I M E N T 21

Archimedes' Principle: Buoyancy and Density

INTRODUCTION AND OBJECTIVES

Some objects float and others sink in a given fluid—a liquid or a gas. The fact that an object floats means it is "buoyed up" by a force equal to its weight. Archimedes (287–212 BCE), a Greek scientist, deduced that the upward buoyant force acting on a floating object is equal to the weight of the fluid it displaces. Thus, an object sinks if its weight exceeds that of the fluid it displaces.

In this experiment, Archimedes' principle will be studied in an application: determining the densities and specific gravities of solid and liquid samples.

After performing this experiment and analyzing the data, you should be able to:

1. Tell whether an object will sink or float in a fluid, knowing the density of each.
2. Distinguish between density and specific gravity.
3. Describe how the densities of objects that sink or float may be determined experimentally.

EQUIPMENT NEEDED

- Triple-beam pan balance with swing platform (or single-beam double-pan balance with swing platform and set of weights).
- Overflow can (or graduated cylinder and eye dropper)
- 2 beakers
- Metal cylinder, irregularly shaped metal object, or metal sinker
- Waxed block of wood
- Saltwater solution and alcohol
- String
- Hydrometer and cylinder

THEORY

When placed in a fluid, an object either floats or sinks. This is most commonly observed in liquids, particularly water, in which "light" objects float and "heavy" objects sink. But the same effect occurs for gases. A falling object is sinking in the atmosphere, whereas other objects float (● Fig. 21.1).

Objects float because they are buoyant, or are buoyed up. That is, when submerged, there must be an upward

Figure 21.1 Gas buoyancy. Archimedes' principle applies to fluids—liquids *or* gases. Here, a helium-filled blimp floats in air. (Bill Aron/PhotoEdit.)

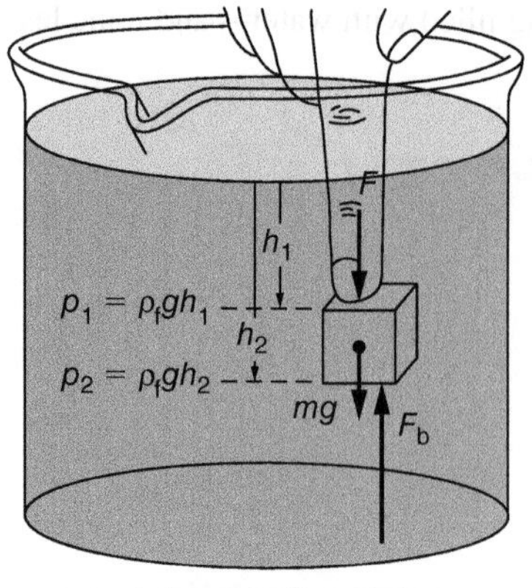

Figure 21.2 Buoyancy. A buoyant force arises from the difference in pressure at different depths. The pressure at the bottom of the submerged block (p_2) is greater than that at the top (p_1), so there is a (buoyant) force directed upward (the arrow is shifted for clarity).

force that is greater than the downward force of the object's weight, and on release the object will be buoyed up and float. When floating, the upward buoyant force is equal to the object's weight. The upward force resulting from an object being wholly or partially immersed in a fluid is called the **buoyant force**. How the buoyant force arises can be understood by considering a buoyant object being held under the surface of a liquid (● Fig. 21.2).

The pressures on the upper and lower surfaces of the block are given by the pressure-depth equations $p_1 = \rho_f g h_1$ and $p_2 = \rho_f g h_2$, respectively, where ρ_f is the density of the fluid. Thus there is a pressure difference, $\Delta p = p_2 - p_1 = \rho_f g(h_2 - h_1)$, which gives an upward force (the buoyant force). In this case (Fig. 21.2), the buoyant force is balanced by the downward applied force and the weight of the block.

It is not difficult to derive an expression for the magnitude of the buoyant force. If both the top and bottom areas of the block are A, the buoyant force (F_b) is given by $F_b = \Delta pA = \rho_f g V_f$, where V_f is the volume of the fluid displaced. But $\rho_f V_f$ is simply the mass of the fluid displaced by the block (recall that $\rho = m/V$). Hence the magnitude of the buoyant force is equal to the weight of the fluid displaced by the block. This general result is known as **Archimedes' principle**:

> *An object immersed wholly or partially in a fluid experiences a buoyant force equal in magnitude to the weight of the volume of fluid that it displaces.*

Thus the magnitude of the buoyant force depends only on the weight of the fluid displaced by the object, *not* on the weight of the object.*

Whether an object will float or sink can be shown mathematically as follows. The weight of an object is $w_o = m_o g = \rho_o g V_o$, where V_o is the volume of the object and $\rho_o = m_o/V_o$. Similarly, the weight of the fluid displaced by the object, or the buoyant force, is $F_b = w_f = m_f g = \rho_f g V_f$. If the object is completely submersed in the fluid, then $V_o = V_f$, and dividing one equation by the other yields

$$\frac{F_b}{w_o} = \frac{\rho_f}{\rho_o} \quad \text{or} \quad F_b = \left(\frac{\rho_f}{\rho_o}\right) w_o \qquad \textbf{(21.1)}$$

Hence, whether the buoyant force (F_b) or the weight of the object (w_o) is greater depends on the densities, and

1. An object **will float** in a fluid if the density of the object, ρ_o, is less than the density of the fluid, ρ_f, that is, $\rho_o < \rho_f$. ($F_b >$ weight of object);
2. An object **will sink** if the object's density is greater than that of the fluid, $\rho_o > \rho_f$. ($F_b <$ weight of object);
3. An object **will float in equilibrium** at any submerged depth where it is placed if its density is equal to that of the fluid, $\rho_o = \rho_f$. ($F_b =$ weight of object).

Specific Gravity and Density

Specific gravity will be used in the study and determination of density. The **specific gravity** *of a solid or liquid is defined as the ratio of the weight of a given volume of the substance* (w_s) *to the weight of an equal volume of water* (w_w):

$$\text{specific gravity (sp. gr.)} = \frac{w_s}{w_w} = \frac{\text{weight of a substance (of given volume)}}{\text{weight of an equal volume of water}} \qquad \textbf{(21.2)}$$

where the subscripts s and w refer to the substance and water, respectively.

Specific gravity is a density-type designation that uses water as a comparison standard. Since it is a weight ratio, specific gravity has no units. The specific gravity can also be expressed as a ratio of densities.

$$\text{sp. gr.} = \frac{\rho_s}{\rho_w} \qquad \textbf{(21.3)}$$

For practical purposes, the density of water is 1 g/cm^3 over the temperature range in which water is liquid,

$$\text{sp. gr.} = \frac{\rho_s}{\rho_w} = \frac{\rho_s\,(\text{g/cm}^3)}{1\,(\text{g/cm}^3)} = \rho_s$$

*Archimedes (287–212 BCE) was a Greek scientist with many accomplishments. He is probably best known from the legend of determining whether a gold crown made for the king was pure gold or whether the craftsman had substituted a quantity of silver for an equivalent amount of gold. According to a Roman account, while pondering the question Archimedes went to the baths, and on immersing himself in a full bath noticed that water flowed out, presumably equal to his body volume. Archimedes recognized a solution to the problem and excitedly jumped out of the bath, running home (unclothed) through the streets of Syracuse (ancient Greece) shouting "Eureka! Eureka!" (Greek for "I have found it!").

Supposedly, he then put quantities of pure gold and silver equal in weight to the king's crown in basins full of water. More water overflowed for the silver than the gold. Testing the crown, more overflowed than for pure gold, implying some silver content. Although Archimedes' solution to the problem involved density and volume, it may have gotten him thinking about buoyancy.

so

$$\boxed{\text{sp. gr.} = \rho_s} \quad (21.4)$$

That is, *the specific gravity is equal to the numerical value of the density of a substance when expressed in g/cm³.*

For example, the density of mercury is 13.6 g/cm³, and mercury has a specific gravity of 13.6. A specific gravity of 13.6 indicates that mercury is 13.6 times more dense than water, $\rho_s = (\text{sp. gr.})\rho_w$, or that a sample of mercury will weigh 13.6 times as much as an equal volume of water.

Archimedes' principle can be used to determine the specific gravity and density of a *submerged* object. By Eq. (21.2),

$$\boxed{\text{sp. gr.} = \frac{\omega_o}{\omega_w} = \frac{\omega_o}{F_b}} \quad (21.5)$$

where ω_o is the weight of the object, ω_w is the weight of the water it displaces, and, by Archimedes' principle, $\omega_w = F_b$.

For a heavy object that sinks, the net force as it does so is equal to $\omega_o - F_b$. (Why?) If attached to a scale while submerged, it would have a measured *apparent* weight ω_o' and $\omega_o' = \omega_o - F_b$. Thus $F_b = \omega_o - \omega_o'$, and Eq. 21.5 may be written

$$\text{sp. gr.} = \frac{\omega_o}{\omega_w} = \frac{\omega_o}{\omega_o - \omega_o'}$$

or, in terms of mass measured on a balance ($m = \omega/g$),

$$\boxed{\text{sp. gr.} = \frac{m_o}{m_o - m_o'} = \rho_o} \quad (21.6)$$

(of a heavy object that sinks)

where ρ_o is the density of the object in g/cm³. This provides us with an experimental method to determine the specific gravity and density of an object that sinks.

To measure the specific gravity and density of an object that floats, or is less dense than water, using Archimedes' principle, it is necessary to use another object of sufficient weight and density to submerge the light object completely.

Letting ω' indicate a submerged weight, $\omega_1 = \omega_o + \omega_s'$ is the measured weight of the object (ω_o) and the sinker (ω_s') with *only* the sinker submerged, and $\omega_2 = \omega_o' + \omega_s'$ is the measured weight when both object and sinker are submerged. Then,

$$\omega_1 - \omega_2 = (\omega_o + \omega_s') - (\omega_o' + \omega_s') = \omega_o - \omega_o'$$

or, in terms of mass,

$$m_1 - m_2 = m_o - m_o'$$

and the specific gravity and density can be found from Eq. 21.6. That is,

$$\boxed{\text{sp. gr.} = \frac{m_o}{m_1 - m_2} = \rho_o} \quad (21.7)$$

(of a heavy object that sinks)

EXPERIMENTAL PROCEDURE

A. Direct Proof of Archimedes' Principle

1. Weigh the metal sample and record its mass, m_o, and the type of metal in the Laboratory Report. Also, determine the mass of an empty beaker, m_b, and record. Fill the overflow can with water, and place it on the balance platform. Attach a string to the sample and suspend it from the balance arm, as illustrated in ● Fig. 21.3.*

2. The overflow from the can when the sample is immersed is caught in the beaker. Take a mass reading m_o' of the submerged object. Make certain that no bubbles adhere to the object. (It is instructive to place the overflow can on a second balance, if available, and note that the "weight" of the overflow does not change as the sample is submerged.)

   Next weigh the beaker and water so as to determine the mass of the displaced water, m_w. (If the can does not fit on the balance platform, first suspend and immerse the object in the full overflow can, and catch the overflow in the beaker and find m_w. Then attach the sample to the balance arm and suspend it in a beaker of water that will fit on the balance platform to find m_o'.)

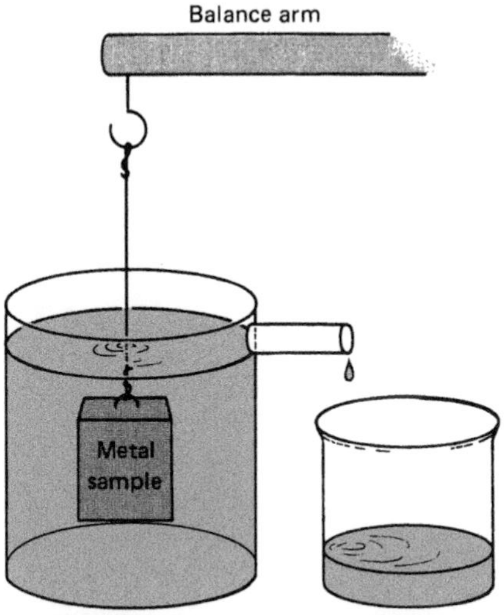

Figure 21.3 Archimedes' principle. The arrangement for proving Archimedes' principle. The weight of the displaced liquid that overflows into the beaker is equal to the reduction in weight of the metal sample when it is submerged, which is equal to the buoyant force.

*You may use an alternative method if no overflow can is available. Attach a string to the sample and place it in a graduated cylinder. Fill the cylinder with water until the sample is completely submerged. Add water (with an eyedropper) until the water level is at a specific reference mark on the cylinder (for example, 35 mL). Remove the sample, shaking any drops of water back into the cylinder, and weigh the cylinder and water (m_b). Refill the cylinder to the reference mark and weigh it again ($m_w + m_b$). The mass of the "overflow" water is then the difference between these measurements.

3. The buoyant force is then the difference between the object's true weight and its submerged weight, $F_b = m_o g - m_o' g$. According to Archimedes' principle, the magnitude of the buoyant force F_b should equal the weight of the displaced water:

$$F_b + \omega_w + m_w g$$

or

$$F_b = (m_o - m_o')g = m_w g$$

Compute the buoyant force, and compare it with the weight of the displaced water by finding the percent difference.

B. Density of a Heavy Solid Object ($\rho_o > \rho_w$)

4. Determine the specific gravity and density of the metal sample. This can be computed using the data from Part A.

C. Density of a Light Solid Object ($\rho_o < \rho_w$)

5. Determine the specific gravity and density of the wooden block by the procedure described in the Theory section. First, measure the mass of the wooden block alone (in air). Then set up as in Fig. 21.3.

Tie the sinker to the wood block, and tie the block to the lower hook of the balance. With the beaker empty, check that the sinker does not touch the bottom of the beaker and that the top of the wooden block is below the top of the beaker. Pour enough water into the beaker to cover the sinker, weigh, add more water until the wooden block is submerged, and then weigh again. Make certain that no air bubbles adhere to the objects during the submerged weighing procedures. The block is waxed so that it does not become waterlogged.

D. Density of a Liquid (ρ_ℓ)

A convenient way to measure the density of a liquid is with a hydrometer, which is a weighted glass bulb and a calibrated stem that floats in liquid. The higher the bulb floats, the greater the density of a liquid. To gain familiarity with a hydrometer, measure the densities of water, saltwater, and alcohol, and record them in the Data Table for Part D. Comment on their relative densities.

Name ______________________ Section ______________ Date ______________
Lab Partner(s) ____

EXPERIMENT 21

Archimedes' Principle: Buoyancy and Density

TI Laboratory Report

A. Direct Proof of Archimedes' Principle

Type of metal ______________

Mass of metal m_o in air ______________

Mass of beaker m_b ______________

Mass of metal m_o' submerged in water ______________

Mass of beaker and displaced water $m_w + m_b$ ______________

Mass of displaced water m_w ______________

Buoyant force (in newtons) ______________

Weight of displaced water (in newtons) ______________

Percent difference ______________

Calculations (show work)

Don't forget units

(continued)

B. Density of a Heavy Solid ($\rho_o > \rho_w$)

Calculations
(show work)

Specific gravity ____________

Density ____________

C. Density of a Light Solid ($\rho_o < \rho_w$)

Mass of block in air ____________

Mass of block and sinker with only sinker submerged ____________

Mass of block and sinker with both submerged ____________

Specific gravity ____________

Density ____________

Calculations
(show work)

Name ______________________ Section ______________ Date ______________

Lab Partner(s) ____

D. Density of a Liquid (ρ_ℓ)

Measured density of water ______________

Measured density of saltwater ______________

Measured density of alcohol ______________

Comments:

TI QUESTIONS

1. Look up the density of the metal of the object used in Parts A and B of the procedure (Appendix A, Table A1), and compare it with the experimental value. If there is any difference, comment on the reason(s).

(continued)

2. In Part B, the string will cause error. When does it lead to an experimental density that is too high? Too low?

3. Discuss the situation that occurs when an object is immersed in a fluid that has the same density as the object.

4. (a) Explain how a submarine is caused to submerge and surface without the use of its propulsion propeller and fins.

   (b) Which has the greater density: (1) ice or water, (2) milk or cream?

5. A block of wood floats in a beaker of water. According to Archimedes' principle, the block experiences an upward buoyant force. If the beaker with the water and floating block were weighed, would the measured weight be less than the sum of the weights of the individual components? Explain.

6. A person can lift 45 kg ($\approx$ 100 lb). Using the experimental value of the specific gravity for the metal object in Part B, how many cubic meters of the metal could the person lift (a) in air, (b) in water? How many actual kilograms of metal is this in air and in water?

7. Explain the principle and construction of a hydrometer. What is the purpose of the common measurements of the specific gravities of an automobile's radiator coolant and battery electrolyte?

Name ____________ Section ____________ Date ____________

Lab Partner(s) ____________

EXPERIMENT 22

Fields and Equipotentials

TI Advance Study Assignment

Read the experiment and answer the following questions.

A. Electric Field

1. What is an electric field, and what does it tell you?

2. What are "lines of force," and what force is it?

3. What are equipotentials, and how are they experimentally determined? What is their relationship to the electric field lines?

(continued)

B. Magnetic Field

4. What is a magnetic field, how is it defined, and what does it tell you?

5. Does the magnetic field $\vec{\mathbf{B}}$ have the same relationship to electric charge as the electric field $\vec{\mathbf{E}}$? Explain.

6. How may a magnetic pole be moved in a magnetic field without doing work?

PRE-LAB DEMONSTRATION | Several good E&M demonstrations—Instructor's Choice

Electrostatics in Action (levitation)
Magnetism through the Hand
Is Money Magnetic?
Eating Magnetic Iron for Breakfast
Which Rod Is Magnetic?

EXPERIMENT 22

Fields and Equipotentials

INTRODUCTION AND OBJECTIVES

When buying groceries, we are often interested in the price per pound. Knowing this, the price for a given amount of an item can be determined. Analogously, it is convenient to know the electric force per unit charge at points in space due to an electric charge configuration; or for the magnetic case, the magnetic force per unit pole or "moving charge." Knowing these, the electric force or magnetic force an interacting object would experience at different locations can easily be calculated.

The electric force per unit charge is a vector quantity called the electric field intensity, or simply the **electric field ($\vec{\mathbf{E}}$)**. By determining the electric force on a test charge at various points in the vicinity of a charge configuration, the electric field may be "mapped," or represented graphically, by lines of force. The English scientist Michael Faraday (1791–1867) introduced the concept of lines of force as an aid in visualizing the magnitude and direction of an electric field.

Similarly, the magnetic force per unit pole is a vector quantity called the magnetic field intensity, or **magnetic field ($\vec{\mathbf{B}}$)**. In this case, the field is mapped out by using the pole of a magnetic compass.

In this experiment, the concept of fields will be investigated and some electric and magnetic field configurations will be determined experimentally.

After performing this experiment and analyzing the data, you should be able to:

1. Describe clearly the concept of a force field.
2. Explain lines of force and the associated physical interpretations.
3. Distinguish between lines of force and equipotentials, and describe their relationships to work.

EQUIPMENT NEEDED

A. Electric Field

- Field mapping board and probes
- Conducting sheets with grids
- Conducting paint
- Connecting wires
- 1.5-V battery (or 10-V dc source)
- Galvanometer [or high-resistance voltmeter or multimeter, or vacuum-tube voltmeter (VTVM) with two-point contact field probe*]
- Single-throw switch
- 3 sheets of Cartesian graph paper

B. Magnetic Field

- 2 bar magnets and 1 horseshoe magnet
- Iron filings
- 3 sheets of paper or overhead transparency material
- Small compass
- 3 sheets of Cartesian graph paper or regular paper

*Leads from the dc input of an oscilloscope work nicely.

THEORY

A. Electric Field

The magnitude of the electrostatic force between two point charges q_1 and q_2 is given by Coulomb's law:

$$F = \frac{kq_1q_2}{r^2} \tag{22.1}$$

where r is the distance between the charges and the constant $k = 9.0 \times 10^9$ N-m^2/C^2. The direction of the force on a charge may be determined by the **law of charges** or **charge-force law**:

Like charges repel, and unlike charges attract.

The magnitude E of the **electric field ($\vec{\mathbf{E}}$)** is defined as the electrical force per unit charge, or $E = F/q_o$ (N/C). By convention, the electric field is determined by using a *positive* test charge q_o. In the case of the electric field

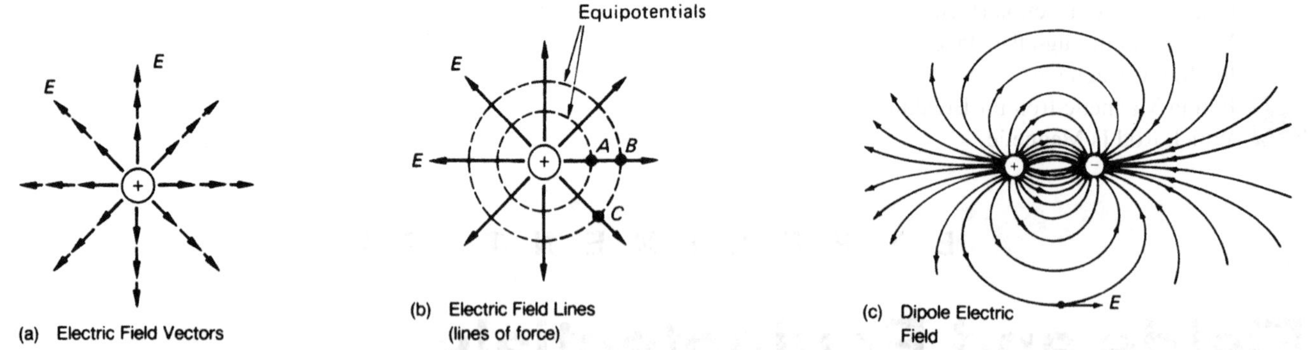

Figure 22.1 Electric field. (a) Electric field vectors near a positive charge. (b) Lines of force with equipotentials for a positive charge. (c) An electric dipole and its electric field. The direction of the electric field at a particular location is tangent to the line of force through that point, as illustrated on the bottom line of force.

associated with a single-source charge q, the magnitude of the electric field a distance r away from the charge is

$$E = \frac{F}{q_o} = \frac{kq_oq}{q_or^2} = \frac{kq}{r^2} \qquad \textbf{(22.2)}$$

(electric field)

The direction of the electric field may be determined by the law of charges—that is, in the direction of the force experienced by the positive test charge.

The electric field vectors for several series of radial points from a positive source charge are illustrated in ● Fig. 22.1a. Notice that the smaller the lengths (magnitudes) of the vectors, the greater the distance from the charge. (Why?)

By drawing lines through the points in the direction of the field vectors, the lines of force are formed (Fig. 22.1b), which give a graphical representation of the electric field. The direction of the electric field at a particular location is tangent to the line of force through that point (Fig. 22.1c). The magnitudes of the electric field are not customarily listed, only the direction of the field lines. However, the closer together the lines of force, the stronger the field.

If a positive charge were released in the vicinity of a stationary positive source charge, it would move along a line of force in the direction indicated (away from the source charge). A negative charge would move along the line of force in the opposite direction. Once the electric field for a particular charge configuration is known, we tend to neglect the charge configuration itself, since the effect of the configuration is given by the field.

Since a free charge moves in an electric field by the action of the electric force, work ($W = Fd$) is done by the field in moving charges from one point to another (for example, from A to B in Fig. 22.1b).

To move a positive charge from B to A would require work supplied by an external force to move the charge against the electric field (force). The work W per charge q_o in moving the charge between two points in an electric field is called the **potential difference**, ΔV, between the points:

$$\boxed{\Delta V_{BA} = V_B - V_A = \frac{W}{q_o}} \qquad \textbf{(22.3)}$$

(It can be shown that the potential at a particular point a distance r from the source charge q is $V = kq/r$. See your textbook.)

If a charge is moved along a path at right angles or perpendicular to the field lines, no work is done ($W = 0$), since there is no force component along the path. Then along such a path (dashed-line paths in Fig. 22.1b), $\Delta V = V_B - V_C = W/q_o = 0$, and $V_C = V_B$. Hence, the potential is constant along paths perpendicular to the field lines. Such paths are called **equipotentials**. (In three dimensions, the path is along an equipotential surface.)

An electric field may be mapped experimentally by determining either the field lines (of force) or the equipotential lines. Static electric fields are difficult to measure, and field lines are more easily determined by measuring small electric currents (flow of charges) maintained in a conducting medium between charge configurations in the form of metal electrodes.

The steady-state electric field lines closely resemble the static field that a like configuration of static charges would produce. The current is measured in terms of the voltage (potential) difference by a high-resistance voltmeter or multimeter (or VTVM).

In other instances, equipotentials are determined, and hence the field lines, using a simple galvanometer as a detector. When no current flows between two probe points, as indicated by a zero deflection on the galvanometer, there is no potential difference between the points ($\Delta V = 0$), and the points are on an equipotential.

B. Magnetic Field

Analogous to an electric field, a **magnetic field ($\vec{\mathbf{B}}$)** was originally defined as the magnetic force per unit pole. The direction of the force at a particular location is that of the force experienced by a north magnetic pole.

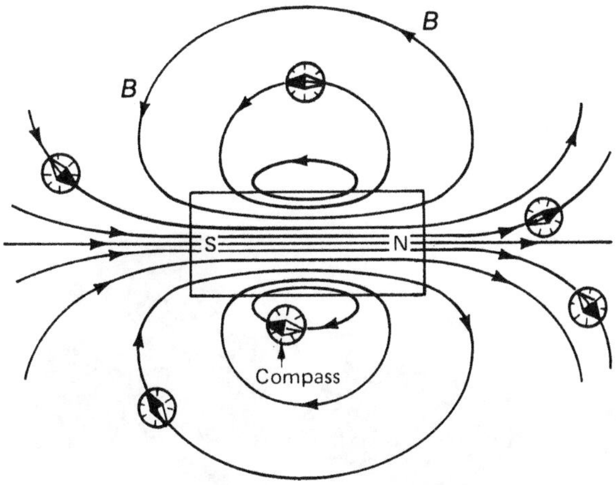

Figure 22.2 Magnetic field. The magnetic force causes a compass needle to line up with the field, and the north pole of the compass points in the direction of the field. If the compass is moved in the direction indicated by the north pole, the path of the compass needle traces out a magnetic field line.

Just as the electric field may be mapped around an electric charge, magnetic lines of force may be mapped around a magnet. A single magnetic pole, or magnetic monopole, has never been observed, so the magnetic field is mapped using the north pole (by convention) of a magnetic dipole, for example, the magnetic needle of a compass. The torque on the compass needle resulting from the magnetic force causes the needle to line up with the field, and the north pole of the compass points in the direction of the field (● Fig. 22.2). If the compass is moved in the direction indicated by the north pole, the path of the compass traces out a field line.

Another observation is that an electric charge q moving nonparallel to a magnetic field experiences a force. For the special case in which the velocity vector $\vec{\mathbf{v}}$ of the charge is perpendicular to the magnetic field $\vec{\mathbf{B}}$, the magnitude of the force is given by

$$F = qvB$$

This gives an expression for the strength (magnitude) of the magnetic field in terms of familiar quantities:

$$\boxed{B = \frac{F}{qv}} \qquad \textbf{(22.4)}$$

(magnetic field)

where the direction of $\vec{\mathbf{B}}$ is perpendicular to the plane of $\vec{\mathbf{v}}$ and $\vec{\mathbf{F}}$. Note that the SI unit of magnetic field is N/A-m, or tesla (T).*

*Other units of magnetic field are the weber/m^2 (Wb/m^2) and the gauss (G). These units are named after early investigators of magnetic phenomena.

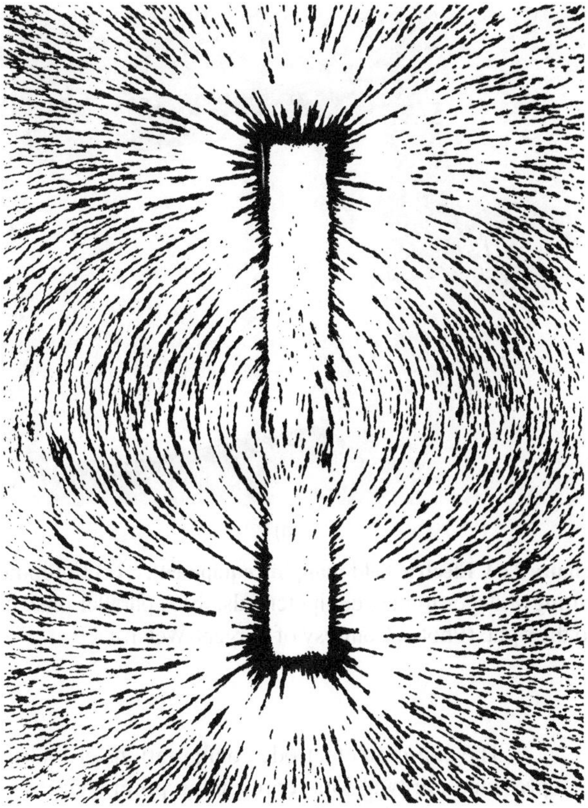

Figure 22.3 Iron filing pattern for a bar magnet. The iron filings become induced magnets and line up with the field, as would a compass needle. (Courtesy of *PSSC Physics,* D.C. Heath and Company with Educational Development Center, Inc., Newton, Massachusetts.)

The magnetic field may then be thought of as the magnetic force "per unit charge" per velocity. The $\vec{\mathbf{B}}$ field has the same form as that mapped out using compass-needle poles.

It is instructive for comparative purposes to draw equipotential lines perpendicular to the field lines, as in the electric field case. No work would be done on a magnetic pole (or electric charge) when it is moved along these equipotential lines. (Why?)

A common method of demonstrating a magnetic field is to sprinkle iron filings over a paper or transparency material covering a magnet (● Fig. 22.3). The iron filings become induced magnets and line up with the field as would a compass needle. This method allows one to visualize the magnetic field configuration quickly.

EXPERIMENTAL PROCEDURE

A. Electric Field

1. An electric field mapping setup is shown in ● Fig. 22.4a. The apparatus consists of a flat board on which is placed a sheet of carbonized conducting paper imprinted with a grid. The sheet has an electrode configuration of conducting silver paint, which provides an electric field when connected to a voltage source (for example, a battery).

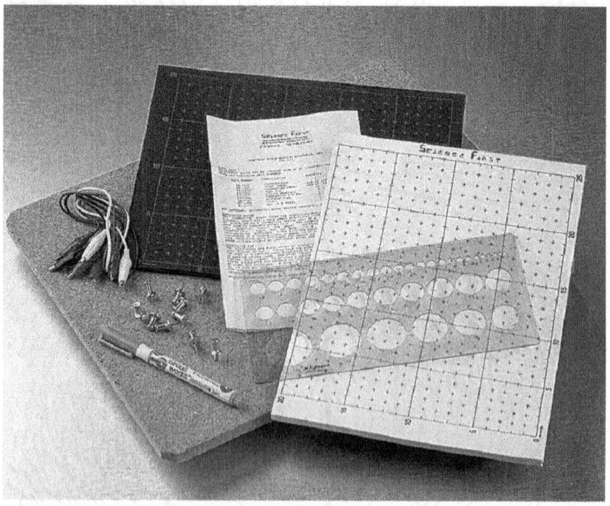

(a)

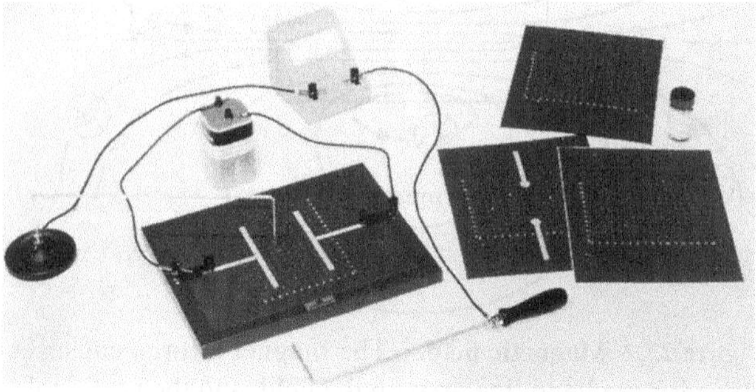

(b)

Figure 22.4 Electric field mapping equipment. (a) Equipment for painting electrodes on conductive paper in preparation for measuring voltages to map equipotentials. (b) A parallel-plate capacitor configuration on the board and an electric dipole configuration to the right. (Photos Courtesy of Sargent-Welch.)

The common electrode configurations ordinarily provided are two dots representing point charges of an electric dipole configuration and two parallel linear electrodes representing a two-dimensional cross section of a parallel-plate capacitor (Fig. 22.4b).

2. Draw the electric dipole configuration on a sheet of graph paper to the same scale and coordinates as those of the painted dipole on the imprinted grid on the conducting sheet. Then place the dipole conducting sheet on the board, and set the contact terminals firmly on the painted electrode connections. If you are using a galvanometer, do Procedures 3 through 7. If you are using a voltmeter, do Procedures 8 through 12.

Galvanometer Measurements

3. Connect the probes to the galvanometer as shown in Fig. 22.4b. The probes are used to locate points in the field that are at equipotential. Connect the voltage source (1.5-V battery) to the board terminals. Place a switch in the circuit (not shown in the figure), and leave it open until you are ready to take measurements.

   Place the stationary probe on the electric dipole sheet at some general point near the edge of the grid area in the region between the electrodes. The potential at this point will serve as a reference potential. Mark the probe position on your graph-paper map.

   The movable probe is then used to determine the location of a series of other points that have the same potential. When the movable probe is at a point with the same potential as that of the stationary reference probe, no deflection will be observed on the galvanometer.

4. Close the switch, and place the movable probe on the conducting paper at some location an appreciable distance away from the stationary probe. Move the probe until the galvanometer shows zero deflection (indicating a point of equipotential), and record this point on the graph-paper map.

   Locate a series of eight or ten points of the same potential across the general field region, and draw a dashed-line curve through these points on the graph-paper map.

5. Choose a new location for the reference probe, 2 cm to 3 cm from the previous reference position, and locate another series of equipotential points. Continue this procedure until you have mapped the field region. Open the switch.

   Draw curves perpendicular to the equipotential lines on the graph-paper map to represent the electric field lines. Do not forget to indicate the field direction on the field lines.

6. Repeat the procedure for the parallel linear (plate) electrode configuration. Be sure to investigate the regions around the ends of the plate electrodes.

7. (*Optional*) Your instructor may wish to have you map the electric field for a nonsymmetric electrode configuration or a configuration of your own choosing. These can be prepared by painting the desired electrode configuration on a conducting sheet with silver paint.

Voltmeter Measurements

8. For the high-resistance voltmeter (or VTVM), the field probe should have two contacts mounted about 2 cm apart. Connect the voltage source (10-V dc) to the board terminals. Place a switch in the circuit (not shown in Fig. 22.4b), and leave it open until you are ready to take measurements.

Close the switch, and with the zeroed voltmeter set on the 10-V scale, position the negative (−) contact of the field probe near the negative electrode. Using the negative probe point as a pivot, rotate the positive (+) contact around the fixed negative contact until the position with the maximum meter reading is found.

Record the positions of the probe contacts on the graph-paper map. (The sensitivity of the voltmeter may be increased by switching to a lower scale. A midscale reading is desirable.)

9. Using the second probe point as a new negative probe point, repeat the procedure to determine another point of maximum meter reading, and record. Continue this procedure until the positive electrode is approached. Draw a smooth curve through these points on the graph-paper map.

Then, starting again at a new position near the negative electrode, repeat these procedures for another field line. Trace out four to six field lines in this manner. Do not forget to indicate the field direction on the lines.

10. Place the negative probe near the center of the field region, and rotate the positive contact until a position is found that gives a *zero* meter reading. Record several of these points on the graph paper with a symbol different from that used for the field lines. Check the zero on the voltmeter frequently, particularly when changing scales.

Use the second point as a new pivot point, as before, and determine a series of null (zero) points. Draw a dashed-line curve through these equipotential points. Determine three to five equipotential lines in this manner.

11. Repeat this procedure for the parallel linear (plate) electrode configuration. Be sure to investigate the regions around the ends of the plate electrodes.

12. (*Optional*) Your instructor may wish to have you map the electric field for a nonsymmetric electrode configuration or a configuration of your own choosing. These can be prepared by painting the desired electrode configuration on a conducting sheet with silver paint.

B. Magnetic Field

13. Covering the magnets with sheets of paper or transparency material, sprinkle iron filings to obtain an iron filing pattern for each of the arrangements shown in ● Fig. 22.5 (Laboratory Report).

For the bar magnet arrangements, the magnets should be separated by several centimeters, depending on the pole strengths of the magnets. Experiment with this distance so that there is enough space between the ends of the magnets to get a good pattern.

14. Sketch the observed magnetic field patterns on Fig. 22.5. After the patterns have been sketched, collect the iron filings on a piece of paper and return them to the filing container (recycling them for someone else's later use). Economy in the laboratory is important.

15. Place the magnets for each arrangement on a piece of graph paper or regular paper. Draw an outline of the magnets for each arrangement on the paper, and label the poles N and S. Using a small compass, trace out (marking on the paper) the magnetic field lines as smooth curves. Draw enough field lines so that the pattern of the magnetic field can be clearly seen. Do not forget to indicate the field direction on the lines.

16. Draw dashed-line curves perpendicular to the field lines.

Name ______________________ Section ______________ Date ______________

Lab Partner(s) ____

EXPERIMENT 22

Fields and Equipotentials

TI Laboratory Report

Attach graphs to Laboratory Report.

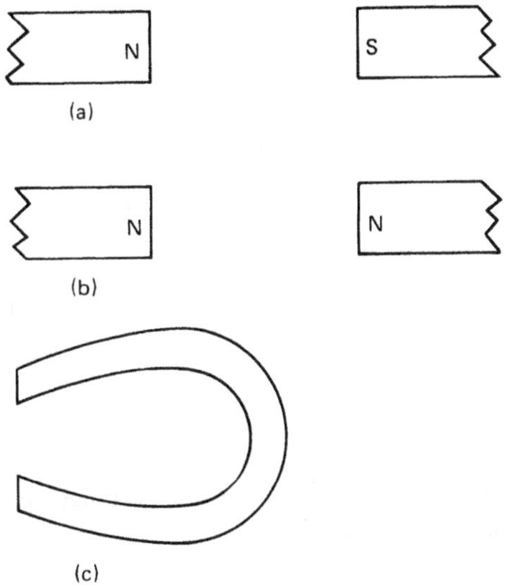

Figure 22.5 See Procedure Section B.

TI QUESTIONS

1. Directions of the fields are indicated on field lines. Why are no directions indicated on equipotential lines?

2. For the dipole configuration, in what region(s) does the electric field have the greatest intensity? Explain how you know from your map, and justify.

Don't forget units

(continued)

3. Comment on the electric field of the parallel plates (a) between the plates, and (b) near the edges of the plates.

4. Sketch the electric field for (a) a negative point charge near a positively charged plate, and (b) two positive point charges.

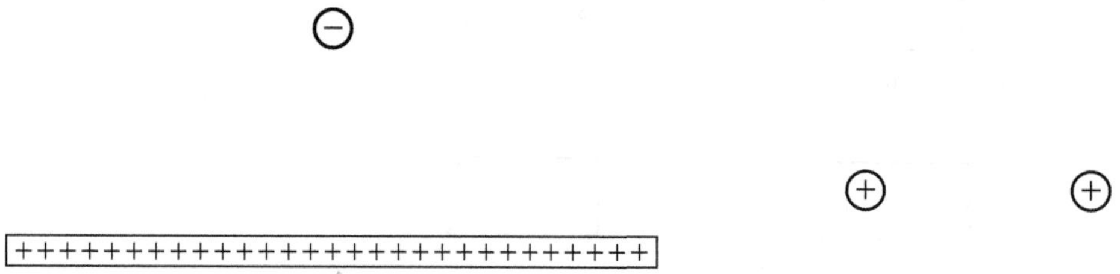

(a) (b)

5. Compare the electric fields and magnetic fields of the experimental arrangements. Comment on any field similarities and differences.

6. Explain how a gravitational field might be mapped. Sketch the gravitational field for two point masses a short distance apart.

Name ______________________ Section ______________ Date ______________

Lab Partner(s) ____

EXPERIMENT 23

Ohm's Law

TI Advance Study Assignment

Read the experiment and answer the following questions.

1. What is the definition of *electrical resistance*?

2. What is an "ohmic" resistance? Are all resistances ohmic in nature?

3. In what ways are liquid and electrical circuits analogous?

4. For a series circuit, what is the terminal voltage of a battery or power supply equal to in terms of the potential differences or voltage drops across circuit components?

(continued)

CI Advance Study Assignment

Read the experiment and answer the following questions.

1. What is a triangle-wave voltage function?

2. What is a nonohmic resistance? How can we distinguish between an ohmic and a nonohmic resistance?

PRE-LAB DEMONSTRATION | **Instructor's Choice**

EXPERIMENT 23

Ohm's Law

OVERVIEW

Experiment 23 examines Ohm's law by TI and CI procedures. In the TI procedure, an experimental circuit makes it possible to investigate (1) the variation of current with voltage, and (2) the variation of current and resistance (constant voltage). The CI procedure looks at the voltage-current relationship not only for an ohmic resistance but also for a nonohmic resistance. Steadily increasing and decreasing voltages are obtained by using a signal generator to produce a triangle-wave voltage.

INTRODUCTION AND OBJECTIVES

One of the most frequently applied relationships in current electricity is that known as **Ohm's law**. This relationship, discovered by the German physicist Georg Ohm (1787–1854), is fundamental to the analysis of electrical circuits. Basically, it relates the voltage (V) and current (I) associated with a resistance (R).

Ohm's law applies to many, but not all, materials. Many materials show a constant resistance over a wide range of applied voltages and are said to be "ohmic." Those that do not are said to be "nonohmic." Common circuit resistors are ohmic, which allows Ohm's law to be used in simple circuit analysis. As will be seen in the theory section, Ohm's law is really a special case of the definition of resistance.

In this experiment, Ohm's law will be investigated as applied to components in a simple circuit.

TI OBJECTIVES

After performing this experiment and analyzing the data, you should be able to:

1. Distinguish between ohmic and nonohmic resistances.
2. Explain current-voltage relationships by Ohm's law.
3. Apply Ohm's law to obtain values of current or voltage in investigating a circuit resistance.

CI OBJECTIVES

1. Verify Ohm's law experimentally.
2. Study the behavior of the current in both an ohmic and a nonohmic resistance.

TI EXPERIMENT 23

Ohm's Law

TI EQUIPMENT NEEDED*

- Ammeter (0 A to 0.5 A)
- Voltmeter (0 V to 10 V dc) or multimeters
- Decade resistance box (0.1 Ω to 99.9 Ω)
- Rheostat (≈ 200 Ω)
- Unknown resistance
- Battery or power supply (6 V)
- Switch
- Connecting wires
- 2 sheets of Cartesian graph paper

*The ranges of the equipment are given as examples. These may be varied to apply to available equipment.

TI THEORY

When a voltage or potential difference (V) is applied across a material, the current (I) in the material is found to be proportional to the voltage, $I \propto V$. The resistance (R) of the material is defined as the ratio of the applied voltage and the resulting current—that is,

$$R = \frac{V}{I} \qquad \text{(TI 23.1)}$$

(definition of electrical resistance)

For many materials, the resistance is constant, or at least approximately so, over a range of voltages. A resistor that has constant resistance is said to obey Ohm's law or to be "ohmic." From TI Eq. (23.1), it can be seen that the unit of resistance is the volt/ampere (V/A). However, the combined unit is called the ohm (Ω), in honor of Georg Ohm (1787–1854), a German physicist, who developed this relationship known as Ohm's law. Note that to avoid confusion with a zero, the ohm is abbreviated with a capital omega (Ω) instead of a capital O.

A plot of V versus I for an ohmic resistance is a straight line (● TI Fig. 23.1). Materials that do not obey Ohm's law are said to be "nonohmic" and have a nonlinear voltage-current relationship. Semiconductors and transistors are nonohmic.

In common practice, **Ohm's law** is written

$$V = IR \qquad \text{(TI 23.2)}$$

where it is understood that R is independent of V. Keep in mind that Ohm's law is not a fundamental law such as Newton's law of gravitation. It is a special case, there being no law that materials must have constant resistance.

To understand the relationships of the quantities in Ohm's law, it is often helpful to consider the analogy of a liquid circuit. (● TI Fig. 23.2).* In a liquid circuit, the force to move the liquid is supplied by a pump. The rate

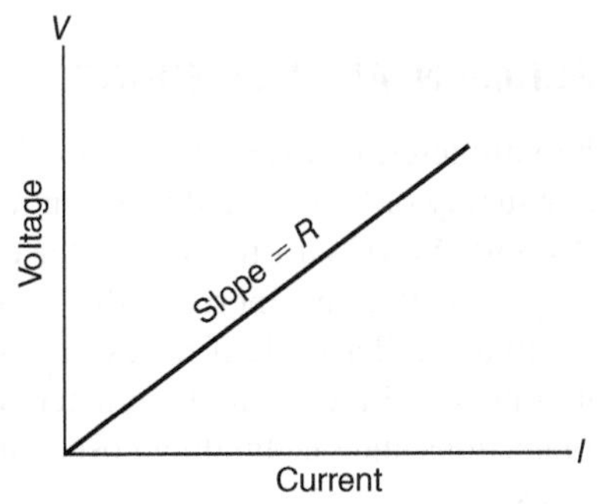

TI Figure 23.1 Ohmic resistance. A voltage-versus-current graph for an ohmic resistance is a straight line, the slope of which is equal to the value of the resistance ($R = V/I$).

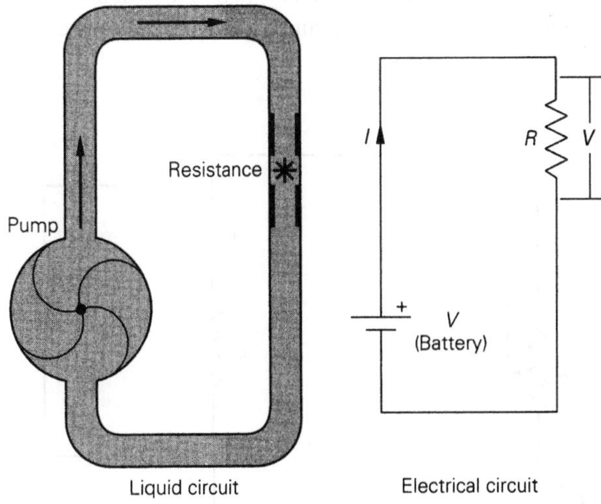

TI Figure 23.2 Analogy to a liquid circuit. In the analogy between a simple electric circuit and a liquid circuit, the pump corresponds to a voltage source, the liquid flow corresponds to electric current, and the paddle wheel hindrance to the flow is analogous to a resistor.

*Keep in mind that an analogy only illustrates a resemblance. Liquid and electrical circuits are physically quite different.

of liquid flow depends on the resistance to the flow (for example, due to some partial obstruction in the circuit pipe, here a paddle wheel)—the greater the resistance, the less the liquid flow.

Analogously, in an electrical circuit, a voltage source (a battery or power supply) supplies the voltage (potential difference) for charge flow, and the magnitude of the current is determined by the resistance R in the circuit. For a given voltage, the greater the resistance, the less the current through the resistance, as may be seen from Ohm's law, $I = V/R$. Notice that the voltage source supplies a voltage "rise" that is equal to the voltage "drop" across the resistance and is given by $V = IR$ (Ohm's law).

In an electrical circuit with two or more resistances and a single voltage source, Ohm's law may be applied to the entire circuit or to any portion of the circuit. When it is applied to the entire circuit, the voltage is the terminal input voltage supplied by the voltage source, and the resistance is the total resistance of the circuit. When Ohm's law is applied to a particular portion of the circuit, the individual voltage drops, currents, and resistances are used for that part of the circuit.

Consider the circuit diagram shown in ● TI Fig. 23.3. This is a *series* circuit. The applied voltage is supplied by a power supply or battery. R_h is a rheostat, a variable resistor that allows the voltage across the resistance R_s to be varied. (This combination is sometimes called a *voltage divider* because the rheostat divides the applied voltage across itself and R_s.)

An ammeter Ⓐ measures the current through the resistor R_s, and a voltmeter Ⓥ registers the voltage drop across both R_s and the ammeter Ⓐ. S is a switch for closing and opening (activating and deactivating) the circuit.

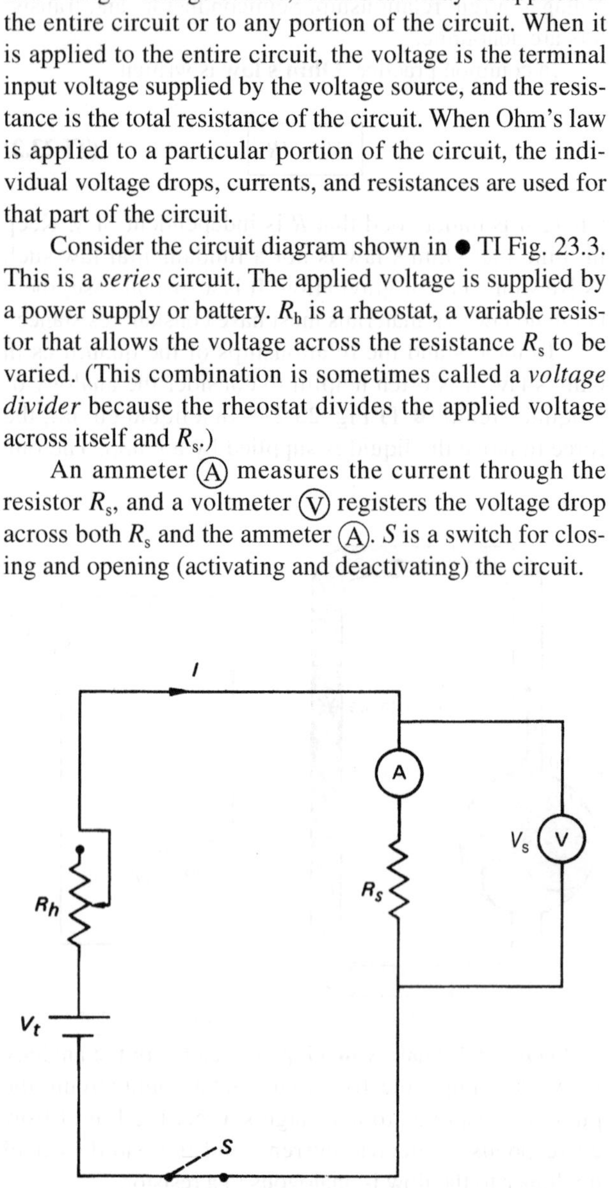

TI Figure 23.3 Circuit diagram. The voltmeter is connected in parallel across the ammeter and the resistance R_s. The other resistance, R_h, is that of the rheostat (continuously variable resistor).

Any component in a circuit that does not generate or supply a voltage acts as a resistance in the circuit. This is true for the connecting wires, the ammeter, and the voltmeter. However, the metallic connecting wires and the ammeter have negligibly small resistances, so they do not greatly affect the current.

A voltmeter has a high resistance, so there is little current through the voltmeter. Hence, to good approximations, the ammeter registers the current in the resistor, and the voltmeter reads the voltage drop across the resistance. These approximations are adequate for most practical applications.

Applying Ohm's law to the portion of the circuit with R_s only,

$$\boxed{V_s = IR_s} \qquad \textbf{(TI 23.3)}$$

where V_s and I are the voltmeter and ammeter readings, respectively. Notice that the same current I flows through the rheostat R_h and the resistance R_s. The voltage drop across R_h is then

$$\boxed{V_h = IR_h} \qquad \textbf{(TI 23.4)}$$

To apply Ohm's law to the entire circuit, we use the fact that the applied voltage "rise" or the terminal voltage V_t of the voltage source must equal the voltage "drops" of the components around the circuit. Then,

$$V_t = V_h + V_s$$

or

$$\boxed{V_t = IR_h + IR_s = I(R_h + R_s)} \qquad \textbf{(TI 23.5)}$$

From TI Eq. (23.5), it can be seen that for a constant R_s, the current through this resistance, and hence its voltage drop V_s, can be varied by varying the rheostat resistance R_h. (The terminal voltage, V_t, is constant.) Similarly, when R_s is varied, the voltage V_s can be maintained constant by adjusting R_h.

TI EXPERIMENTAL PROCEDURE

1. With the voltmeter, measure the terminal voltage of the power supply or battery, and record it in the Laboratory Report. Start with the voltmeter connection to the largest scale, and increase the sensitivity by changing to a smaller scale if necessary. Most common laboratory voltmeters and ammeters have three scale connections and one binding post common to all three scales.

   It is a good practice to take measurements initially with the meters connected to the largest scales. This prevents the instruments from being "pegged" (the needle forced off the scale in galvanometer-type meters) and possibly damaged, should the magnitude of the voltage or current exceed the smaller scale limits. A scale setting may be changed for greater sensitivity by moving the connection (or turning the switch on a multimeter) to a lower scale after the general magnitude and measurement are known.

Also, take care to ensure the proper polarity (+ and −); connect plus (+) to plus (+), and minus (−) to minus (−). Otherwise, the meter will be "pegged" in the opposite direction.

2. Set up the circuit shown in the circuit diagram (TI Fig. 23.3) with the switch open. A standard decade resistance box is used for R_s. Set the rheostat resistance R_h for maximum resistance and the value of the R_s to about 50 Ω. *Be careful with the polarity of the meters. Have the instructor check the circuit before closing the switch.*

A. Variation of Current with Voltage

3. After the instructor has checked the circuit, close the switch and read the voltage and current on the meters. Open the switch after the readings are taken, and record them in TI Data Table 1. Repeat this procedure for a series of four successively lower rheostat settings along the length of the rheostat.

   It is convenient for data analysis to adjust the rheostat (after closing the switch) so that evenly spaced and convenient ammeter readings are obtained. The switch should be closed only long enough to obtain the necessary readings. This prevents unnecessary heating in the circuit and running the battery down.

4. Repeat Procedure 3 for another value of R_s (about 30 Ω).

5. Repeat Procedure 3 for the unknown resistance, and record the data in TI Data Table 2. Relatively low values of voltage may be required. Your instructor will discuss this and the proper connection. *Do not perform this procedure without instructions.*

6. Plot the results for both decade box resistances on a single V_s-versus-I_s graph, and draw straight lines that best fit the sets of data. Determine the slopes of the lines, and compare them with the constant values of R_s of the decade box by computing the percent errors. According to Ohm's law, the corresponding values should be equal.

7. Plot V_s versus I_s for the unknown resistance. What conclusions about the unknown resistance can you draw from the graphs?

B. Variation of Current and Resistance (V_s constant)

8. This portion of the experiment uses the same circuit arrangement as before. In this case, the voltage V_s is maintained constant by adjusting the rheostat resistance R_h when the R_s is varied.

   Initially, set the rheostat near maximum resistance and the resistance R_s of the decade box to about 100 Ω. Record the value of R_s in TI Data Table 3.

   Close the circuit and adjust the rheostat for a convenient voltmeter reading (about 4 V). Record the voltmeter reading as the constant voltage V_s in TI Data Table 3. Record the current and resistance in the table. Open the circuit after making the readings.

9. Repeat this procedure for four more successive steps of current by reducing the value of R_s of the decade box. Keep the voltage across R_s constant for each setting by adjusting the rheostat resistance R_h. Do not reduce R_s below 30 Ω.

10. Plot the results on an I_s-versus-$1/R_s$ graph, and draw a straight line that best fits the data. (Reciprocal ohms, $1/R$, is commonly given the unit name "mhos.") Determine the slope of the line, and compare it with the constant value of V_s by computing the percent error. According to Ohm's law, these values should be equal.

Name ______________________ Section ______________ Date ______________
Lab Partner(s) ____

TI EXPERIMENT 23

Ohm's Law

TI *Laboratory Report*

A. Variation of Current with Voltage

TI DATA TABLE 1

Terminal voltage V_t ______________

Reading	Constant R_s ______		Constant R_s ______	
	Voltage V_s ( )	Current I_s ( )	Voltage V_s ( )	Current I_s ( )
1				
2				
3				
4				
5				

Calculations
(show work)

Slope of lines	Percent error from R_s
______________	______________
______________	______________

Don't forget units

(continued)

TI **DATA TABLE 2** Unknown Resistance

Reading	Voltage V_s ( )	Current I_s ( )
1		
2		
3		
4		
5		
6		
7		

Conclusions from graph:

Name ______________ Section ______________ Date ______________

Lab Partner(s) ______________

E X P E R I M E N T 2 3 **Ohm's Law** *Laboratory Report*

B. Variation of Current with Resistance (V_s constant)

TI DATA TABLE 3

Constant voltage V_s ______________

Reading	Current I_s ( )	Resistance R_s ( )	$1/R_s$ ( )
1			
2			
3			
4			
5			

Calculations (show work)

Slope of lines ______________

Percent error from V_s ______________

TI QUESTIONS

1. If the switch were kept closed during the procedures and the circuit components heated up, how would this affect the measurements? (*Hint*: See Experiment 25.)

(continued)

2. Devise and draw a circuit using a long, straight wire resistor, instead of a decade box, that would allow the study of the variation of voltage with resistance (I_s constant). According to Ohm's law, what would a graph of the data from this circuit show?

3. Compute the values of R_h and the voltage drops across this resistance for the two situations in TI Data Table 1, reading 1. How do the values compare?

C I EXPERIMENT 23

Ohm's Law

EQUIPMENT NEEDED

This activity is designed for the Science Workshop 750 Interface, which has a built-in function generator. It is easily adapted to use with an external wave function generator. Just substitute the available triangle-function generator for the signal generator in the procedure.

- 100-Ω resistor
- 2 cables with alligator clips
- 6-V lightbulb
- Voltage sensor (PASCO CI-6503)
- Science Workshop 750 Interface

THEORY

As discussed in the TI Theory section, for many materials the resistance remains constant over a range of voltages. Such materials are called "ohmic" and they obey **Ohm's law**:

$$V = IR \quad \textbf{(CI 23.1)}$$

For such a material, a graph of voltage versus current is a straight line, the slope of which is the value of the resistance, as shown in TI Fig. 23.1.

In this CI part of the experiment, the relationship between current and voltage for both an ohmic and a nonohmic component of a circuit will be investigated. The current will be measured as the voltage across a component is steadily increased and decreased. If the component is ohmic, the current should be directly proportional to the voltage.

To achieve a steadily increasing and decreasing voltage, a signal generator is used, which can produce what is called a triangle-wave voltage. ● CI Fig. 23.1 shows how the voltage from such a source varies with time. Notice that it increases up to a maximum value, then drops steadily back to zero, and then, with a change of polarity, increases in the opposite direction. This repeats with a certain fixed frequency.

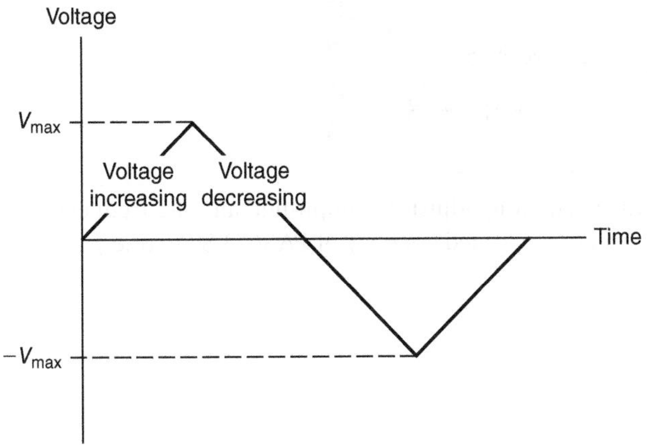

CI Figure 23.1 A triangle-wave voltage function. With a triangle-wave voltage function, the voltage will increase up to a maximum value, drop steadily back to zero, and then change the polarity and increase in the opposite direction. This will repeat with a certain fixed frequency.

SETTING UP DATA STUDIO

1. Open Data Studio and choose "Create Experiment."
2. The Experiment Setup window will open and you will see a picture of the Science Workshop interface. There are seven channels to choose from, and a signal generator. (Digital Channels 1, 2, 3 and 4 are the small buttons on the left; analog Channels A, B and C are the larger buttons on the right; the signal generator is all the way to the right, as shown in ● CI Fig. 23.2).
3. Click on the Channel A button in the picture. A window with a list of sensors will open.
4. Choose the Voltage Sensor from the list and press OK.
5. Connect the sensor to Channel A of the interface, as shown on the computer screen.
6. Click on the picture of the signal generator. The Signal Generator window will open (See ● CI Fig. 23.3).
7. The default form of the signal generator function is a sine wave. Change it to a triangle wave by selecting from the drop menu.
8. Set the amplitude to 5.00 V.
9. Set the frequency to 0.500 Hz. This will produce a triangle wave with a period of 2 seconds.
10. Click on the Measurement and Sample Rate button on the Signal Generator window. A list of measurements will open. Choose to measure the output current, and deselect all others.
11. Do not close the Signal Generator window. Move it toward the bottom of the screen.
12. The Data list should now have two icons: one for the voltage reading of the sensor and one for the output current of the source.
13. Create a graph by dragging the "Voltage" icon from the Data list and dropping it on the "Graph" icon in the Displays list. A graph of voltage versus time will open. The graph window will be called Graph 1.
14. Drag the "Output Current" icon from the Data list, and drop it on top of the x-axis of the graph. The time axis should change to a current axis. Graph 1 is now a graph of voltage versus current.

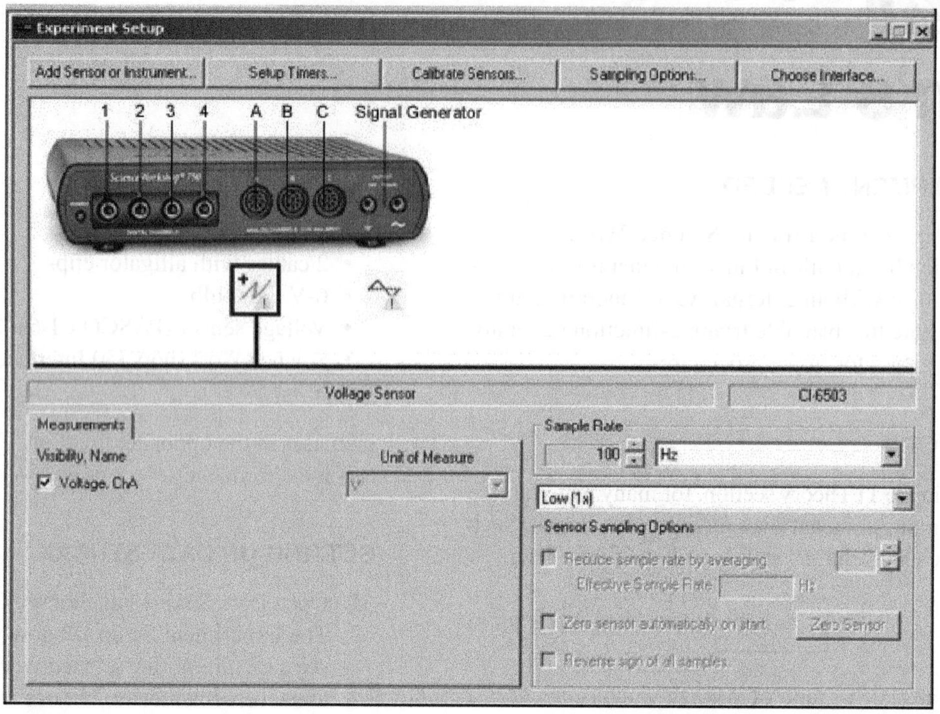

CI Figure 23.2 The Experiment Setup window. The voltage sensor is connected to Channel A and works as a multimeter. The signal generator of the Science Workshop interface is used as the voltage source that produces a triangle-wave function. (Reprinted courtesy of PASCO Scientific.)

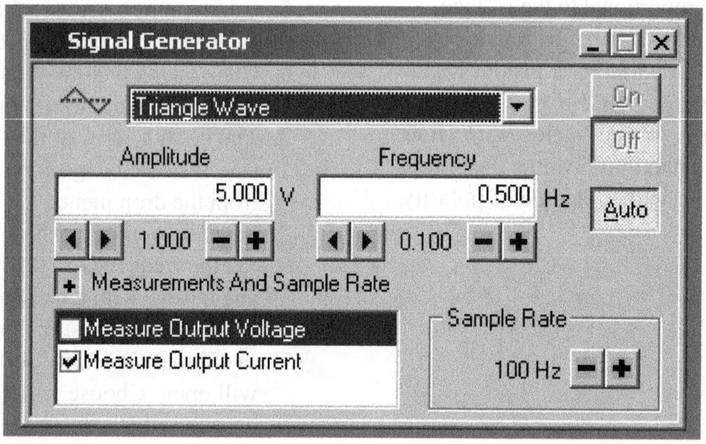

CI Figure 23.3 The Signal Generator window. Choose a triangle-wave function, adjust the amplitude and the frequency as specified in the setup procedure, and choose to measure the output current. (Reprinted courtesy of PASCO Scientific.)

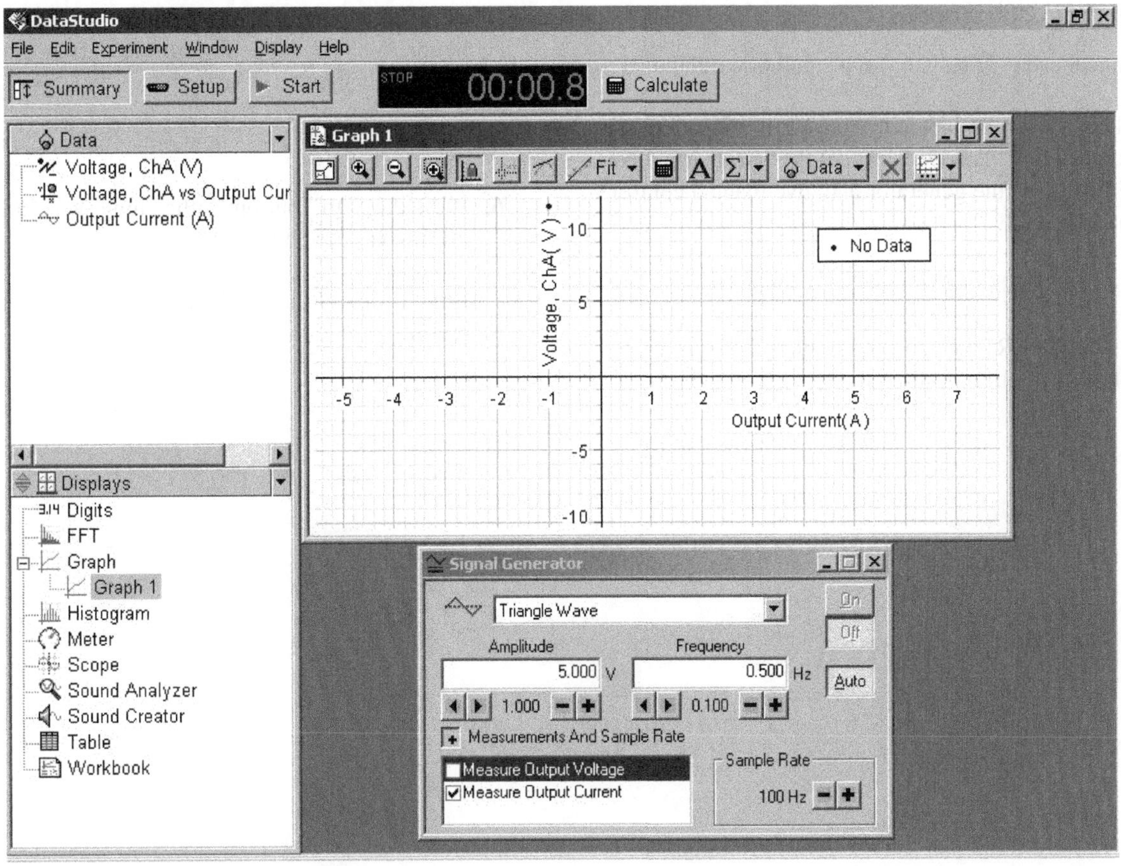

CI Figure 23.4 Data Studio setup. A graph of voltage versus current will show the variations for an ohmic and a nonohmic resistor. The Signal Generator window remains active to manually control the output during experimental Procedure B. (Reprinted courtesy of PASCO Scientific.)

15. ● CI Fig. 23.4 shows what the screen should look like once the setup is complete. The size of the graph window can be changed, if needed. The Signal Generator window will need to stay visible for Procedure B of the experiment, where the output voltage will be manually controlled.

CI EXPERIMENTAL PROCEDURE

A. Ohmic Component

1. Connect the signal generator to the 100-Ω resistor. The voltage sensor will measure the voltage drop across the resistor, as shown in the circuit of ● CI Fig. 23.5.

2. Press the START button and click on the Scale-to-Fit button of the graph toolbar. (That is the leftmost button of the graph toolbar.) After a few seconds, press the STOP button. A cycle is complete after 2 seconds, but it will not affect the experiment if it runs longer than that. In fact, let it run longer and follow the plot on the screen as it appears. What is happening to the current as the voltage changes?

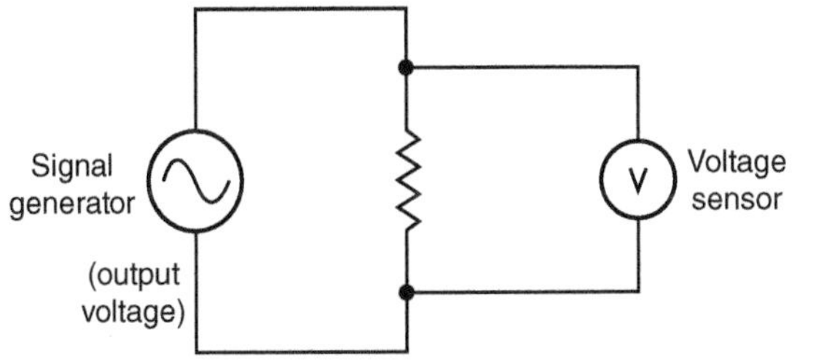

CI Figure 23.5 The experimental setup. The resistor (ohmic or nonohmic) is connected to the signal generator. The voltage sensor measures the voltage drop across the resistor.

3. Print a copy of the graph and paste it to the laboratory report.

4. As expected, the graph for the ohmic resistor is a straight line. Use the Fit drop menu (on the graph toolbar) to select a "Linear Fit" for the data. Record the slope of the line, and compare it to the known value of the resistance by calculating a percent error.

B. Nonohmic Component

1. Change the 100-Ω resistor for a small 6-V lightbulb.

2. Click on the button labeled "Auto" in the Signal Generator window. This will cancel the automatic ON/OFF feature of the generator and give manual control of the signal.

3. Press the "On" button of the signal generator.

4. Press the START button, and collect data for a few seconds, enough to observe the pattern on the screen. Press the Scale-to-Fit button if needed to see the data better. What is happening to the current now, as the voltage changes?

5. Press the STOP button to end the data collection.

6. Press the "Off" button of the signal generator to turn off the output voltage.

7. Print a copy of the graph and paste it to the laboratory report.

Name ______________________ Section ______________ Date ______________

Lab Partner(s) ____

CI EXPERIMENT 23

Ohm's Law

CI Laboratory Report

A. Ohmic Component

Don't forget to attach the graph to the laboratory report.

Calculations
(show work)

Slope of line ______________

Percent error from *R* ______________

B. Nonohmic Component

Don't forget to attach the graph to the laboratory report.

CI QUESTIONS

1. The graph of voltage versus current for the nonohmic resistor was not a straight line. Describe what happened to the current as the voltage increased, compared to what happened for the ohmic resistor.

2. Why does the graph for the nonohmic resistor "loop"? (*Hint*: What happens to the lightbulb filament as the current increases?)

Don't forget units

(continued)

3. Describe what is happening to the resistance of the lightbulb as the voltage increases. (*Hint*: Look at the graph in segments, and treat each segment as though it were a straight line with slope equal to the resistance.)

Name ______________________ Section ______________ Date ______________

Lab Partner(s) ____

EXPERIMENT 24

The Measurement of Resistance: Ammeter-Voltmeter Methods and Wheatstone Bridge Method

TI Advance Study Assignment

Read the experiment and answer the following questions.

A. Ammeter-Voltmeter Method

1. When one is measuring a resistance with an ammeter and a voltmeter, is the resistance given exactly by $R = V/I$? Explain.

2. Comment on the relative magnitudes of the resistances of an ammeter and a voltmeter.

3. Is (a) an ammeter and (b) a voltmeter connected in series or parallel with a circuit component (a resistance)? Explain.

(continued)

B. Wheatstone Bridge Method

4. Why is the Wheatstone bridge called a "null" instrument?

5. When the galvanometer in a Wheatstone bridge circuit shows no deflection, why are the voltages across opposite branches on each side of the galvanometer necessarily equal?

6. For a slide-wire Wheatstone bridge, why should the sliding key not be moved with the key depressed?

EXPERIMENT 24

The Measurement of Resistance: Ammeter-Voltmeter Methods and Wheatstone Bridge Method

INTRODUCTION AND OBJECTIVES

The magnitude of a resistance can be measured by several methods. One common method is to measure the voltage drop V across a resistance in a circuit with a voltmeter and the current I through the resistance with an ammeter. By Ohm's law, then, $R = V/I$. However, the ratio of the measured voltage and current does not give an exact value of the resistance because of the resistances of the meters.

This problem is eliminated when one measures a resistance, or, more properly, compares a resistance with a standard resistance in a Wheatstone bridge circuit [named after the British physicist Sir Charles Wheatstone (1802–1875)].* In this experiment, the ammeter-voltmeter and the Wheatstone bridge methods of measuring resistances will be investigated.

After performing this experiment and analyzing the data, you should be able to:

1. Describe the two ways to measure resistance with an ammeter and a voltmeter, and explain how they differ.
2. Describe the basic principle and operation of the Wheatstone bridge.
3. Discuss the relative accuracy of the ammeter-voltmeter methods and the Wheatstone bridge method of measuring resistance.

*The Wheatstone bridge was popularized and promoted by Sir Charles Wheatstone; however, the British mathematician Samuel Christie invented it.

EQUIPMENT NEEDED†

A. Ammeter-Voltmeter Methods

- Ammeter (0 A to 0.5 A)
- Voltmeter (0 V to 3 V)
- Rheostat (10 Ω)
- Resistors (for example, 10 Ω and 25 Ω)
- Battery or power supply (3 V)
- Connecting wires

B. Wheatstone Bridge Method

- Slide-wire Wheatstone bridge
- Galvanometer
- Standard decade resistance box (0.1 Ω to 99.9 Ω)
- Single-pole, single-throw switch

†The ranges of the equipment are given as examples. These may be varied to apply to available equipment.

THEORY

A. Ammeter-Voltmeter Methods

There are two basic arrangements by which resistance is measured with an ammeter and a voltmeter. One circuit is shown in ● Fig. 24.1. The current I through the resistance R is measured with an ammeter, and the potential difference or voltage drop V across the resistance is measured with a voltmeter. Then, by Ohm's law, $R = V/I$.

Strictly speaking, however, this value of the resistance is not altogether correct, since the current registered on the ammeter divides between the resistance R and the voltmeter in parallel. A voltmeter is a high-resistance instrument and draws relatively little current, provided that voltmeter resistance R_v is much greater than R. Hence, it is more appropriate to write

$$\boxed{R \simeq \frac{V}{I} \quad \text{if} \quad R_v \gg R} \qquad \textbf{(24.1)}$$

For more accurate resistance measurement, one must take the resistance of the voltmeter into account. The current drawn by the voltmeter is $I_v = V/R_v$. Since the total current I divides between the resistance and the voltmeter in the parallel branch,

$$I = I_R + I_v$$

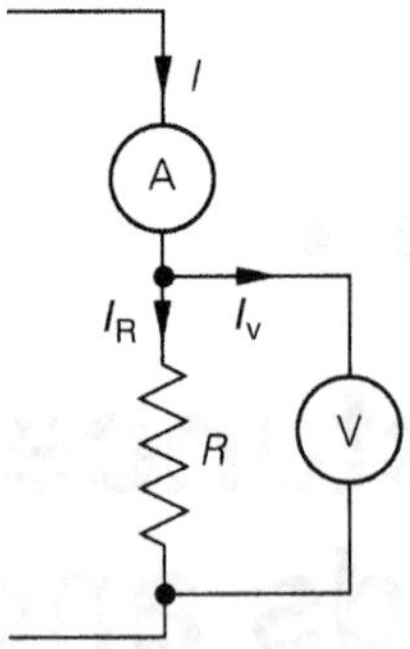

Figure 24.1 Resistance measurement. One of the basic arrangements for measuring resistance with an ammeter and a voltmeter. The ammeter measures the sum of the currents through the resistance and the voltmeter. Therefore, the true value of R is *greater* than the measured value, if the measured value is taken to be V/I.

or

$$I_R = I - I_v \tag{24.2}$$

where I_R is the true current through the resistance. Then, by Ohm's law,

$$R = \frac{V}{I_R} = \frac{V}{I - I_v} = \frac{V}{I - (V/R_v)} \tag{24.3}$$

Another possible arrangement for measuring R is shown in the circuit diagram in ● Fig. 24.2. In this case, the ammeter measures the current through R alone, but now the voltmeter reads the voltage drop across *both* the ammeter *and* the resistance. Since the ammeter is a low-resistance instrument, to a good approximation

$$R \simeq \frac{V}{I} \quad \text{if} \quad R_a \ll R \tag{24.4}$$

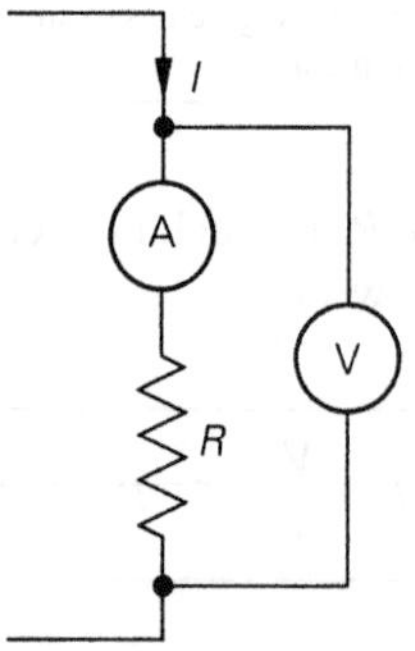

Figure 24.2 Resistance measurement. Another basic arrangement for measuring resistance with an ammeter and a voltmeter. The ammeter measures the current through R, but the voltmeter is across R and the ammeter. Therefore, the true value of R is *less* than the measured value, if the measured value is taken to be V/I.

where R_a is the resistance of the ammeter. When $R_a << R$, the voltage drop across R_a, that is, $V_a = IR_a$, is small compared to that across R, which is $V_R = IR$.

Taking the voltage drop or the resistance of the ammeter into account,

$$\begin{aligned} V = V_R + V_a &= IR + IR_a \\ &= I(R + R_a) = IR' \end{aligned}$$

and

$$R' = R + R_a \tag{24.5}$$

Solving for R and substituting for R' from the first equation:

$$R = \frac{V}{I} - R_a \tag{24.6}$$

B. Wheatstone Bridge Method

The basic diagram of a Wheatstone bridge circuit is shown in ● Fig. 24.3. In its simplest form, the bridge circuit consists of four resistors, a battery or voltage source, and a sensitive galvanometer. The values of R_1, R_2, and R_s are all known, and R_x is the unknown resistance.

Switch S is closed, and the bridge is balanced by adjusting the standard resistance R_s until the galvanometer shows no deflection (indicating no current flow through the galvanometer branch). As a result, the Wheatstone bridge is called a "null" instrument. This is analogous to an ordinary double-pan beam balance, which shows a null or zero reading when there are equal masses on its pans.

Assume that the Wheatstone bridge is balanced so that the galvanometer registers no current. Then points b and c in the circuit are at the same potential; current I_1 flows through both R_s and R_x, and current I_2 flows through both R_1 and R_2.

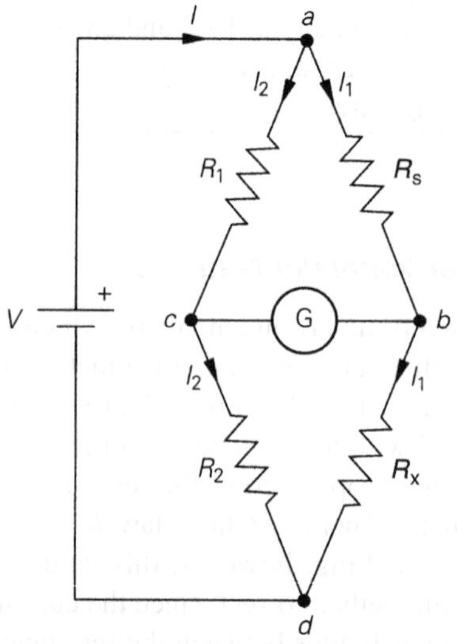

Figure 24.3 Wheatstone bridge circuit diagram. See text for description.

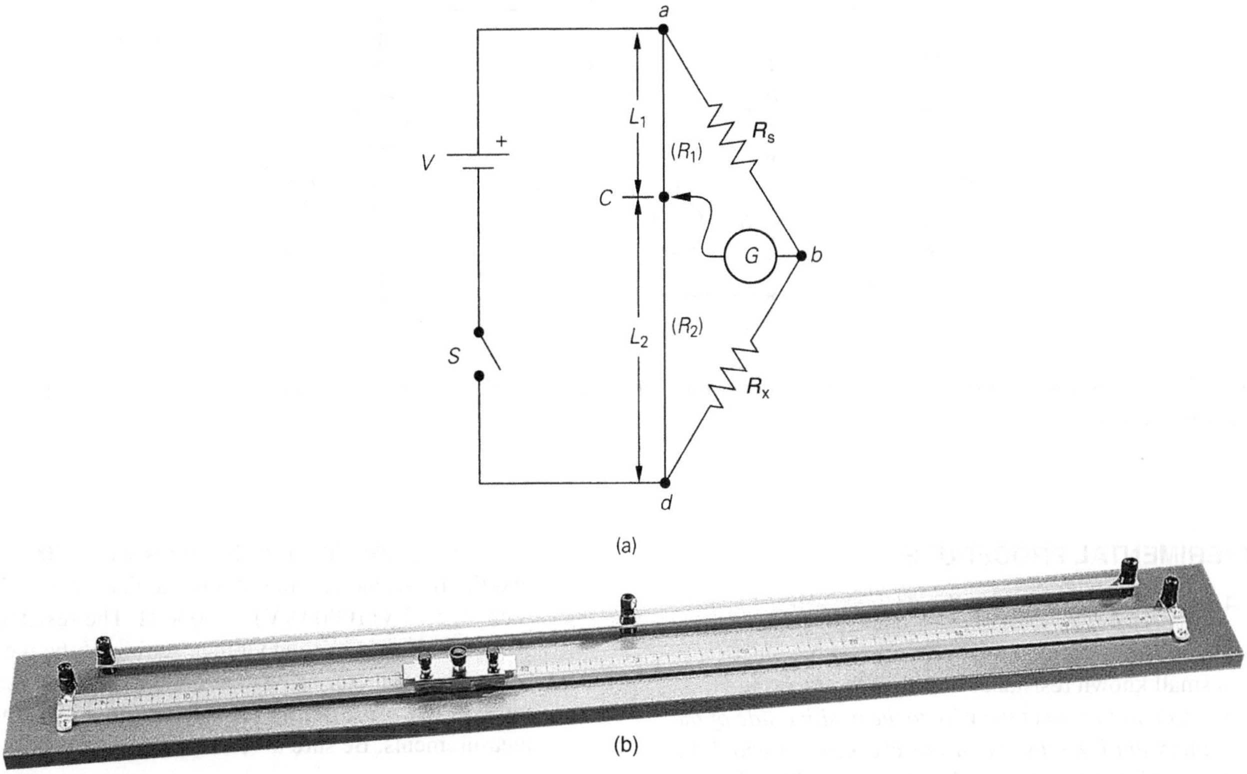

Figure 24.4 Slide-wire Wheatstone bridge. (a) Circuit diagram for resistance measurements. The resistances R_1 and R_2 are varied by sliding the contact C along the wire. The galvanometer G is used to indicate when the bridge is "balanced." (b) The contact slides over the wire on a meterstick. When contact is made, the lengths L_1 and L_2 on either side of C are easily read. (Photo Courtesy of Sargent-Welch.)

Also, the voltage drop V_{ab} across R_s is equal to the voltage drop across R_1, V_{ac}, for a zero galvanometer deflection:

$$\boxed{V_{ab} = V_{ac}}$$

Similarly,

$$\boxed{V_{bd} = V_{cd}} \qquad \textbf{(24.7)}$$

(Why?)

Writing these equations in terms of currents and resistances, by Ohm's law,

$$\boxed{\begin{aligned} I_1 R_x &= I_2 R_2 \\ I_1 R_s &= I_2 R_1 \end{aligned}} \qquad \textbf{(24.8)}$$

Then, dividing one equation by the other and solving for R_x yields

$$\boxed{R_x = \left(\frac{R_2}{R_1}\right) R_s} \qquad \textbf{(24.9)}$$

Hence, when the bridge is balanced, the unknown resistance R_x can be found in terms of the standard resistance R_s and the ratio R_2/R_1.

Notice that the difficulties of the ammeter-voltmeter methods are eliminated. The Wheatstone bridge in effect compares the unknown resistance R with a standard resistance R_s. Should $R_1 = R_2$, then $R_x = R_s$.

The circuit diagram for a slide-wire form of the Wheatstone bridge is shown in ● Fig. 24.4 along with a photo of an actual bridge. The line from a to d represents a wire, and C is a contact key that slides along the wire so as to divide the wire into different-length segments.

The resistances of the segments are proportional to their lengths, so the resistance ratio may be expressed in terms of a length ratio:

$$\boxed{\frac{R_2}{R_1} = \frac{L_2}{L_1}} \qquad \textbf{(24.10)}$$

Equation 24.9 can then be written in terms of the length ratio:

$$\boxed{R_x = \left(\frac{L_2}{L_1}\right) R_s} \qquad \textbf{(24.10)}$$

This type of bridge is convenient since the length segments can be measured easily. The resistances R_1 and R_2 of the length segments may be quite small relative to R_x and R_s because the bridge equation depends only on the ratio R_2/R_1 or L_2/L_1. This fact makes it possible to use a wire as one side of the bridge.

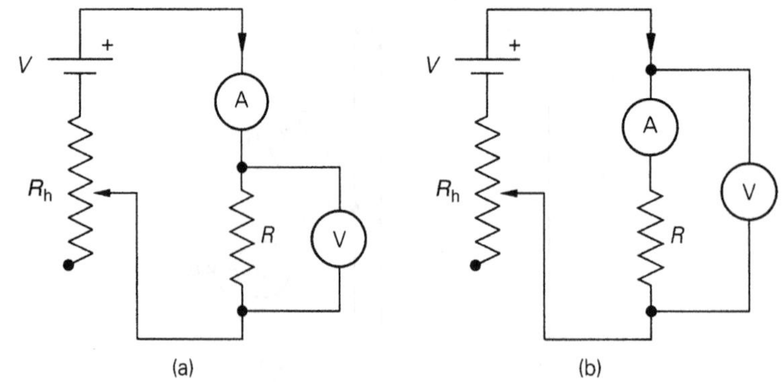

Figure 24.5 Resistance measurement. Circuit diagrams for experimental procedures for ammeter-voltmeter methods of measuring resistance.

EXPERIMENTAL PROCEDURE

A. Ammeter-Voltmeter Methods

1. Set up a circuit as shown in ● Fig. 24.5a, where R is a small known resistance and R_h is the rheostat. However, *do not connect the wire to the positive side of the battery until the instructor has checked it.* Record the value of R in the first of the spaces provided for this purpose in Part A of the Laboratory Report.

   Most common meters have three scale connections, with a binding post common to all three scales. It is a good practice initially to make connections to the largest scale. This prevents the instruments from being "pegged" (and possibly damaged) should the magnitudes of the current and voltage exceed the smaller scales' limits.

   The scale setting may be changed for greater sensitivity by moving the connection to a lower scale after the general magnitude of a measurement is known.

   Also, attention should be given to the proper polarity (+ and −). Otherwise, the meter will be "pegged" in the opposite direction. Connect + to + and – to –. However, *do not activate the circuit until your laboratory instructor has checked it.*

2. The current in the circuit is varied by varying the rheostat resistance R_h. Activate the circuit and take three different readings of the ammeter and voltmeter for three different currents. Adjust R_h so that the three currents differ as much as possible. Record the data in Data Table 1, and deactivate the circuit after each of the three readings until the rheostat is set for the next reading.

   Also, record the resistance of the voltmeter. The resistance of the meter will be found on the meter face or will be supplied by the instructor. The voltmeter resistance is commonly given as so many ohms per volt, which is the total resistance of the meter divided by the full-scale reading.

   For example, if the meter has a resistance of 1000 Ω/V and the full-scale reading of a particular range is 3 V, then $R_v = 3\text{ V}(1000\ \Omega/\text{V}) = 3000\ \Omega$. The resistance in ohm/volt applies to any range setting of the meter. (*Note*: If voltmeter scales are changed during readings, R_v will be different for different sets of V and I measurements. Be sure to record this if it occurs.)

3. Using Eq. 24.3, compute the value of R for each current setting and find the average value. Compare this with the accepted value by finding the percent error.

4. Set up a circuit as shown in Fig. 24.5b. This is accomplished by changing only one wire in the previous circuit. Repeat the measurements as in Procedure 2 for this circuit, recording your findings in Data Table 2.

5. **(a)** Compute the resistance $R' = V/I$ directly from each set of current and voltage measurements, and find the average value.
   (b) When one is not taking into account the ammeter resistance, R' is taken to be the value of the resistance R. Compare the average experimental value of R' with the accepted value of R by finding the percent error.
   (c) Using the values of R and R', compute R_a [Eq. 24.5]. Mentally compare the magnitudes of the ammeter and voltmeter resistances.

6. Repeat the previous procedures with a large known resistance. Record its accepted value in the space provided in Data Table 3, and use Data Tables 3 and 4 for your findings.

B. Wheatstone Bridge Method

7. Set up a slide-wire Wheatstone bridge circuit as in Fig. 24.4a, using the previous small known resistance R as R_x. *Leave the switch open until the instructor checks the circuit.* The wires connecting the resistances and

the bridge should be as short as practically possible. The decade resistance box is used for R_s. This should be initially set for a value about equal to R_x.

Contact is made to the wire by sliding contact key *C*. *Do not slide the key along the wire while it is pressed down.* This will scrape the wire, causing it to be nonuniform. *Have the instructor check your setup before activating the circuit.*

8. Activate the circuit by closing the switch *S*, and balance the bridge by moving the slide-wire contact. Open the switch, and record R_s, L_1, and L_2 in Data Table 5. Leave the switch open except when you are actually making measurements.

9. Repeat Procedures 7 and 8 for R_s settings of (a) $R_s \approx 3R_x$, and (b) $R_s \approx 0.3R_x$.

10. Compute the value of R_x for each case, and find the average value. Compare this value to the accepted value of *R* by finding the percent error.

11. Repeat the previous procedures with a large known resistance. Record your findings in Data Table 6.

Name ______________________ Section ______________ Date ______________

Lab Partner(s) ____

EXPERIMENT 24

The Measurement of Resistance: Ammeter-Voltmeter Methods and Wheatstone Bridge Method

TI Laboratory Report

A. Ammeter-Voltmeter Methods

DATA TABLE 1

Purpose: To measure resistance values.

Accepted value of R ______________

Voltmeter resistance R_v ______________

Rheostat setting R_h	Current I ( )	Voltage V ( )	Resistance R ( )
1			
2			
3			
		Average R	

Percent error of R ______________

DATA TABLE 2

Purpose: To measure resistance values.

Rheostat setting R_h	Current I ( )	Voltage V ( )	$R' = V/I$ ( )
1			
2			
3			
		Average R'	

Percent error of R' ______________

Ammeter resistance R_a ______________

Calculations
(show work)

Don't forget units

(continued)

DATA TABLE 3

Purpose: To measure resistance values.

Voltmeter resistance R_v ______________

Rheostat setting R_h	Current I ( )	Voltage V ( )	Resistance R ( )
1			
2			
3			
		Average R	

Percent error of R ______________

Accepted value of R ______________

DATA TABLE 4

Purpose: To measure resistance values.

Rheostat setting R_h	Current I ( )	Voltage V ( )	$R' = V/I$ ( )
1			
2			
3			
		Average R'	

Percent error of R' ______________

Ammeter resistance R_a ______________

Calculations
(show work)

Name ______________________ Section ______________ Date ______________

Lab Partner(s) ____

B. Wheatstone Bridge Method

DATA TABLE 5

Purpose: To measure resistance values.

Accepted value of R ______________

R_s ( )	L_1 ( )	L_2 ( )	R ( )
1			
2			
3			
		Average R	

Percent error ______________

DATA TABLE 6

Purpose: To measure resistance values.

Accepted value of R ______________

R_s ( )	L_1 ( )	L_2 ( )	R ( )
1			
2			
3			
		Average R	

Percent error ______________

Calculations
(show work)

TI QUESTIONS

A. Ammeter-Voltmeter Methods

1. An ideal ammeter would have zero resistance, and an ideal voltmeter would have an infinite resistance. Explain why we would desire these ideal cases when using the meters.

(continued)

2. If, in general, R were calculated as $R = V/I$, which circuit arrangement in Part A of the experiment would have the smallest error? Explain.

3. (a) Prove that the true resistance R is given by

$$R = R'\left(1 - \frac{R_a}{R'}\right)$$

where $R' = V/I$ is the measured resistance as given by the voltmeter and ammeter readings for measurements done by the arrangement in Fig. 24.2 or Fig. 24.5b. Is the true resistance larger or smaller than the apparent resistance?

(b) Prove that the true resistance R is given approximately by

$$R = R'\left(1 + \frac{R'}{R_v}\right)$$

where $R' = V/I$ is the measured resistance as given by the voltmeter and ammeter readings for measurements done by the arrangement in Fig. 24.1 or Fig. 24.5a. (*Hint*: Use the binomial theorem.)

$$\frac{1}{1 - \dfrac{R'}{R_v}} \simeq 1 + \frac{R'}{R_v}$$

Is the true resistance larger or smaller than the apparent resistance? Explain.

Name ______________________ Section ______________ Date ______________

Lab Partner(s) ____

4. For each of the circuits used in the preceding question, for what values of R (large or small) does the error in taking R as equal to V/I become large enough to be important?

B. Wheatstone Bridge Method

5. Why should the wires connecting the resistances and the bridge be as short as possible?

6. Suppose that the slide-wire on the bridge did not have a uniform cross section. How would this affect your measurements? Was there any experimental evidence of this?

Name ______________________ Section ______________ Date ______________

Lab Partner(s) ____

EXPERIMENT 25

The Temperature Dependence of Resistance

TI Advance Study Assignment

Read the experiment and answer the following questions.

1. Does the resistance of all substances increase with temperature? Explain.

2. What is the temperature coefficient of resistance, and what are its units?

3. Distinguish between a positive and a negative temperature coefficient of resistance.

4. Are the α of a metal conductor and the β of a thermistor the same? Explain.

(continued)

5. What are the circuit conditions when a Wheatstone bridge is “balanced”?

6. In using the equations to determine the temperature dependence of resistance, what temperature scale is used?

EXPERIMENT 25

The Temperature Dependence of Resistance

INTRODUCTION AND OBJECTIVES

The electrical resistance of all substances varies somewhat with temperature. For pure metals and most alloys, the resistance increases with increasing temperature. However, for some substances, such as carbon and many electrolytes (conducting solutions), the resistance decreases with increasing temperature. Then, too, for some special alloys [for example, constantan (55% Cu–45% Ni)], the resistance is virtually independent of temperature over a limited range.

The temperature dependence of resistance for a substance is commonly expressed in terms of its **temperature coefficient of resistance**, which is the fractional change in the resistance per degree change in temperature. For many electrical applications, it is important to know temperature coefficients and to take into account the temperature dependence of resistances. In this experiment, this temperature dependence will be investigated and the temperature coefficients of some materials determined.

After performing this experiment and analyzing the data, you should be able to:

1. Explain how the resistances of common metallic conductors vary with temperature.
2. Discuss the temperature coefficient of resistance for various materials and the differences among them.
3. Describe what is meant by the exponential temperature coefficient of a thermistor.

EQUIPMENT NEEDED

- Slide-wire Wheatstone bridge assembly (with a 3-V battery and a single-pole, single-throw switch)
- Standard decade resistance box
- Copper coil (and optional constantan or manganese coil)
- Thermistor
- Immersion vessel and stirrer
- Thermometer
- Immersion heater and power source (or Bunsen burner and stand or hot plate)
- 2 sheets of Cartesian graph paper

THEORY

The change in resistance, ΔR, of a substance is proportional to the change in temperature, ΔT. This change in resistance is commonly expressed in terms of the fractional change $\Delta R/R_o$, where R_o is the initial resistance. For many substances, for example metals, the change in resistance is to a good approximation a linear function of temperature:

$$\boxed{\frac{\Delta R}{R_o} = \alpha \Delta T} \qquad (25.1)$$

where the constant of proportionality α is called the **temperature coefficient of resistance** and has the units of inverse temperature, 1/°C, or °C^{-1}.

For the change in temperature $\Delta T = T - T_o$, it is convenient to take the initial temperature T_o as 0 °C and with $\Delta R = R - R_o$, Eq. 25.1 can be written

$$\frac{R - R_o}{R_o} = \alpha T$$

or

$$\boxed{R = R_o + R_o \alpha T = R_o(1 + \alpha T)} \qquad (25.2)$$

where R is then the resistance of the conductor at some temperature T (°C), and R_o is the resistance at $T_o = 0$ °C. The linearity of the temperature dependence is only approximate, but Eq. 25.2 can be used over moderate temperature ranges for all but the most accurate work.

In contrast to pure metals, which have positive temperature coefficients of resistance (increase in resistance with increase in temperature), some materials have negative temperature coefficients (decrease in resistance with an increase in temperature). Carbon is an example, and negative temperature coefficients of resistance generally occur in materials of intermediate conductivity, or *semiconducting* materials.

Carbon has a relatively small negative temperature coefficient of resistance compared to other semiconducting materials. Such materials with large negative temperature coefficients are used in commercial components called thermistors. A **thermistor** is a thermally sensitive resistor made of semiconducting materials such as oxides of manganese, nickel, and cobalt.

Because of relatively large (negative) temperature coefficients, thermistors are very sensitive to small temperature changes and are used for temperature measurements in a variety of temperature-sensing applications, such as for voltage regulation and time-delay switches.

Unlike common metal conductors, for a thermistor, the change of resistance with a change of temperature is nonlinear, and the α in Eq. (25.1) is not constant. The temperature dependence of a thermistor is given by an exponential function,

$$\boxed{R = R_a e^{\beta(1/T - 1/T_a)}} \quad (25.3)$$

where R = resistance at a temperature T (in kelvins, K)
R_a = resistance at temperature T_a (K)
T_a = initial temperature (K), near ambient room temperature in the experiment
e = 2.718, the base of natural logarithms
β = *exponential* temperature coefficient of resistance, which has Kelvin temperature units (K).

In this case, as T increases, the exponential function, and hence the resistance R, becomes smaller. This expression can be written in terms of the natural logarithm (to the base e) as

$$\boxed{\ln\left(\frac{R}{R_a}\right) = \beta\left(\frac{1}{T} - \frac{1}{T_a}\right)} \quad (25.4)$$

Hence, when $y = \ln(R/R_a)$ versus $x = (1/T - 1/T_a)$ is plotted on a Cartesian graph, β is the slope of the line. This, too, is an approximation, but β is reasonably constant for moderate temperature ranges.

The temperature coefficient of resistance of a material can be determined by using an experimental arrangement with a slide-wire Wheatstone bridge circuit, as illustrated in ● Fig. 25.1. The resistance, R_c, of a material (coil of wire) when the bridge circuit is balanced is given by

$$R_c = \left(\frac{R_2}{R_1}\right) R_s$$

or

$$\boxed{R_c = \left(\frac{L_2}{L_1}\right) R_s} \quad (25.5)$$

where R_s is a standard resistance and R_2/R_1 and L_2/L_1 are the ratios of the resistances and lengths of the slide-wire segments, respectively. (See Experiment 24 for the theory

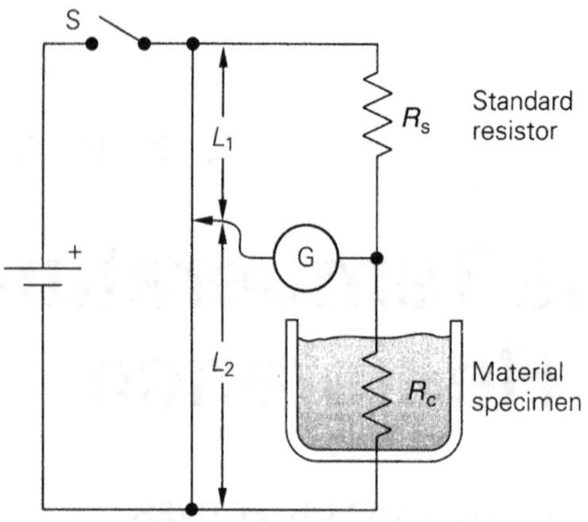

Figure 25.1 Temperature dependence of resistance. The circuit diagram for the experimental procedure to measure the temperature dependence of resistance. See text for description.

of the Wheatstone bridge.) By measuring the resistance of a material at various temperatures, the temperature coefficient can be determined.

EXPERIMENTAL PROCEDURE

A. Metal Conductor(s)

1. Set up the circuit as in Fig. 25.1 with the copper coil in the container of water (near room temperature) and the heating arrangement for the water (immersion heater or other heat source). Place the thermometer in the water. *Have the instructor check your setup.*

2. After your setup has been checked, close the switch and balance the bridge circuit to measure the resistance R_c of the coil at the initial water temperature. The value of the standard resistance R_s should be selected so that the bridge is balanced with the contact key C as near the center of the slide-wire as possible. Then, with $L_1 \simeq L_2$, it follows that $R_c \simeq R_s$ [Eq. 25.5].

   Record in Data Table 1 the initial temperature of the water, the magnitude of R_s, and the lengths of the wire segments of the bridge.

3. *Slowly*, raise the temperature of the water by about 10 °C. Stir the water while heating, and discontinue heating when the temperature is about 2 °C below the desired temperature. Continue stirring until a maximum steady temperature is reached. Balance the bridge, and record the measurements. Adjust R_s, if necessary. Record the measurements of temperature and bridge length in the data table.

4. Repeat Procedure 3, taking a series of measurements at approximately 10 °C temperature intervals until a final temperature of about 90 °C is reached.

5. (*Optional*) Repeat the foregoing procedures using the constantan wire coil, starting near room temperature. (Use Data Table 1A.)

6. Compute R_c of the coil(s) at the various temperatures and plot a graph of R_c versus T with a temperature range of 0 °C to 100 °C. Draw the straight line(s) that best fit(s) the data, and extrapolate the line(s) to the y-axis. Determine the slope and y-intercept of the line(s).

   From the slope, find the temperature coefficient of resistance for the specimen(s) and compare with the accepted value found in Appendix A, Table A6, by computing the percent error.

B. Thermistor

7. Replace the coil with the thermistor in the bridge circuit, and repeat the previous measurement Procedures 1 through 4, starting at a temperature near room temperature. In this portion of the experiment, exercise great care in order to have temperatures as constant as possible when making resistance measurements, since a thermistor shows considerable variation in resistance with temperature.

8. (a) Find the quantities listed in the second part of Data Table 2.
   (b) Plot a graph of $y = \ln(R/R_a)$ versus $x = (1/T - 1/T_a)\ \mathrm{K}^{-1}$, and draw the straight line that best fits the data.
   (c) Determine the slope of the line, which is the value of β. Compare this to the accepted value provided by the instructor by computing the percent error.

Name ______________________ Section ______________ Date ______________

Lab Partner(s) ____

EXPERIMENT 25

The Temperature Dependence of Resistance

TI *Laboratory Report*

A. Metal Conductor(s)

DATA TABLE 1

Purpose: To determine the temperature coefficient of resistance.

Material ______________

Temperature ( )	Decade box resistance R_s ( )	L_1 ( )	L_2 ( )	Coil resistance $R = (L_2/L_1)R_s$ ( )

Calculations
(show work)

Slope $R_o\alpha$ ______________

Intercept R_o ______________

Experimental α ______________

Accepted α ______________

Percent error ______________

Don't forget units

(continued)

DATA TABLE 1A (Optional)

Purpose: To determine the temperature coefficient of resistance.

Material ____________

Temperature ( )	Decade box resistance R_s ( )	L_1 ( )	L_2 ( )	Coil resistance R_c ( )

Calculations
(show work)

Slope $R_o\alpha$ ____________

Intercept R_o ____________

Experimental α ____________

Accepted α ____________

Percent error ____________

Name ________________ Section ________________ Date ________________

Lab Partner(s) ________________

B. Thermistor

DATA TABLE 2

Purpose: To determine the exponential temperature coefficient of resistance.

Temperature T ( )	Decade box resistance R_s ( )	L_1 ( )	L_2 ( )	Thermistor resistance $R = (L_2/L_1)R_s$ ( )
(T_a)				(R_a)

Temperature ( ) $T_K = T_C + 273$	$1/T$	$1/T - 1/T_a$	R/R_a	$\ln(R/R_a)$

Calculations (show work)

Slope β ________________

Accepted β ________________

Percent error ________________

(continued)

TI QUESTIONS

A. Metal Conductor

1. What is the value of α for copper in terms of Fahrenheit degrees? If the resistance is a linear function on the Celsius scale, will it be a linear function on the Fahrenheit scale? Explain.

2. Replot the copper data for R_c versus T with a smaller temperature scale extending to -300 °C, and extrapolate the line to the temperature axis. At what temperature would the resistance go to zero? What are the practical electrical implications for a conductor with zero resistance?

   [It is interesting to note that the value of α is roughly the same for many pure metals: approximately $\frac{1}{273}$, or 0.004 °C^{-1}. This is the same as the value of the coefficient of expansion of an ideal gas. Also, some metals and alloys do become "superconductors," or have zero resistance, at low temperatures. Some "high-temperature" ceramic materials show superconductivity at liquid nitrogen temperatures (77 K, or -196 °C, or -321 °F).]

3. A coil of copper wire has a resistance of 10.0 Ω, and a coil of silver wire has a resistance of 10.1 Ω, both at 0 °C. At what temperature would the resistance of the coils be equal?

Name ______________________ Section ______________ Date ______________

Lab Partner(s) ____

B. Thermistor

4. Explain why the ambient temperature T_a for the thermistor cannot be taken as $T_a = 0$ °C, and why the expression for the resistance is written $R = R_o e^{\beta/T}$, where T is in degrees Celsius.

5. Assuming that β remained constant, what would be the resistance of the thermistor in the experiment as the temperature approached absolute zero?

6. Assume the temperature coefficient of resistance α to be defined over the temperature range $\Delta T = T - T_a$, where $T_a > 273$ K (0 °C), by $R - R_a = -R_a\alpha(T - T_a)$. Show that for a thermistor, α is a function of temperature given by:

$$\alpha = \frac{1 - e^{\beta(1/T - 1/T_a)}}{T - T_a}$$

Name ______________________ Section ______________ Date ______________

Lab Partner(s) ____

EXPERIMENT 26

Resistances in Series and Parallel

TI/CI Advance Study Assignment

TI/CI QUESTIONS

Read the experiment and answer the following questions.

TI-CI 1. Explain the difference between series and parallel connections.

TI-CI 2. Consider resistors connected in series.

a. How are the voltage drops across the individual resistors related to the voltage supplied by the battery?

b. How are the currents through the individual resistors related to the current supplied by the battery?

TI-CI 3. Consider resistors connected in parallel.

a. How are the voltage drops across the individual resistors related to the voltage supplied by the battery?

b. How are the currents through the individual resistors related to the current supplied by the battery?

(continued)

TI 4. Give (draw and explain) an analogy to liquid flow for the series–parallel circuit in Part C of the experiment.

CI 4. In a plot of voltage versus current, what physical quantity is represented by the slope of the graph?

TI 5. How would the current divide in a parallel branch of a circuit containing two resistors R_1 and R_2 if (a) $R_1 = R_2$ and (b) $R_1 = 4R_2$?

EXPERIMENT 26

Resistances in Series and Parallel

OVERVIEW

Experiment 26 examines resistances in parallel and series combinations with both TI and CI procedures. In the TI procedure, the resistances are measured using a voltmeter and an ammeter. In the CI procedure, measurements are made with a voltage (and current) sensor, and graphs of V versus I are plotted, from which the resistances are given by the slopes.

INTRODUCTION AND OBJECTIVES

The components of simple circuits are connected in series and/or parallel arrangements. Each component may be represented as a resistance to the current in the circuit. In computing the voltage and current requirements of the circuit (or part of the circuit), it is necessary to know the equivalent resistances of the series and parallel arrangements.

In this experiment, the circuit characteristics of resistors in series and parallel will be investigated. A particular circuit will first be analyzed theoretically, and then those predictions will be checked experimentally.

After performing this experiment and analyzing the data, you should be able to:

TI-CI 1. Describe the current-voltage relationships for resistances in series.

TI-CI 2. Describe the current-voltage relationships for resistances in parallel.

TI 3. Reduce a simple series–parallel resistance circuit to a single equivalent resistance, and compute the voltage drops across and the currents through each resistance in the circuit.

CI 3. Describe the changes in the slopes of V-versus-I graphs as more resistors are connected in (a) series and (b) parallel.

T I EXPERIMENT 26

Resistances in Series and Parallel

TI EQUIPMENT NEEDED*

- Battery or power supply (3 V)
- Ammeter (0 mA to 500 mA)
- Voltmeter (0 V to 3 V)
- Single-pole, single-throw (SPST) switch
- 4 resistors (10 Ω, 20 Ω, 100 Ω, and 10 kΩ; composition type, 1 W)
- Connecting wires

*The ranges of the equipment are given as examples. These may be varied to apply to available equipment.

TI THEORY

A. Resistances in Series

Resistors are said to be **connected in series** when connected as in ● TI Fig. 26.1a. (The resistors are connected in line, or "head to tail" so to speak, although there is no distinction between the connecting ends of a resistor.) When connected to a voltage source V and the switch is closed, the source supplies a current I to the circuit.

By the conservation of charge, this current I flows through each resistor. The voltage drop across each resistor is not equal to V, but the *sum* of the voltage drops is:

$$V = V_1 + V_2 + V_3 \qquad \text{(TI 26.1)}$$

In an analogous liquid-gravity circuit (TI Fig. 26.1b), a pump, corresponding to the voltage source, raises the liquid a distance h. The liquid then falls or "drops" through three series paddle wheel "resistors" and the distances h_1, h_2, and h_3. The liquid "rise" supplied by the pump is equal to the sum of the liquid "drops," $h = h_1 + h_2 + h_3$. Analogously, the voltage "rise" supplied by the source is equal to the sum of the voltage drops across the resistors [TI Eq. 26.1].†

The voltage drop across each resistor is given by Ohm's law (for example, $V_1 = IR_1$). TI Eq. 26.1 may be written

$$\begin{aligned} V &= V_1 + V_2 + V_3 \\ &= IR_1 + IR_2 + IR_3 \\ &= I(R_1 + R_2 + R_3) \end{aligned} \qquad \text{(TI 26.2)}$$

For a voltage across a single resistance R_s in a circuit, $V = IR_s$, and by comparison,

$$R_s = R_1 + R_2 + R_3 \qquad \text{(TI 26.3)}$$

(resistances in series)

†Keep in mind that an analogy represents only a resemblance. Liquid and electrical circuits are quite different physically.

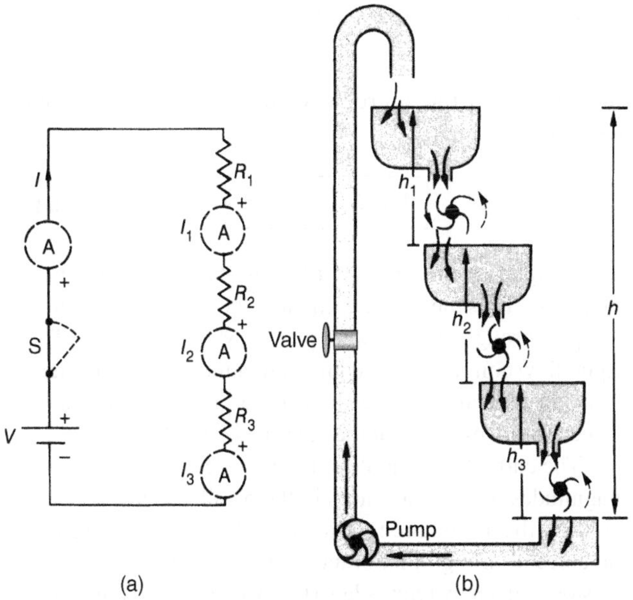

TI Figure 26.1 Resistances in series. (a) A circuit diagram for resistors connected in series. (The resistors are connected "head" to "tail.") The Ⓐ symbol represents an ammeter that will be used in experimental setups. (b) A liquid analogy on the left for the circuit diagram of resistors in series on the right. The analogies are pump–voltage source, valve–switch, liquid flow–current, and paddle wheels–resistors. See text for more description.

where R_s *is the equivalent resistance of the resistors in series*. That is, the three resistors in series could be replaced by a single resistor with a value of R_s, with the same current I in the circuit.

B. Resistances in Parallel

Resistors are said to be **connected in parallel** when connected as in ● TI Fig. 26.2a. (In this arrangement, all the "heads" are connected together, as are all of the "tails.") The voltage drops across all the resistors are the same and equal to the voltage V of the source. However, the

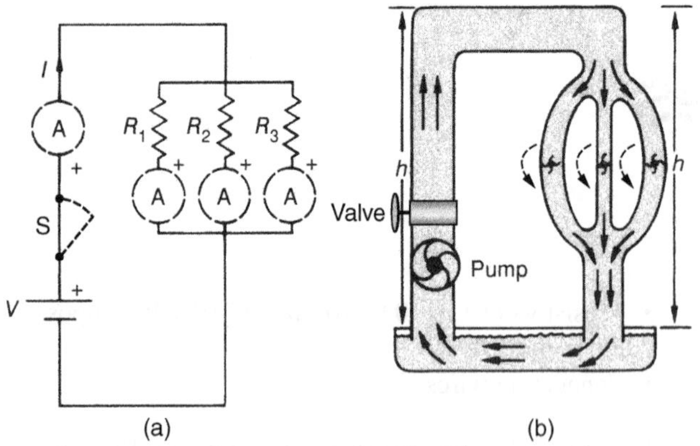

TI Figure 26.2 Resistances in parallel. (a) A circuit diagram for resistors in parallel. (All the "heads" are connected together, as are all of the "tails.") (b) A liquid analogy on the left for the circuit diagram of resistors in parallel on the right. See text for description.

current I from the source divides among the resistors such that

$$I = I_1 + I_2 + I_3 \quad \textbf{(TI 26.4)}$$

In the liquid circuit analogy (TI Fig. 26.2b), the height h the pump raises the liquid is equal to the distance the liquid "drops" through each parallel paddle wheel "resistor." The liquid flow coming into the junction of the parallel arrangement divides among the three pipe paths, analogously to the current dividing in the electrical circuit.

The current in a parallel circuit divides according to the magnitudes of the resistances in the parallel branches—the smaller the resistance of a given branch, the greater the current through that branch. The current through each resistor is given by Ohm's law (for example, $I_1 = V/R_1$), and TI Eq. (26.4) may be written

$$\begin{aligned} I = I_1 + I_2 + I_3 &= \frac{V}{R_1} + \frac{V}{R_2} + \frac{V}{R_3} \\ &= V\left(\frac{1}{R_1} + \frac{1}{R_2} + \frac{1}{R_3}\right) \end{aligned} \quad \textbf{(TI 26.5)}$$

For the current through a single resistance R_p in a circuit, $I = V/R_p$, and by comparison,

$$\frac{1}{R_p} = \frac{1}{R_1} + \frac{1}{R_2} + \frac{1}{R_3} \quad \textbf{(TI 26.6)}$$

(three resistances in parallel)

where R_p *is the equivalent resistance of the resistors in parallel.* That is, the three resistors in parallel could be replaced by a single resistor with a value of R_p, and the same current I would be drawn from the battery.

The previous developments for equivalent resistances may be extended to any number of resistors (that is, $R_s = R_1 + R_2 + R_3 + R_4 + \cdots$ and $1/R_p = 1/R_1 + 1/R_2 + 1/R_3 + 1/R_4 + \cdots$).

In many instances, two resistors are connected in parallel in a circuit, and

$$\frac{1}{R_p} = \frac{1}{R_1} + \frac{1}{R_2}$$

or

$$R_p = \frac{R_1 R_2}{R_1 + R_2} \quad \textbf{(TI 26.7)}$$

(two resistances in parallel)

This particular form of R_p for two resistors may be more convenient for calculations than the reciprocal form.

Also, in a circuit with three resistors in parallel, the equivalent resistance of two of the resistors can be found by TI Eq. (26.7), and then the equation may be applied again to the equivalent resistance and the other resistance in parallel to find the total equivalent resistance of the three parallel resistors. However, if your calculator has a $1/x$ function, the reciprocal form may be easier to use.

Note that the voltage drops across R_1 and R_2 in parallel are the same, and by Ohm's law,

$$I_1 R_1 = I_2 R_2$$

or

$$\frac{I_1}{I_2} = \frac{R_2}{R_1} \quad \textbf{(TI 26.8)}$$

TI Example 26.1 Given two resistors R_1 and R_2, with $R_2 = 2R_1$, in parallel in a circuit. What fraction of the current I from the voltage source goes through each resistor?

Solution With $R_2 = 2R_1$, or $R_2/R_1 = 2$, by TI Eq. (26.8)

$$I_1 = \left(\frac{R_2}{R_1}\right) I_2 = 2I_2$$

Since $I = I_1 + I_2$,

$$I = I_1 + I_2 = 2I_2 + I_2 = 3I_2$$

or

$$I_2 = \frac{I}{3}$$

Hence, the current divides, with one-third going through R_2 and two-thirds going through R_1.

Thus, the ratio of the resistances gives the relative magnitudes of the currents in the resistors.

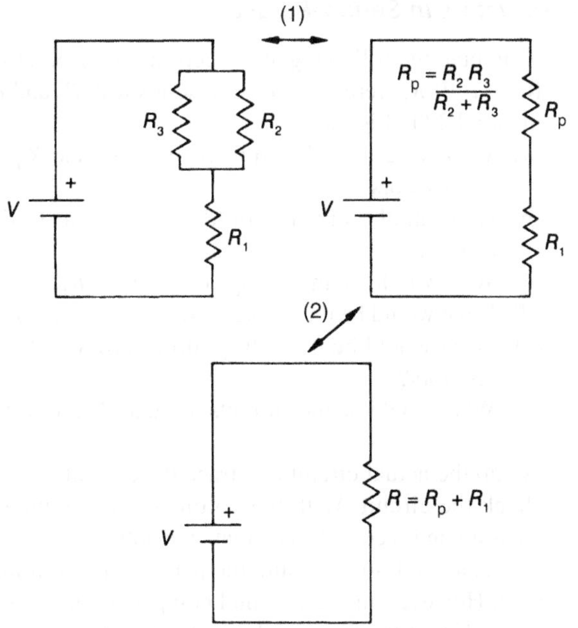

TI Figure 26.3 Circuit reduction. Series and parallel resistances are combined to find the equivalent resistance of a series–parallel circuit. See text for description.

Consider the circuit in ● TI Fig. 26.3. To find the equivalent resistance of this series–parallel circuit, one first "collapses" the parallel branch into a single equivalent resistance, which is given by TI Eq. 26.7. This equivalent resistance is in series with R_1, and the total equivalent resistance R of the circuit is $R = R_1 + R_p$.

TI EXPERIMENTAL PROCEDURE

1. Examine the resistors. The colored bands conform to a color code that gives the value of a resistor. Look up the color code in Appendix A, Table A5, read the value of each resistor, and record in the Laboratory Report. Designate the smallest resistance as R_1 and consecutively larger values as R_2, R_3, and R_4.

2. In the following procedures, you will be asked to compute theoretically various quantities for a given circuit arrangement. The quantities are then determined by actual circuit measurements, and the calculated and experimental results are compared. Before initially activating each circuit arrangement, *have the circuit checked by the instructor, unless otherwise instructed.*

A. Resistors in Series

3. Set up a series circuit with R_1, R_2, and R_3, as in TI Fig. 26.1a, with a switch and ammeter in the circuit next to the voltage source. A convenient way to check a circuit to see whether it is properly connected is to trace the path of the current (with your finger) through the circuit. Do this making sure that the current goes through each circuit component in series. Remember, an ammeter is *always* connected in series, and for proper polarity, + is connected to +.

   Connect the voltmeter across (in parallel with) the voltage source. After having the circuit checked by the instructor, close the switch. If using a variable power supply, adjust the voltage, if necessary, to the suggested value (3.0 V). Read and record the voltmeter value (V). This is the voltage "rise" of the source.

   (*Note*: If the needle of the ammeter goes in the wrong direction, reverse the polarity, that is, reverse the hook-up of the leads of the ammeter.)

   Open the circuit after completing the reading.

4. Using the resistor values and the measured voltage, *compute* (a) the equivalent resistance R_s of the circuit, (b) the current in the circuit, and (c) the voltage drop across each resistor. Show your calculations in the Laboratory Report.

5. Returning to the experimental circuit, close the switch and read the current I. Compare this with the computed value by finding the percent difference. Open the switch, and move the ammeter in the circuit to the position "after" the first resistor [that is, on the opposite side of the resistor from the voltage source so as to measure the current through (coming from) the resistor]. Record this as I_1.

   Carry out this procedure for each resistor, and record the currents in the Laboratory Report. Leave the switch closed only while readings are being taken.

6. Remove the ammeter from the circuit, and with the voltmeter, measure and record the voltage drop across each resistor and across all three resistors as a group. Remember, a voltmeter is *always* connected in parallel or "across" a circuit element to measure its voltage drop.

7. Compare the experimentally measured values with the theoretically computed values by finding the percent error. (Use the theoretical values as the accepted values.)

B. Resistors in Parallel

8. Set up a parallel circuit with R_1, R_2, and R_3, as in TI Fig. 26.2a, with the ammeter and voltmeter connected as before in Procedure 3. Check the circuit arrangement by tracing the current from the source through the circuit to see that it divides into three parallel branches at the junction of the resistors and comes together again at the opposite junction.

   Close the circuit (after it has been checked), and record the voltage and current readings in the Laboratory Report. (If using a variable power supply, adjust the voltage if necessary.)

   Open the circuit after taking the reading.

9. Using the resistor values and the measured voltage, *compute* (a) the equivalent resistance R_p of the circuit, (b) the current supplied by the source, and (c) the current through each resistor. Show your calculations in the Laboratory Report.

10. Returning to the experimental circuit, measure and record the voltage drops across each resistor and across all three resistors as a group.

Remove the voltmeter, and connect the ammeter so as to measure the current I supplied by the source. Then move the ammeter to measure the current through each resistor by connecting the meter between a given resistor and one of the common junctions. The ammeter positions are shown in TI Fig. 26.2. Leave the switch closed only while readings are being taken.

11. Compare the theoretical and experimental values by computing the percent errors.

12. (*Optional*) Repeat Procedures 8 through 11 with R_2 replaced by R_4.

C. Resistors in Series–Parallel

13. (Compute the following and record in the Laboratory Report.) If R_1 were connected in series with R_2 and R_3 in parallel (TI Fig. 26.3):

(a) What would be the equivalent resistance R_{sp} of the resistors?

(b) How much current would be supplied by the source?

(c) What would be the voltage drop across R_1?

(d) What would be the voltage drop across R_2 and R_3?

(e) What would be the voltage drop across all three resistors?

(f) What would be the currents through R_2 and R_3?

14. Set up the actual circuit and trace the current flow to check the circuit. With the voltmeter and ammeter, measure and record the calculated quantities.

You need not compute the percent errors in this case. However, make a mental comparison to satisfy yourself that the measured quantities agree with the computed values within experimental error.

Name ____________ Section ____________ Date ____________

Lab Partner(s) ____________

TI EXPERIMENT 26

Resistances in Series and Parallel

TI Laboratory Report

Resistor values R_1 ____________ R_3 ____________

R_2 ____________ R_4 ____________

A. Resistors in Series

Calculations (show work)

Source voltage V ____________

Equivalent resistance R_s ____________

Current I ____________

Voltage drops across resistors V_1 ____________

V_2 ____________

V_3 ____________

Don't forget units

(continued)

Experimental measurements

Percent error

I	________			________
I_1	________	V_1	________	________
I_2	________	V_2	________	________
I_3	________	V_3	________	________

$V_1 + V_2 + V_3$ ________

V across resistors as a group ________

B. Resistors in Parallel

Calculations
(show work)

Source voltage V ________

Equivalent resistance R_p ________

Current I ________

Current through resistors I_1 ________

I_2 ________

I_3 ________

Experimental measurements

Percent error

		I	________	________
V_1	________	I_1	________	________
V_2	________	I_2	________	________
V_3	________	I_3	________	________

$I_1 + I_2 + I_3$ ________

Name ______________________ Section ______________ Date ______________
Lab Partner(s) ____

EXPERIMENT 26 Resistances in Series and Parallel — *Laboratory Report*

(Optional Procedure)

Calculations
(show work)

Source voltage V ______________

Equivalent resistance R_p ______________

Current I ______________

Current through resistors I_1 ______________

I_3 ______________

I_4 ______________

Experimental measurements

				Percent error
		I	______	______
V_1	______	I_1	______	______
V_3	______	I_3	______	______
V_4	______	I_4	______	______

(continued)

C. Resistors in Series–Parallel

Calculations
(show work)

Source voltage V ____________

Equivalent resistance R_{sp} ____________

Current I ____________

Voltage drops

V_1 ____________

$V_2 = V_3$ ____________

Experimental measurements

Currents

I_2 ____________

I_3 ____________

I ____________

V_1 ____________

$V_2 = V_3$ ____________

I_2 ____________

I_3 ____________

TI QUESTIONS

1. Discuss the sources of error in the experiment.

2. Suppose that the resistors in the various circuit diagrams represented the resistances of lightbulbs. When a lightbulb "burns out," the circuit is open through that particular component, that is, *R* is infinite. Would the remaining bulbs continue to burn for the following conditions? If so, would the bulbs burn more brightly (draw more current) or burn more dimly (draw less current), if:
   (a) R_2 burned out in the circuit in Part A?

   (b) R_1 burned out in the circuit in Part B?

   (c) R_3 also burned out in the circuit in Part B?

   (d) R_3 burned out in the circuit in Part C?

(continued)

(e) R_1 also burned out in the circuit in Part C?

3. Explain the effect of replacing R_2 with R_4 in Procedure 12. (Explain theoretically even if Procedure 12 of the experiment was not done.)

4. For the circuit in Fig. 26.3, $V = 12$ V, $R_1 = 4\ \Omega$, $R_2 = 6\ \Omega$, and $R_3 = 3\ \Omega$. Show that the power supplied by the battery ($P = IV$) is equal to that dissipated in the resistors (I^2R). What principle does this illustrate? Use the accompanying table. (Consider values significant to two decimal places.)

Circuit element	Current I	Power dissipated P
$R_1 = 4\ \Omega$		
$R_2 = 6\ \Omega$		
$R_3 = 3\ \Omega$		
		(total)
		Power supplied
Battery $V = 12$ V		

(Show calculations)

5. Given three resistors of different values, how many possible resistance values could be obtained by using one or more of the resistors? (List the specific combinations, for example, R_1 and R_2 in series.)

CI EXPERIMENT 26

Resistances in Series and Parallel

CI EQUIPMENT NEEDED

This activity is designed for the Science Workshop 750 Interface, which has a built-in function generator.

- Voltage sensor (PASCO CI-6503)
- Science Workshop 750 Interface
- Cables and alligator clips
- 3 1000-Ω resistors

THEORY

According to Ohm's law, the current through a resistor is proportional to the voltage but inversely proportional to the resistance:

$$I = \frac{V}{R} \qquad \text{(CI 26.1)}$$

Thus, if the resistance of a circuit increases, the current decreases; and if the resistance of a circuit decreases, the current increases. On the other hand, the larger the voltage, the larger the current. The overall current in a circuit thus depends on the interplay between the amount of voltage and the amount of resistance.

In this experiment, the total amount of resistance in a circuit will be varied by connecting resistors in series and then in parallel. An increasing voltage will be applied, and the overall current in the circuit (through the voltage source) will be measured.

Rewriting Ohm's law as $V = IR$, notice that a plot of voltage (in the y-axis) versus current (in the x-axis) must result in a straight line, with the slope equal to the overall resistance in the circuit:

$$\begin{array}{ccccc} V & = & R & I \\ \downarrow & & \downarrow & \downarrow \\ y & = & m & x \end{array}$$

Using voltage (and current) sensors, we will find the resistances of the circuits by measuring the slope of a voltage-versus-current plot.

CI SETTING UP DATA STUDIO

1. Open Data Studio, and choose "Create Experiment."
2. The Experiment Setup window will open, and you will see a picture of the Science Workshop interface. There are seven channels to choose from and a signal generator. (Digital Channels 1, 2, 3, and 4 are the small buttons on the left; analog Channels A, B, and C are the larger buttons on the right; the signal generator is all the way to the right, as shown in ● CI Fig. 26.1.)
3. Click on the Channel A button in the picture. A window with a list of sensors will open.
4. Choose the Voltage Sensor from the list, and press OK.
5. Connect the sensor to Channel A of the interface, as shown on the computer screen.
6. The screen now shows you the properties of the Voltage Sensor directly under the picture of the interface. Adjust the sample rate to 20 Hz.
7. Click on the picture of the Signal Generator. The Signal Generator window will open. (See ● CI Fig. 26.2.)
8. The default form of the signal generator function is a sine wave. Change it to a "Positive Up Ramp Wave" by selecting from the drop menu.
9. Set the amplitude to 2.0 V and the frequency to 0.20 Hz.
10. Click on the Measurements and Sample Rate button on the Signal Generator window. A list of measurements will open. Choose to measure the output current. Deselect the measurement of the output voltage.
11. Press the Sampling Options button on the top toolbar of the Experiment Setup window. The Sampling Options window will open. Under "Automatic Stop," set the time to 4.5 seconds. Click OK.
12. Click on the Calculate button on the main toolbar. The calculator will open. Follow the next steps:
    a. Clear the definition box at the top, and enter the following formula in it:

       Voltage = smooth (20, x)

    b. Press the top Accept button after entering the formula. Notice that the variable x will appear, waiting to be defined.
    c. To define the variable, click on the drop menu button on the side of the variable. Define x as a Data Measurement and when prompted choose Voltage (ChA).

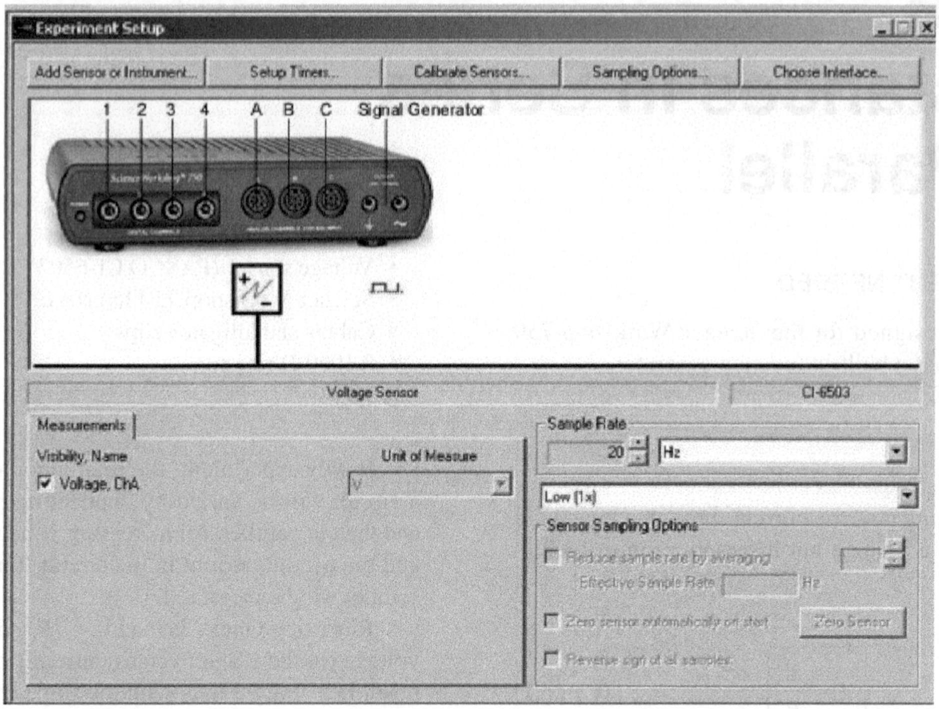

CI Figure 26.1 The Experiment Setup window. The voltage sensor is connected to Channel A and works as a multimeter. The signal generator of the Science Workshop interface is used as the voltage source that produces a positive ramp-up function. (Reprinted courtesy of PASCO Scientific.)

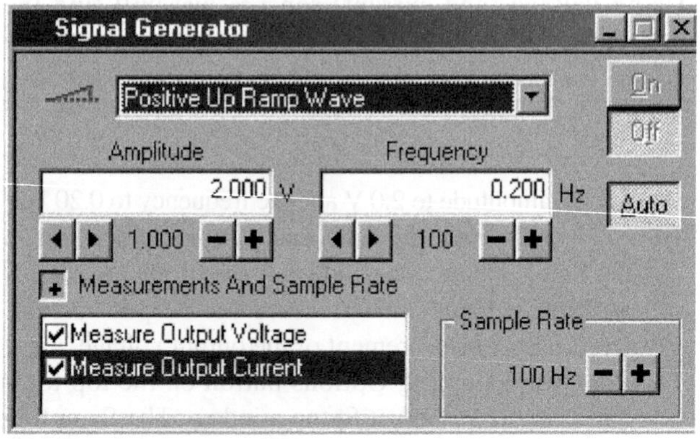

CI Figure 26.2 The Signal Generator window. Choose a positive up ramp wave function, adjust the amplitude and the frequency as specified in the setup procedure, and choose to measure the output current. (Reprinted courtesy of PASCO Scientific.)

d. Press the Accept button.

e. Click on the New button again to define another calculation.

f. Clear the definition box, and enter the following formula in it:

Current = smooth (20, x)

g. Press the Accept button after entering the formula. Notice that the variable x will again appear, waiting to be defined.

h. This time define x as a Data Measurement and, when prompted, choose Output Current.

i. Press Accept again.

13. The data list on the top left of the screen should now have the following items: Voltage ChA, Output Current, Voltage, and Current, where a small calculator icon identifies the quantities that are calculated, not measured.

14. Drag the "Voltage" (calculator) icon from the data list and drop it on the "Graph" icon of the displays list. A graph of voltage versus time will open in a window called Graph 1.

15. Drag the "Current" (calculator) icon from the data list and drop it on top of the time axis of Graph 1. The time axis will change into a current axis. The graph should now be of voltage versus current. ● CI Fig. 26.3 shows what the screen should look like at this point.

16. Double-click anywhere on the graph. The graph settings window will open. Make the following selections:

Under the tab Appearance:

Data:

Connect data points in bold.

Deselect the buttons marked "Show Data Points" and "Show Legend Symbols."

Click OK to accept the changes and exit the graph settings window.

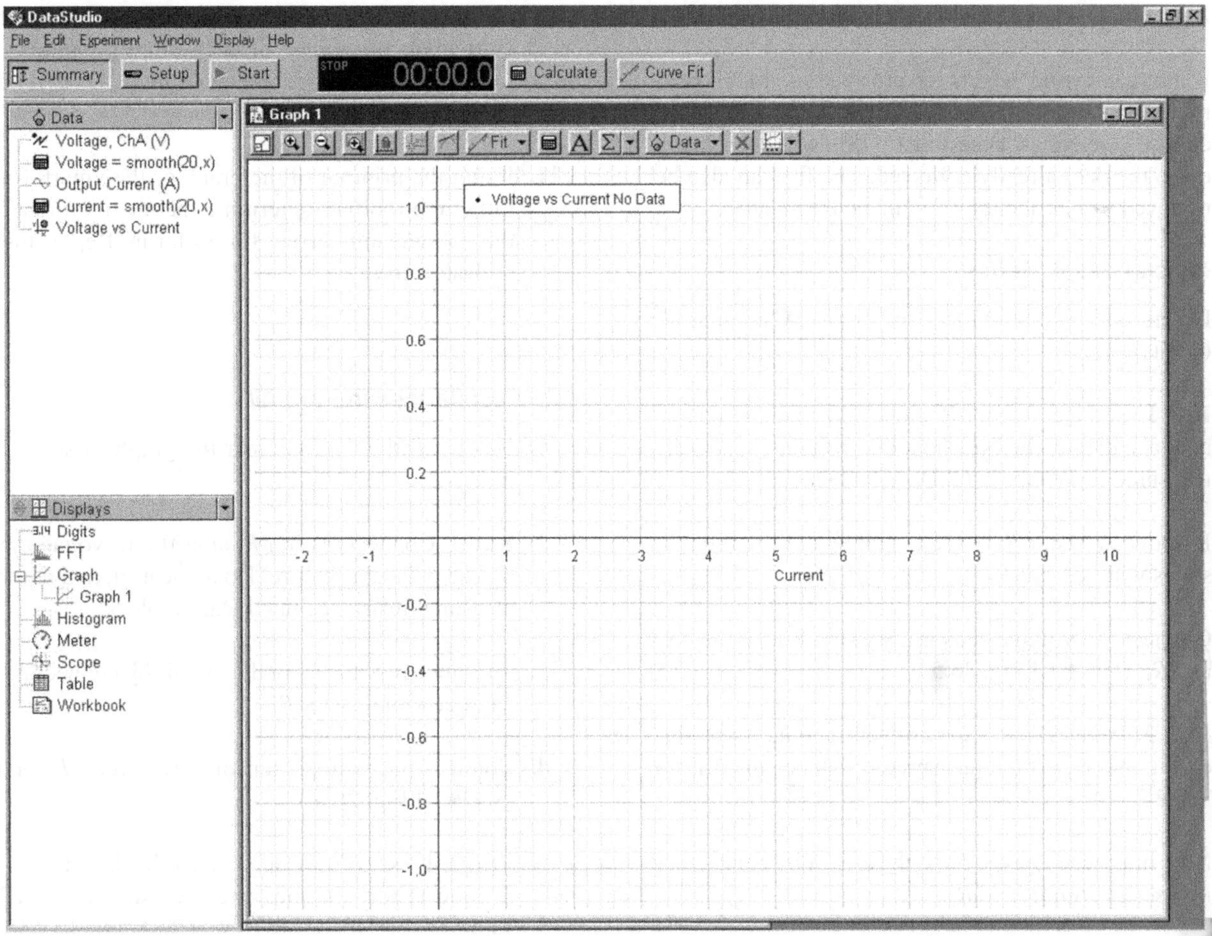

CI Figure 26.3 Data Studio setup. A graph of voltage versus current will be used to examine different simple circuits. The slope of the graph will represent the resistance of the circuit. (Reprinted courtesy of PASCO Scientific.)

CI EXPERIMENTAL PROCEDURE

A. Measuring Resistance

1. Get three resistors and label them R_1, R_2, and R_3.
2. Connect R_1 to the output source of the 750 Interface, using cables and alligator clips, if needed. A circuit diagram for this setup is shown in ● CI Fig. 26.4.
3. Put alligator clips on the prongs of the voltage sensor, and connect the voltage sensor across the resistor. Make sure that the positive of the voltage sensor (red lead) is connected to the positive lead of the resistor.
4. Press the START button. Data collection will stop automatically after 4.5 seconds.
5. Press the Scale-to-Fit button on the graph toolbar. The Scale-to-Fit button is the leftmost button on the graph toolbar. This will scale all data to fit the full screen.

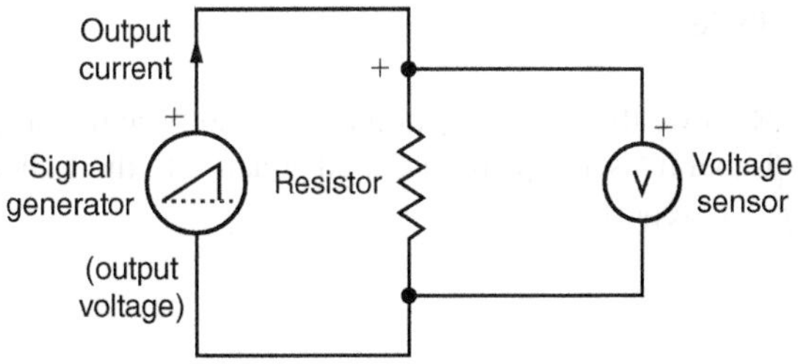

CI Figure 26.4 The experimental setup. A single resistor is connected to the source, with the voltage sensor connected across the resistor. The positive (red) lead of the voltage sensor must connect to the positive lead of the resistor.

6. Use the Fit menu (on the graph toolbar) to do a "Linear Fit" of the data. A box with information about the fit will appear. Report the slope of the line in CI Data Table 1 as the value of R_1. Do not forget units.
7. Repeat the experiment two more times, and determine an average value for R_1.

8. Repeat the process individually with R_2 and R_3.

9. If the graph window gets too crowded, go to "Experiment" (in the main menu, top of the screen) and choose "Delete all Data Runs." This will completely erase the data already collected. The fits can also be removed by deselecting them in the Fit menu.

B. Resistances in Series

1. Delete all the data to clear the graph. Also clear all the fits.

2. **Run 1**: Connect resistor R_1 alone to the voltage source and take data, as before. Do a linear fit, and report the measured resistance in CI Data Table 2.

3. Introduce resistor R_2 to the circuit by connecting it in series with resistor R_1.

4. Connect the voltage sensor across both resistors, R_2 and R_1. (See ● CI Fig. 26.5.)

5. **Run 2**: Press START and collect the data. Do a linear fit, and report the measured resistance in CI Data Table 2.

6. Now introduce resistor R_3 to the circuit by connecting it in series with R_2 and R_1.

7. Connect the voltage sensor across all three resistors. (See CI Fig. 26.5.)

8. **Run 3**: Press START and collect the data. Do a linear fit, and report the measured resistance in CI Data Table 2.

9. Remove the fit information boxes, and print the graph. Label it "Series Circuits," and attach it to the Laboratory Report.

10. Calculate the theoretical (expected) value of the equivalent resistance of each circuit. Compare the theoretical values with the measured ones by taking a percent difference.

11. Using the printout of the graph or the Smart-Tool on the graph toolbar, determine the maximum value of the voltage and current for each run. Report them in CI Data Table 2.

C. Resistances in Parallel

1. Delete all the data to clear the graph. Also clear all the fits.

2. **Run 1**: Connect resistor R_1 alone to the voltage source and take data, as before. Do a linear fit, and report the measured resistance in CI Data Table 3.

3. Introduce resistor R_2 to the circuit by connecting it in parallel with resistor R_1.

4. Connect the voltage sensor across both R_2 and R_1. (See ● CI Fig. 26.6.)

5. **Run 2**: Press START and collect the data. Do a linear fit, and report the measured resistance in CI Data Table 3.

6. Now introduce resistor R_3 to the circuit by connecting it in parallel with R_2 and R_1.

7. Connect the voltage sensor across the three resistors. (See CI Fig. 26.6.)

8. **Run 3**: Press START and collect the data. Do a linear fit, and report the measured resistance in CI Data Table 3.

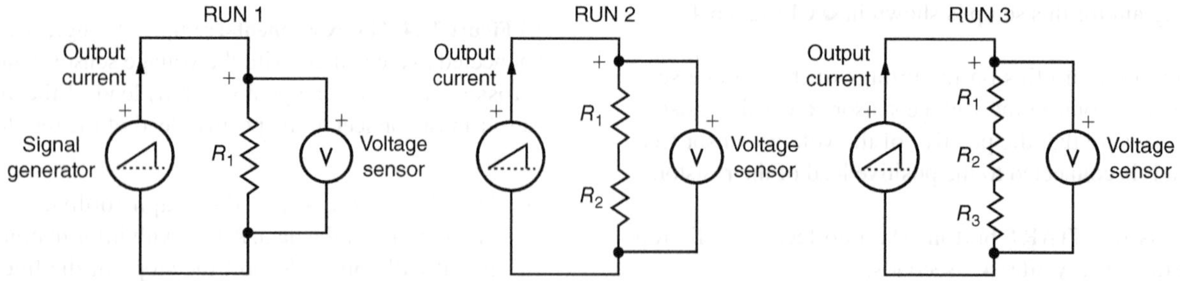

CI Figure 26.5 Resistors connected in series. Three different series circuits will be analyzed, each time adding an extra resistor to the series.

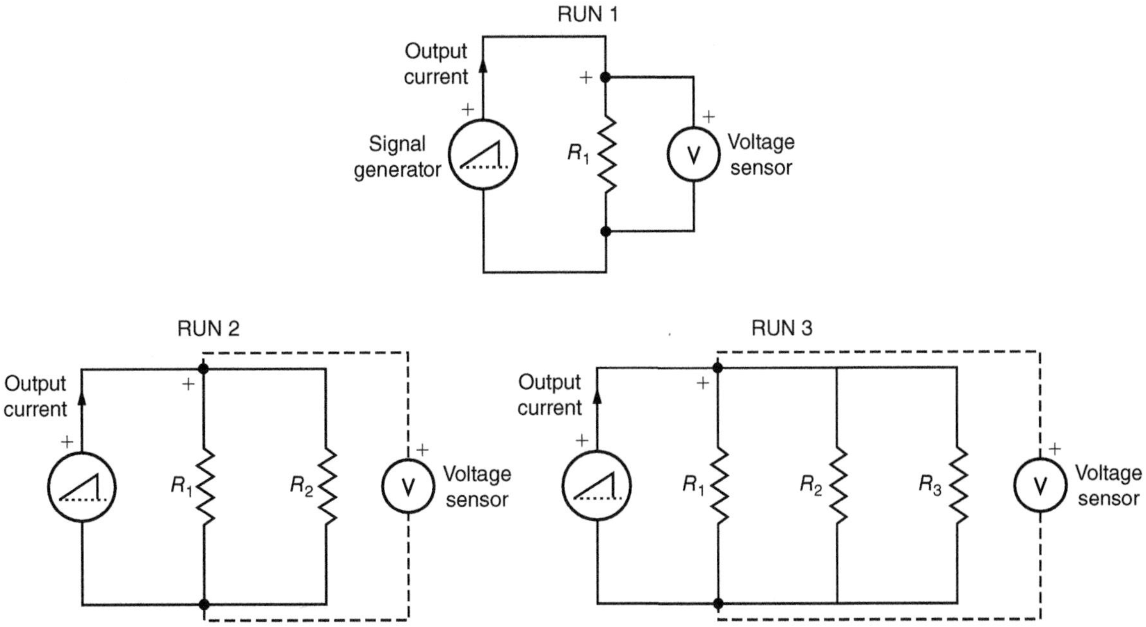

CI Figure 26.6 Resistors connected in parallel. Three different parallel circuits will be analyzed, each time adding an extra branch to the circuit.

9. Remove the fit information boxes, and print the graph. Label it "Parallel Circuits," and attach it to the Laboratory Report.

10. Calculate the theoretical (expected) value of the equivalent resistance of each circuit. Compare the theoretical values with the measured ones by taking a percent difference.

11. Using the printout of the graph or the Smart-Tool on the graph toolbar, determine the maximum value of the voltage and the maximum value of the current for each run. Report them in CI Data Table 3.

Name ______________________ Section ______________ Date ______________
Lab Partner(s) ____

CI EXPERIMENT 26

Resistances in Series and Parallel

CI *Laboratory Report*

A. Measuring Resistance

CI DATA TABLE 1

Purpose: To measure the actual resistance of each of the three resistors.

Resistor	Slope measurements	Average resistance
R_1	1.	
	2.	
	3.	
R_2	1.	
	2.	
	3.	
R_3	1.	
	2.	
	3.	

Don't forget units

(continued)

B. Resistances in Series

CI DATA TABLE 2

Purpose: To experimentally measure the equivalent resistance of series circuits.

	Run 1 R_1 alone	Run 2 R_1 and R_2 in series	Run 3 R_1, R_2, and R_3 in series
Measured equivalent resistance			
Theoretical equivalent resistance $R_s = R_1 + R_2 + \cdots$			
Percent difference			
Maximum voltage			
Maximum current			

Name ________________ Section ________________ Date ________________

Lab Partner(s) ________________

C. Resistances in Parallel

CI DATA TABLE 3

Purpose: To experimentally measure the equivalent resistance of parallel circuits.

	Run 1 R_1 alone	Run 2 R_1 and R_2 in parallel	Run 3 R_1, R_2, and R_3 in parallel
Measured equivalent resistance			
Theoretical equivalent resistance $R_p = \left(\frac{1}{R_1} + \frac{1}{R_2} + \cdots\right)^{-1}$			
Percent difference			
Maximum voltage			
Maximum current			

Don't forget units

(continued)

CI QUESTIONS

1. As more resistors were added to the series circuit, what happened to the total resistance of the circuit?

2. For approximately the same maximum voltage, what happened to the maximum current as more resistors were added to the series circuit?

3. As more resistors were added to the parallel circuit, what happened to the total resistance of the circuit?

4. For approximately the same maximum voltage, what happened to the maximum current as more resistors were added to the parallel circuit?

Name ______________________ Section ______________ Date ______________

Lab Partner(s) ____

EXPERIMENT 27

Joule Heat

TI Advance Study Assignment

Read the experiment and answer the following questions.

1. Explain (a) joule heat and (b) I^2R losses.

2. What is the difference between joule heat and power?

3. Given two different resistances, how does the total joule heat vary if they are connected to a battery of fixed voltage V_o (a) in series and (b) in parallel?

(continued)

4. In the experiment, why isn't the final temperature of the system read at the same time that the power supply is unplugged and the timer stopped?

5. Suppose that oil were used in the experiment instead of water. Would (a) the joule heat and (b) the temperature rise be the same? Explain.

PRE-LAB DEMONSTRATION | **Instructor's Choice**

EXPERIMENT 27

Joule Heat

INTRODUCTION AND OBJECTIVES

Whenever there is an electrical current in a conductor, some electrical energy is converted into heat energy. For a given current I, the energy conversion is greater in a conductor of greater resistance. This is analogous to the conversion of mechanical energy into heat energy due to frictional resistance.

The heat generated (or power dissipated) in an electrical circuit is commonly referred to as **joule heat**, after James Prescott Joule (1818–1889), the English scientist who investigated the conversion of electrical energy into heat (and also the mechanical equivalent of heat).

In many electrical applications, such as electrical motors, joule heat is an undesirable loss of energy. However, in other applications, such as toasters and electrical heaters, electrical energy is purposefully converted into heat energy. In this experiment, the heating effect of an electrical current and the "electrical equivalent of heat" will be investigated.

After performing this experiment and analyzing the data, you should be able to:

1. Describe what is meant by joule heat.
2. Explain the factors on which joule heat depends.
3. Tell how joule heat may be measured experimentally.

EQUIPMENT NEEDED

- Electrocalorimeter (immersion heater and calorimeter)
- Power supply or battery (12 V)
- Ammeter (0 A to 3 A)
- Voltmeter (30 V)
- Rheostat (40 Ω)
- Connecting wires
- Thermometer
- Stopwatch or laboratory timer
- Laboratory balance
- Ice

THEORY

The work W done (or energy expended) per unit charge in moving a charge q from one point to another is the potential difference or voltage V, that is,

$$\boxed{V = \frac{W}{q} \text{ or } W = qV} \quad \textbf{(27.1)}$$

The time rate of flow of charge is described in terms of current I, and

$$\boxed{I = \frac{q}{t}} \quad \textbf{(27.2)}$$

Hence, Eq. 27.1 may be written

$$\boxed{W = qV = IVt} \quad \textbf{(27.3)}$$

which represents the work done or the energy expended in a circuit in a time t. Dividing this by t gives the work or energy per time, or power P:

$$\boxed{P = \frac{W}{t} = IV} \quad \textbf{(27.4)}$$

When this general expression is applied to a resistance R, for which $V = IR$ (Ohm's law), the expanded energy or work in Eq. 27.4 can be written

$$\boxed{W = IVt = I^2Rt = \frac{V^2t}{R}} \quad \textbf{(27.5)}$$

The electrical energy expended is manifested as heat energy and is commonly called **joule heat** or **I^2R losses,** I^2R being the power or energy expended per time.

$$\boxed{\frac{W}{t} = IV = I^2R = \frac{V^2}{R}} \quad \textbf{(27.6)}$$

Equation 27.6 shows how the joule heat varies with resistance:

1. For a constant current I, the joule heat is directly proportional to the resistance, I^2R.

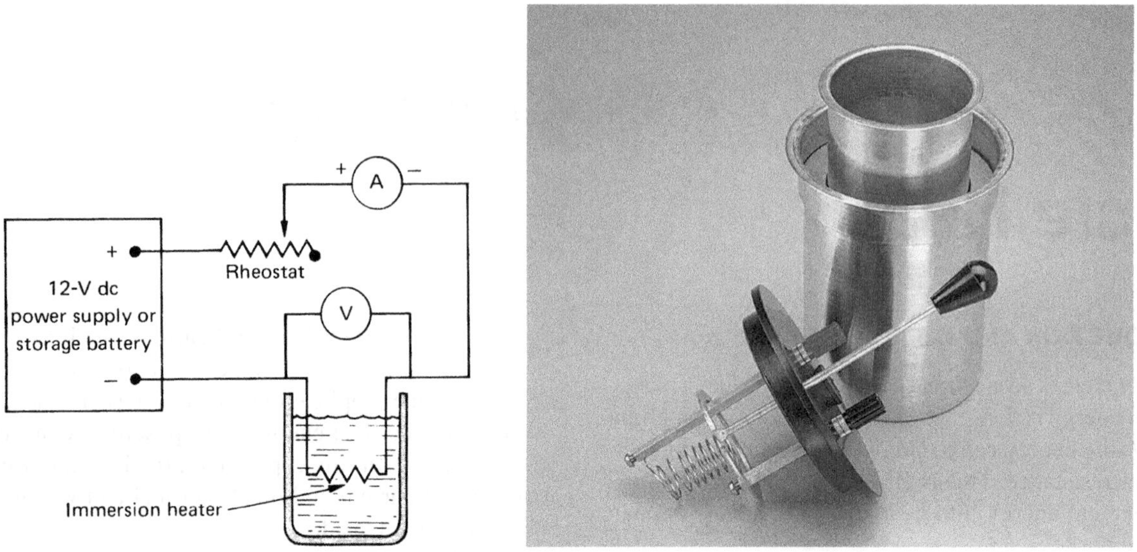

(a) (b)

Figure 27.1 Joule heat determination. (a) The circuit diagram for the experimental procedure to measure joule heat. See text for description. (b) An electric calorimeter. (Photo Courtesy of Sargent-Welch.)

2. For a constant voltage V, the joule heat is inversely proportional to the resistance, V^2/R.

The energy expended in an electrical circuit, as given by Eq. 27.6, has the unit of joule (J). The relationship (conversion factor) between joules and heat units in calories was established by James Joule from mechanical considerations—the **mechanical equivalent of heat**.

You may have learned that in his mechanical experiment, Joule had a descending weight turn a paddle wheel in a liquid. He then correlated the mechanical (gravitational) potential energy lost by the descending weight to the heat generated in the liquid. The result was 1 cal = 4.186 J, or 1 kcal = 4186 J. A similar electrical experiment may be done to determine the electrical equivalent of heat. By the conservation of energy, the heat equivalents of mechanical and electrical energy are the same (that is, 1 cal = 4.186 J).

Experimentally, the amount of electrical joule heat generated in a circuit element of resistance R is measured by calorimetry methods. If a current is passed through a resistance (immersion heater) in a calorimeter with water in an arrangement as illustrated in ● Fig. 27.1, then by the conservation of energy, the electrical energy expended in the resistance is equal to the heat energy (joule heat) Q gained by the system:

$$\text{electrical energy expended} = \text{heat gained}$$
$$W = Q$$
$$IVt = mc\,\Delta T$$

or

$$IVt = (m_w c_w + m_{cal} c_{cal} + m_{coil} c_{coil})(T_f - T_i) \quad \textbf{(27.7)}$$

where the m's and c's are the masses and specific heats of the water, calorimeter cup, and immersion coil, respectively, as indicated by the subscripts. T_f and T_i are the final and initial temperatures of the system, respectively. (See Experiment 20 for a detailed theory of calorimetry procedure.)

EXPERIMENTAL PROCEDURE

1. Determine and record in the Laboratory Report the masses of the inner calorimeter cup (without ring) and the coil of the immersion heater. (The latter may be supplied by the instructor if the coil is permanently mounted.)

Also, record the types of materials and their specific heats in cal/(g-°C). (The type of material and specific heat of the calorimeter cup are usually stamped on the cup. For the coil, usually copper, a table of specific heats is given in Appendix A, Table A4.)

2. Fill the calorimeter cup about two-thirds full of cool tap water several degrees below room temperature. (The cup should be filled high enough that the immersion heater will be completely covered when immersed later.) Determine and record the mass of the calorimeter cup with the water.

3. Place the immersion heater in the calorimeter cup, and set up the circuit as illustrated in Fig. 27.1 with the rheostat set at its maximum resistance. Make certain that the heating coil is completely immersed. If not, add more water and reweigh the cup and water as in Procedure 2. Reweigh the cup and water as in Procedure 2. *Do not plug in the power supply (or connect the battery) until the circuit has been checked by the instructor.*

4. After the circuit has been checked, plug in the power supply set to 10 V to 12 V. Adjust the rheostat until there is a constant current between 2 A and 3 A in the circuit as indicated on the ammeter. (If a variable power supply is used, it may also be used to make fine current adjustments.) Then unplug the power supply. This procedure should be done as quickly as possible to avoid heating the water.

5. Add some ice to the water in the calorimeter cup with the immersion coil or thermometer. When the ice has melted, measure and record the equilibrium temperature T_i. This should be 5 °C to 8 °C below room temperature.

Then plug in the power supply and at the same time start the stopwatch or laboratory timer. Immediately read and record the initial ammeter and voltmeter readings.

As time goes on, keep the current as constant as possible by varying the rheostat (and/or the power supply). Record the voltage and the current every minute. Stir the water frequently.

6. When the temperature of the water (and calorimeter system) is 10 °C to 15 °C above the initial temperature, simultaneously unplug the power supply and stop the timer at the time of a particular minute-interval reading. Continue stirring until a maximum temperature is reached, and record this temperature (T_f).

7. Compute the electrical energy expended in the coil (in joules) from the electrical and time readings. Use the average value of the voltage readings as the effective voltage across the coil.

8. **(a)** Compute the heat energy (in calories) gained by the calorimeter system.
(b) Then take the ratio of the electrical and heat energy results to find the "electrical equivalent of heat" (J/cal or J/kcal). Compare this to the value of the mechanical equivalent of heat by computing the percent error.

9. If time permits (ask your instructor), repeat the experiment and use the average value of the experimental results in determining the percent error. Attach these results to the Laboratory Report.

Name ______________________ Section ______________ Date ______________

Lab Partner(s) ____

EXPERIMENT 27

Joule Heat

TI Laboratory Report

	Mass	Material	Specific heat
Calorimeter cup	______	______	______
Immersion coil	______	______	______
Calorimeter cup and water	______		
Water	______	______	______

DATA TABLE

Purpose: To determine the mechanical equivalent of heat.

Time ( )	Voltage V ( )	Current I ( )	Temperature ( )
0			
			T_i ______
			T_f ______

Calculations
(show work)

Average voltage ______

Average current ______

Electrical energy expended ______

Heat energy gained ______

Ratio of results ______

Percent error ______

Don't forget units

(continued)

TI QUESTIONS

1. What are the major sources of error in the experiment? Why should the initial temperature of the water be several degrees below room temperature?

2. Why was it necessary to make adjustments to maintain a constant current in the circuit?

3. If the cost of electricity is 12 cents per kWh, what was the cost of the electricity used in performing the experiment?

4. (a) Circular metal wires in electrical circuits may have different cross-sectional areas (different diameters) and different lengths. For a given applied voltage, how would the joule heat vary with these parameters?
   (b) Would the wire material make a difference? (*Hint*: See *resistivity* in your textbook.)

5. Do heating appliances such as hair dryers and toasters have high-resistance or low-resistance elements? Explain.

Name ______________________ Section ______________ Date ______________

Lab Partner(s) ____

EXPERIMENT 28

The *RC* Time Constant

(Manual Timing)

TI Advance Study Assignment

Read the experiment and answer the following questions.

1. What is an *RC* time constant?

2. What must the unit(s) of an *RC* time constant be? Show this explicitly. (*Hint*: $Q = CV$ and $V = IR$.)

3. When an *RC* series circuit is connected to a dc source, what is the voltage on a capacitor after one time constant when (a) charging from zero voltage and (b) discharging from a fully charged condition?

4. If the resistance in a capacitor circuit is increased, does the charging time of the capacitor increase or decrease? Explain.

(continued)

5. Can the voltage across a capacitor be measured with a common voltmeter? Explain.

EXPERIMENT 28

The *RC* Time Constant

(Manual Timing)

INTRODUCTION AND OBJECTIVES

When a capacitor is connected to a dc power supply or battery, charge builds up on the capacitor plates, and the potential difference or voltage across the plates increases until it equals the voltage of the source. At any time, the charge Q of the capacitor is related to the voltage across the capacitor plates by $Q = CV$, where C is the capacitance of the capacitor in farads (F).

The rate of voltage rise depends on the capacitance of the capacitor and the resistance in the circuit. Similarly, when a charged capacitor is discharged, the rate of voltage decay depends on the same parameters.

Both the charging time and discharge time of a capacitor are characterized by a quantity called the **time constant** τ, which is the product of the capacitance C and the series resistance R, that is, $\tau = RC$. In this experiment, the time constants and the charging and discharging characteristics of capacitors will be investigated.

After performing this experiment and analyzing the data, you should be able to:

1. Explain the *RC* time constant and what its value means in terms of circuit characteristics.
2. Describe how a capacitor charges and discharges through a resistor as a function of time.
3. Tell how an *RC* time constant may be measured experimentally.

EQUIPMENT NEEDED

- 2 capacitors (for example, 1000 μF and 2200 μF electrolytic)
- 2 resistors (for example, 4.5 kΩ and 10 kΩ)
- Power supply or battery (12 V)
- High-resistance digital readout voltmeter
- Single-pole, double-throw (SPDT) switch
- Connecting wires
- Laboratory timer
- 2 sheets of Cartesian graph paper

THEORY

When a capacitor is charged through a resistor by a dc voltage source (the single-pole, double-throw switch S in position *a* in ● Fig. 28.1), the charge in the capacitor and the voltage across the capacitor increase with time. The voltage V as a function of time is given by

$$V = V_o(1 - e^{-t/RC}) = V_o(1 - e^{-t/\tau}) \quad \textbf{(28.1)}$$

(charging voltage)

where the exponential $e = 2.718$ is the base of natural logarithms and V_o is the voltage of the source.

The quantity $\tau = RC$ is called the **time constant** of the circuit. The curve of the exponential rise of the voltage with time during the charging process is illustrated in ● Fig. 28.2.

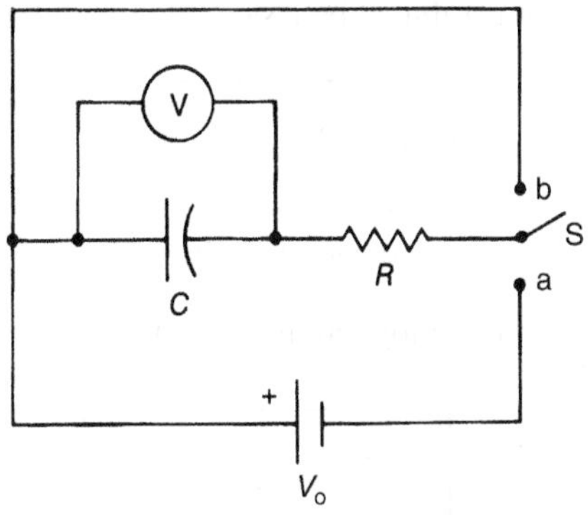

Figure 28.1 Capacitor charging and discharging. The circuit diagram for charging (switch S in position *a*) and discharging (switch S in position *b*) a capacitor through a resistor.

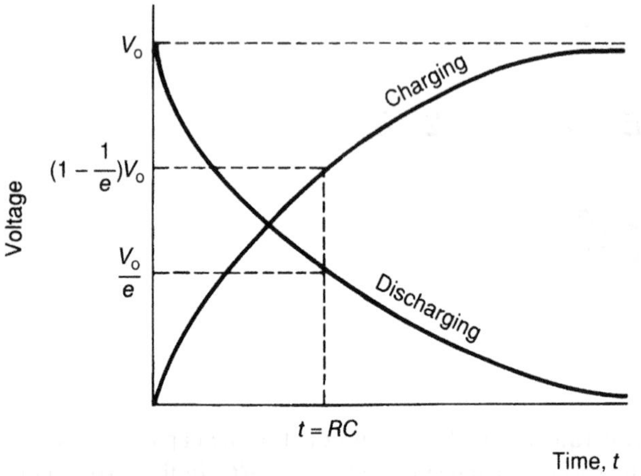

Figure 28.2 Voltage versus time. A graph illustrating voltage as a function of time for capacitor charging and discharging. The "steepness" of the curve depends on the time constant *RC*.

At time $t = \tau = RC$ (one time constant), the voltage across the capacitor has grown to a value of $(1 - \frac{1}{e})$ of V_o; that is, $V = V_o(1 - e^{-RC/RC}) = V_o(1 - e^{-1}) = V_o(1 - \frac{1}{e}) = 0.63\ V_o$.

When the fully charged capacitor is discharged through a resistor (switch S in position b in Fig. 28.1), the voltage across (and the charge on) the capacitor "decays," or decreases with time, according to the equation

$$\boxed{V = V_o e^{-t/RC}} \qquad \textbf{(28.2)}$$

(discharging voltage)

The exponential decay of the voltage with time is also illustrated in Fig. 28.2. After a time $t = \tau = RC$ (one time constant), the voltage across the capacitor has decreased to a value of $\frac{1}{e}$ of V_o; that is, $V = V_o e^{-t/RC} = V_o e^{-RC/RC} = V_o e^{-1} = V_o/e = 0.37\ V_o$. In order to analyze the voltage versus time, it is helpful to put Eqs. 28.1 and 28.2 in the form of a straight line. From Eq. 28.1, $(V_o - V) = V_o e^{-t/RC}$ and taking the natural logarithm of both sides of the equation gives

$$\boxed{\ln(V_o - V) = \frac{t}{RC} + \ln V_o} \qquad \textbf{(28.3)}$$

(charging voltage)

Taking the natural logarithm (base *e*) of Eq. 28.2,

$$\boxed{\ln V = -\frac{t}{RC} + \ln V_o} \qquad \textbf{(28.4)}$$

(discharging voltage)

Both of these equations have the form of the equation of a straight line, $y = mx + b$. (Can you identify the variables and constants?) Both have slopes of magnitude $1/RC$. Hence the time constant of a circuit can be found from the slopes of the graphs of $\ln(V_o - V)$ versus *t* and/or $\ln V$ versus *t*.

EXPERIMENTAL PROCEDURE

1. Set up the circuit as shown in Fig. 28.1 with the capacitor of smaller capacitance and resistor of larger resistance. It is often necessary to use a series combination of resistors to obtain the large resistances required in the experiment.

   The resistance of a resistor may be determined from the colored bands on the resistor. (See Appendix A, Table A5, for the resistor color code.)

   Record the value of the capacitance C_1 and the resistance R_1 in Data Table 1. Also prepare the laboratory timer for time measurements. *Have the instructor check the circuit before closing the switch.*

2. Close the switch to position *a,* and note the voltage rise of the capacitor on the voltmeter. When the capacitor is fully charged, move the switch to position *b*, and note the voltage decrease as the capacitor discharges. In the following procedures, the voltage is read as a function of time. You should try trial time runs to become familiar with the procedures.

3. Simultaneously close the switch to position *a* and start the timer. Read and record the capacitor voltage at small time intervals (for example, 3–5 seconds) until the capacitor is fully charged (V_o). This should be done with two persons working together.

   If necessary, however, the switch may be opened (and the timer stopped) to stop the charging process after a given interval without appreciable error if a high-quality, low-leakage capacitor is used.

4. After the capacitor is fully charged, open the switch to the neutral position and reset the timer. Then, simultaneously close the switch to position *b* and start the timer. Read and record the decreasing voltage at small time intervals. Open the switch when the capacitor is discharged.

5. Replace R_1 and C_1 with R_2 and C_2 (smaller resistance and larger capacitance), and repeat Procedures 3 and 4, using Data Table 2 to record your findings.

6. Compute the quantity $(V_o - V)$ for the charging and discharging processes, respectively. Then find the value of $\ln(V_o - V)$ and $\ln V$.

7. On a Cartesian graph, plot $\ln(V_o - V)$ versus *t* for both sets of data. On the other graph, plot $\ln V$ versus *t* for both sets of data. Draw the straight lines that best fit the data, and determine the slope of each line. Record the slopes in the data tables. Compute the time constants from the average slope values.

8. Compute $\tau_1 = R_1C_1$ and $\tau_2 = R_2C_2$ from the given resistance and capacitance values, and compare with the experimental values by finding the percent errors. (*Note*: The resistors and capacitors may have appreciable tolerances (± %) or vary from the given values.)

Name ______________________ Section ______________ Date ______________

Lab Partner(s) ____

EXPERIMENT 28

The *RC* Time Constant

(Manual Timing)

TI Laboratory Report

DATA TABLE 1

C_1 ______________

Purpose: To determine the *RC* time constant.

R_1 ______________

Charging				Discharging		
V ( )	t ( )	$V_o - V$	$\ln(V_o - V)$	V ( )	t ( )	$\ln V$
				V_o		
V_o						

Calculations
(show work)

Slope (charging) ______________

Slope (discharging) ______________

Average slope ______________

R_1C_1 (from slope) ______________

R_1C_1 (from given values) ______________

Percent error ______________

Don't forget units

(continued)

DATA TABLE 2

C_2 ______________

Purpose: To determine the *RC* time constant.

R_2 ______________

Charging				Discharging		
V ( )	t ( )	$V_o - V$	$\ln (V_o - V)$	V ( )	t ( )	$\ln V$
				V_o		
V_o						

Calculations (show work)

Slope (charging) ______________

Slope (discharging) ______________

Average slope ______________

R_2C_2 (from slope) ______________

R_2C_2 (from given values) ______________

Percent error ______________

Name ______________________ Section ______________ Date ______________

Lab Partner(s) ____

TI QUESTIONS

1. Show that the magnitude of the charge on a capacitor is given by $Q = Q_o(1 - e^{-t/\tau})$ and $Q = Q_o e^{-t/\tau}$ for charging and discharging, respectively.

2. What is the voltage across a capacitor after a time of two constants when (a) charging from zero voltage and (b) discharging from a fully charged condition?

3. With $V = V_o e^{-t/RC}$, it mathematically takes an infinite time for a capacitor in an *RC* circuit to discharge. Practically, how many time constants does it take for a capacitor to discharge to less than 1% of its initial voltage?

(continued)

4. Show that the time for the voltage in the *RC* circuit to rise to $V_o/2$ ("half-max") is $t_{1/2} = \tau \ln 2$.

5. A 2.0-μF capacitor in a circuit in series with a resistance of 1.0 MΩ is charged with a 6.0-V battery. How long would it take to charge the capacitor to three-fourths of its maximum voltage?

Name ____________________ Section ____________ Date ____________

Lab Partner(s) ____________________

EXPERIMENT 29

The *RC* Time Constant

(Electronic Timing)

TI Advance Study Assignment

Read the experiment and answer the following questions.

1. Compare the voltages across a capacitor in dc and ac *RC* circuits.

2. How is the time base of the horizontal oscilloscope trace determined?

3. What is the significance of the *RC* time constant for the circuit?

4. Explain how the time constant of an *RC* circuit is determined from a stationary oscilloscope pattern.

(continued)

CI Advance Study Assignment

Read the experiment and answer the following questions.

1. What is the time constant of an *RC* circuit, and what are the units of measurement?

2. How many time constants will you have to wait before you can consider the capacitor "fully charged"?

EXPERIMENT 29

The *RC* Time Constant

(Electronic Timing)

OVERVIEW

Experiment 29 examines the *RC* time constant using complementary electronic TI and CI approaches. In the TI procedure, the time constant of an *RC* circuit is determined from an oscilloscope trace of voltage versus time. This is done for combinations of *RC* values.

In the CI procedure, a voltage sensor monitors voltage changes for charging and discharging and supplies data to the computer. From computer-drawn graphs of voltage versus time, the time constant is determined—the point of 63% of maximum voltage for charging and 37% of the maximum voltage for discharging. The procedure is done for two resistances.

INTRODUCTION AND OBJECTIVES

The oscilloscope can be used to study many ac circuit characteristics. The screen display of voltage versus time makes it possible to observe a variety of measurements. In particular, in an ***RC* (resistance-capacitance) circuit**, the charging of the capacitor can be visually observed. And using the horizontal time scale, the time constant of the charging process can be readily determined.

In this experiment, the oscilloscope will be used to determine the time constant of an *RC* circuit as the capacitor is continually charged and discharged by an ac signal voltage.

TI OBJECTIVES

After performing this experiment and analyzing the data, you should be able to:

1. Explain the charging characteristics of a capacitor with ac voltage.
2. Appreciate how the oscilloscope can be used to monitor electrical characteristics and to make electrical measurements.
3. Describe how an *RC* time constant may be measured from an oscilloscope trace.

CI OBJECTIVES

The purpose of this experiment is to investigate the charging and discharging of a capacitor in a series *RC* circuit. The time constant of the circuit will be determined experimentally and compared to the theoretical value. After performing this experiment and analyzing the data, you should be able to:

1. Describe the charging and discharging of a capacitor through a resistor.
2. Explain how the time constant can be measured experimentally.
3. Explain what the *RC* time constant means in terms of circuit characteristics.

The *RC* Time Constant

(Electronic Timing)

TI EQUIPMENT NEEDED

- Function generator (square wave)
- Oscilloscope
- 3 capacitors (0.05 μF, 0.1 μF, and 0.2 μF, or capacitor substitution box)
- 3 resistors (5 kΩ, 10 kΩ, and 20 kΩ or resistance box)
- Connecting wires
- 2 sheets of Cartesian graph paper
- (*Optional*) Unknown resistor wrapped in masking tape to conceal value

TI THEORY

When an *RC* circuit is connected to a dc voltage source, charge must flow into the capacitor before the voltage across the capacitor can change. This takes time. As the voltage across the capacitor becomes closer to that of the source, the flow of charge becomes slower and slower. The capacitor voltage approaches the supply voltage as an asymptote—coming ever closer, but never getting there.

When the capacitor starts with no voltage across it, $V = 0$ at $t = 0$, the subsequent changing voltage is given by the equation

$$V = V_o(1 - e^{-t/RC}) = V_o(1 - e^{-t/\tau}) \qquad \text{(TI 29.1)}$$

where e is the base of the natural logarithms ($e = 2.718\ldots$), V_o is the voltage of the dc source, R the resistance in the circuit, and C the capacitance. The quantity $\tau = RC$ is the **time constant** of the circuit. (See the Theory section in Experiment 28.)

After a time of one time constant, $t = \tau = RC$, the voltage is

$$V = V_o(1 - e^{-RC/RC}) = V_o(1 - e^{-1}) = V_o(0.63)$$

or

$$\frac{V}{V_o} = 0.63 \qquad \text{(TI 29.2)}$$

That is, the voltage across the capacitor is 0.63 (or 63%) of its maximum value (● TI Fig. 29.1). For a dc voltage source, the capacitor voltage further increases to V_o and maintains this voltage unless discharged.

However, for an ac voltage source, the capacitor voltage increases and decreases as the voltage of the applied signal alternately increases and decreases. For example, suppose that a square-wave ac signal as illustrated in ● TI Fig. 29.2 is applied to the circuit. This has the effect of continuously charging and discharging the capacitor.* The voltage across the capacitor increases according to TI Eq. (29.1) and then decreases according to the relationship†

$$V = V_o e^{-t/RC} \qquad \text{(TI 29.3)}$$

On an oscilloscope, the time base or the magnitude of the horizontal time axis is determined by the SWEEP TIME/DIV. From this control setting, you can determine time functions for traces on the screen. For example, suppose two complete wave cycles of a stationary sinusoidal pattern cover 6.66 horizontal divisions with a SWEEP TIME/DIV

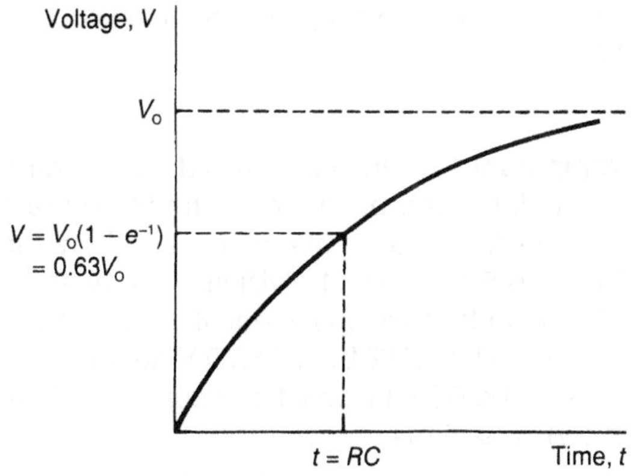

TI Figure 29.1 Voltage rise. A typical graph of voltage versus time for a capacitor charging in an *RC* circuit. In a time $t = RC$, the capacitor charges to 63% of its maximum value.

*The square-wave generator actually is constantly reversing the charge on the capacitor, but the trace has the same rise time as though it were charging and discharging.

†It should be noted that the high point on the charging curve and the low point on the decay curve in Fig. 29.2 are not $V = V_o$ and $V = 0$, respectively, since it takes infinite times for the capacitor to charge and discharge to these values. However, if the time constant is several times smaller than one-half the period T of the square wave, $T = 1/f$, then to a good approximation the high and low points of the curve may be taken to correspond to $V = V_o$ and $V = 0$, respectively.

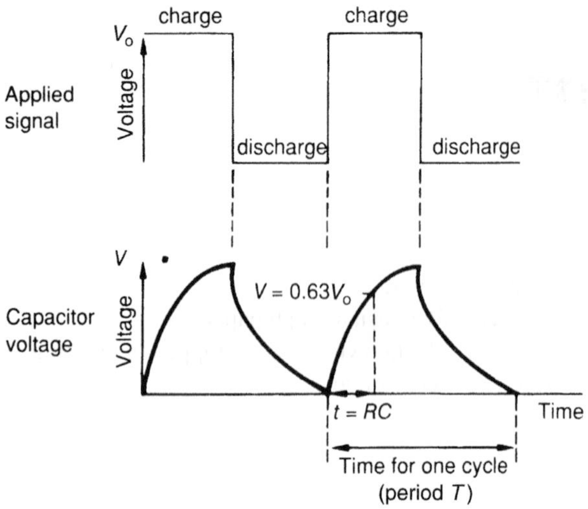

TI Figure 29.2 Charging and discharging. When a square-wave signal is applied to a capacitor in an *RC* circuit, the capacitor periodically charges and discharges, as shown here on a voltage-versus-time graph.

setting of 5 ms/div. Then, the time for these two cycles is time = ST/div × div = 5 ms/div × 6.66 div = 33.3 ms, so the time for one cycle or the period of the wave is $T =$ 33.3 ms/2 = 16.7 ms. (What is the frequency of the wave?)

The time constant of an *RC* circuit can be determined from a stationary oscilloscope pattern of the capacitor voltage versus time. This is done by finding the horizontal distance (time) needed for the trace to reach $0.63V_o$.

On an oscilloscope, time is measured as a horizontal distance. The scale is set by the knob marked SWEEP TIME/DIV.

TI Example 29.1 If the horizontal distance from the starting point to the point where the trace reaches 63% of the maximum voltage V_o, as shown in TI Fig. 29.1, is 6.5 divisions (1 division ≈ 1 cm), the time for 6.5 horizontal divisions is equal to one time constant (τ). With the SWEEP TIME/DIV set at 5 ms/div, the value of the *RC* time constant would be (6.5 div) × (5 ms/div) = 32.5 ms.

TI EXPERIMENTAL PROCEDURE

1. Turn on the oscilloscope and function generator. Set the function generator frequency to 100 Hz and the wave amplitude near maximum. Connect the square-wave output of the function generator directly to the vertical input terminals of the oscilloscope.

   Set the oscilloscope as follows. (*Note:* Different oscilloscopes differ somewhat in the names and locations of controls. **Vertical**: CH A DC, VOLTS/DIV 0.5, MODE CH A, POSITION Center Trace, CH B GND. **Horizontal**: TIME/DIV 2 mSEC, POSITION Center the trace. **Triggering**: LEVEL 12:00 position, COUPLING/SYNC AC SLOW, SOURCE INT, SLOPE +). The Vertical VOLTS/DIV and Horizontal TIME/DIV will be used here.

   Check that the small red knobs in the center of the VOLTS/DIV and TIME/DIV controls are in the calibrated position. Adjust the FOCUS and INTENSITY controls for a sharp, clear trace. ***Caution:*** Intensity should be kept low to protect the phosphor on the screen. If time permits, experiment with the controls to see how they affect the display.

   Obtain a stationary trace of one or two cycles of the square-wave pattern on the screen. Adjust the vertical VOLTS/DIV and the function generator amplitude until the pattern is exactly 8 divisions high. (This is about 8 cm high.) If V_o is 8 divisions, the 5-division horizontal line will be very close to the $0.63V_o$ criterion for measuring the time constant (since 5/8 = 0.625, actually $0.625V_o$).

2. Then set up the circuit as shown in ● TI Fig. 29.3, with $R = R_1 = 10\ \text{k}\Omega$ and $C = C_1 = 0.1\ \mu\text{F}$. *Have the instructor check the circuit before attaching the final lead to the oscilloscope.*

3. Close the oscilloscope circuit by connecting the wire to the circuit, and note the pattern. Carefully adjust the trigger controls so that the curve starts upward at the left end of the trace. The exponential rise time can be observed in greater detail by increasing the sweep rate (decreasing the TIME/DIV).

   Adjust the time (TIME/DIV) until the rising curve extends well across the screen. Be sure that the variable TIME/DIV remains in the calibrated position.

4. With the total pattern 8 divisions high, the time constant is represented by the horizontal distance from the point where the trace starts to move up to the point where it crosses the horizontal line 5 divisions up. The time is found by multiplying the horizontal distance by the TIME/DIV setting (see TI Example 29.1). Record in TI Data Table 1.

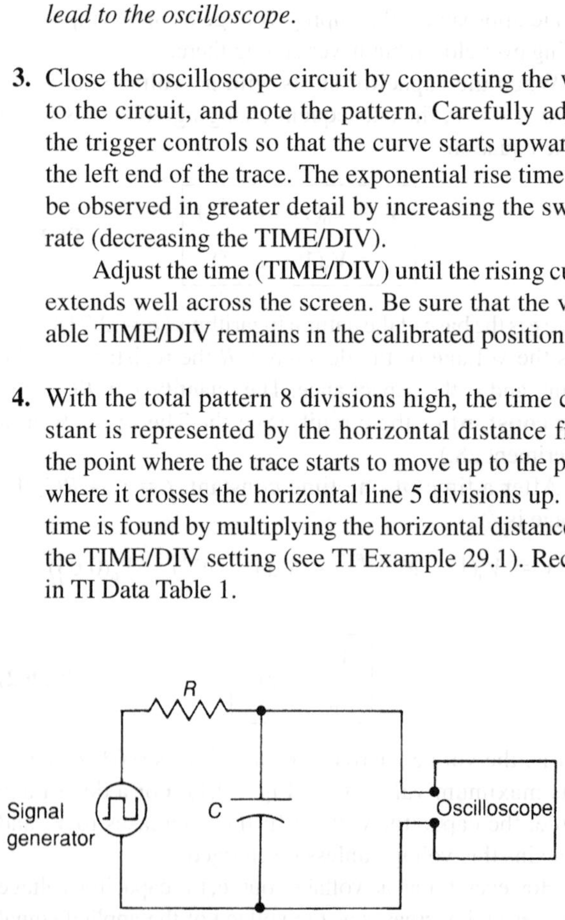

TI Figure 29.3 *RC* circuit. Circuit diagram for the experimental procedure for studying *RC* circuits. See text for description.

5. Open the circuit, and repeat Procedures 3 and 4 with $R = R_2 = 5\ \text{k}\Omega$ and $R = R_3 = 20\ \text{k}\Omega$. Record in TI Data Table 2.

6. On a Cartesian graph, plot the experimental τ versus R. Determine the slope of the straight line that best fits the data. To what does the value of the slope correspond?

7. Replace R with $R_1 = 10\ \text{k}\Omega$, and repeat Procedures 3 and 4 with $C = C_2 = 0.05\ \mu\text{F}$ and $C = C_3 = 0.2\ \mu\text{F}$.

8. On a Cartesian graph, plot the experimental τ versus C. (You should have three data points for τ with R_1. Why?) Determine the slope of the straight line that best fits the data. To what does the value of the slope correspond?

9. Compute the time constants for each of the *RC* combinations using the known R and C values, and compare with the experimentally determined values by finding the percent errors.

10. (*Optional*) Use your knowledge gained in this experiment to determine experimentally the value of the unknown resistor. Remove the masking tape after doing so, and compute the percent error.

Name ____________________ Section ____________ Date ____________

Lab Partner(s) ____

TI EXPERIMENT 29

The *RC* Time Constant

(Electronic Timing)

TI *Laboratory Report*

TI DATA TABLE 1

Purpose: To determine the effect of *R* on the time constant.

	R ( )	C ( )	Divisions for 0.63 rise	Sweep time = div	Exp. time constant	Computed RC	Percent error
Case 1 R_1C_1							
Case 2 R_2C_1							
Case 3 R_3C_1							

Calculations
(show work)

Slope of the τ-versus-R plot ____________

Percent difference between slope and C_1 ____________

Don't forget units

(continued)

TI DATA TABLE 2

Purpose: To determine the effect of C on the time constant.

	R ( )	C ( )	Divisions for 0.63 rise	Sweep time = div	Exp. time constant	Computed RC	Percent error
Case 4 R_1C_2							
Case 5 R_1C_3							

Slope of the τ-versus-C plot __________

Percent difference between slope and R_1 __________

Experimental RC time constant __________

Capacitance C __________

Computed R __________

Marked value of R __________

Percent error __________

Name ______________________ Section ______________ Date ______________

Lab Partner(s) ______________________

TI QUESTIONS

1. Judging on the basis of your experimental results, under what conditions are the charging times of different *RC* circuits the same?

2. In the form $V = V_o(1 - e^{-t/\tau})$, the $\tau = RC$ in the exponential must have units of time. (Why?) Show that this is the case.

3. How could the value of an unknown capacitance be determined using the experimental procedures? Show explicitly by assuming a value for an experimentally determined time constant.

CI EXPERIMENT 29

The *RC* Time Constant

(Electronic Timing)

EQUIPMENT NEEDED

This activity is designed for the Science Workshop 750 Interface, which has a built-in function generator.

- 1000-Ω resistor
- 330-μF capacitor
- Voltage sensor (PASCO CI-6503)
- Cables and alligator clips
- Multimeter (that can measure resistance and capacitance)
- Second resistor of different value

THEORY

A. Charging a Capacitor

● CI Fig. 29.1 shows a series *RC* circuit: a resistor connected in series with a capacitor and a power source of voltage V_o. As soon as the voltage source is turned on, the capacitor starts charging. As the charge in the capacitor increases exponentially with time, so does the voltage across its plates. The voltage across the capacitor at any time t is given by

$$\boxed{V = V_o(1 - e^{-t/RC})} \qquad \textbf{(CI 29.1)}$$

The quantity *RC* is called the time constant τ of the circuit, that is,

$$\boxed{\tau = RC} \qquad \textbf{(CI 29.2)}$$

With the resistance measured in ohms and the capacitance in farads, it is easy to show that the time constant has units of seconds. (See Question 1.) In terms of the time constant, CI Eq. 29.1 can be written as

$$\boxed{V = V_o(1 - e^{-t/\tau})} \qquad \textbf{(CI 29.3)}$$

The voltage across the capacitor will increase exponentially with time until it matches the voltage of the source.

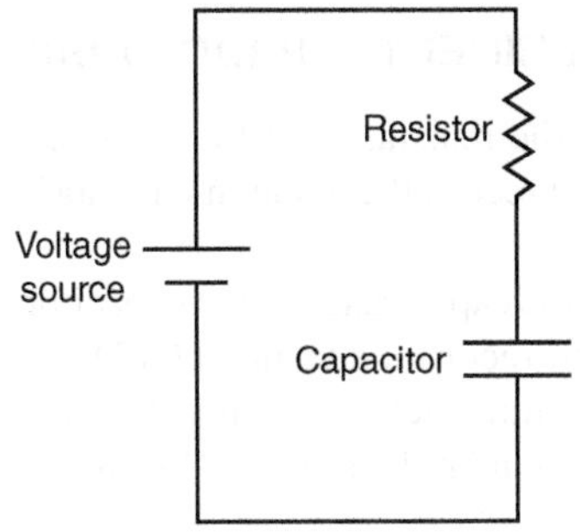

CI Figure 29.1 A series *RC* circuit. A capacitor and a resistor are connected in series to a voltage source.

The capacitor is fully charged when $V = V_o$, which theoretically requires an infinite amount of time, $t \rightarrow \infty$. In practice, however, it is said that the capacitor is fully charged if we wait long enough. But how long is "long enough"? Let's say until the voltage across the capacitor is 99.9% of the voltage of the source. The time it takes for this to happen can be calculated as follows:

$$\begin{aligned} V &= V_o(1 - e^{-t/\tau}) \\ 0.999\,V_o &= V_o(1 - e^{-t/\tau}) \\ 0.999 &= 1 - e^{-t/\tau} \\ e^{-t/\tau} &= 1 - 0.999 \\ e^{-t/\tau} &= 0.001 \\ \frac{-t}{\tau} &= \ln(0.001) \end{aligned}$$

Thus the time needed is

$$\boxed{t = -\tau \ln(0.001) = 6.9\tau \approx 7\tau} \qquad \textbf{(CI 29.4)}$$

For experimental purposes, for a time of about seven time constants, the capacitor is considered to be fully charged.

Another time that is of special interest is the time constant itself. Notice that at a time $t = \tau = RC$, one time constant after starting the charging process, the voltage across the capacitor has increased to 63% of the voltage of the source, as shown here:

$$\boxed{\begin{aligned} V &= V_o(1 - e^{-t/\tau}) \\ &= V_o(1 - e^{-\tau/\tau}) \\ &= V_o(1 - e^{-1}) \\ &= 0.63V_o \end{aligned}} \qquad \textbf{(CI 29.5)}$$

Notice that if you experimentally find at what time the voltage is 63% of the maximum, you are finding the time constant of the circuit.

In this experiment, the voltage source will be a signal generator that will produce a positive square wave.

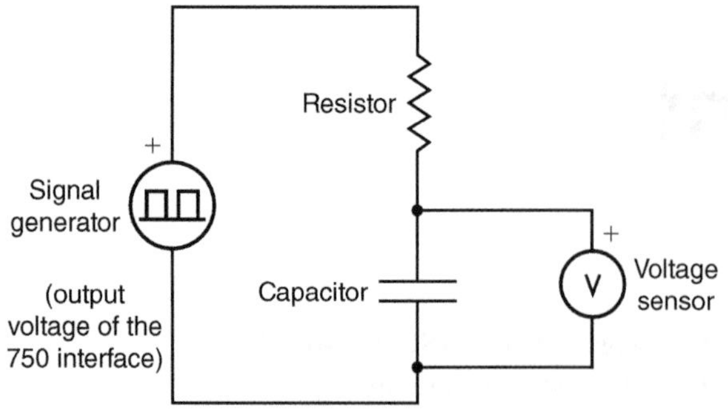

CI Figure 29.2 The experimental setup. The signal generator of the 750 Interface will be the voltage source for this experiment. A positive-square-wave voltage function will be used to periodically charge and discharge the capacitor. A voltage sensor will keep track of the voltage across the capacitor.

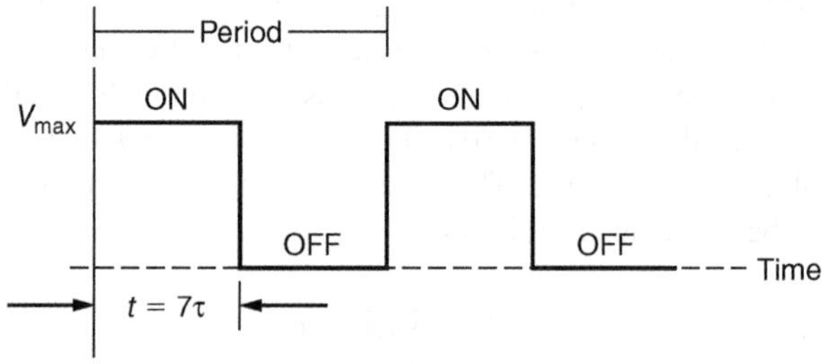

CI Figure 29.3 A positive square wave. The voltage periodically turns ON and OFF. To make sure the time it remains ON is enough to charge the capacitor fully, the time needed will be approximated to seven time constants (7τ), and the frequency of the signal will be adjusted accordingly.

The circuit is shown in ● CI Fig. 29.2. The voltage source is the signal generator of the PASCO Science Workshop 750 Interface. A voltage sensor will keep track of the voltage across the capacitor. A positive square wave is shown in ● CI Fig. 29.3. The voltage source will periodically turn on and off, charging and discharging the capacitor. To make sure that the capacitor gets fully charged before the source turns off, it will be necessary to set up the square wave so that the time it remains "ON" is at least seven time constants, as explained by CI Eq. 29.4. The experimental procedure contains detailed instructions on how to do this.

B. Discharging a Capacitor

When the voltage source is turned off, the charge in the capacitor flows back through the resistor. As the charge in the capacitor decreases, the voltage across the capacitor also decreases. The decrease is exponential, and as a function of time, it is described by the equation

$$V = V_o e^{-t/\tau} \qquad \textbf{(CI 29.6)}$$

In this case, notice that one time constant after the discharge begins, the voltage across the capacitor will be 37% of the original fully charged voltage of V_o:

$$\begin{aligned} V &= V_o e^{-t/\tau} \\ &= V_o e^{-\tau/\tau} \\ &= V_o e^{-1} \\ &= 0.37 V_o \end{aligned} \qquad \textbf{(CI 29.7)}$$

Thus the discharging of the capacitor can also be used to find the time constant experimentally by determining how long it takes for the voltage to decrease to 37% of the initial maximum value.

In this experiment, the charging and discharging of the capacitor will be observed in a plot of voltage versus time. The time constant of the circuit will be directly measured from the plot.

SETTING UP DATA STUDIO

1. Open Data Studio, and choose "Create Experiment."
2. The Experiment Setup window will open, and you will see a picture of the Science Workshop interface. There are seven channels to choose from and a signal generator. (Digital channels 1, 2, 3, and 4 are the small buttons on the left; analog channels A, B, and C are the larger buttons on the right; the signal generator is all the way to the right, as shown in ● CI Fig. 29.4.)
4. Click on the channel A button in the picture. A window with a list of sensors will open.
5. Choose the Voltage Sensor from the list, and press OK.
6. Connect the sensor to channel A of the interface, as shown on the computer screen.
7. Click on the picture of the signal generator. The Signal Generator window will open.
8. The default form of the signal generator function is a sine wave. Change it to a positive square wave of amplitude 3.0 V. (*Note:* Be sure to choose the "Positive Square Wave," not the one that says just "Square Wave." Scrolling down the list may be needed.) The frequency of the signal will depend on the values of *R* and *C* and will be entered later on.

CI EXPERIMENTAL PROCEDURE

1. Measure the resistance of the resistor using a multimeter, and record the value in CI Data Table 1.

2. Measure the capacitance of the capacitor using a multimeter, and record the value in CI Data Table 1. If the available multimeter does not measure capacitance, then use the manufacturer's value as the capacitance.

3. Calculate the theoretical time constant, and enter the value in CI Data Table 1.

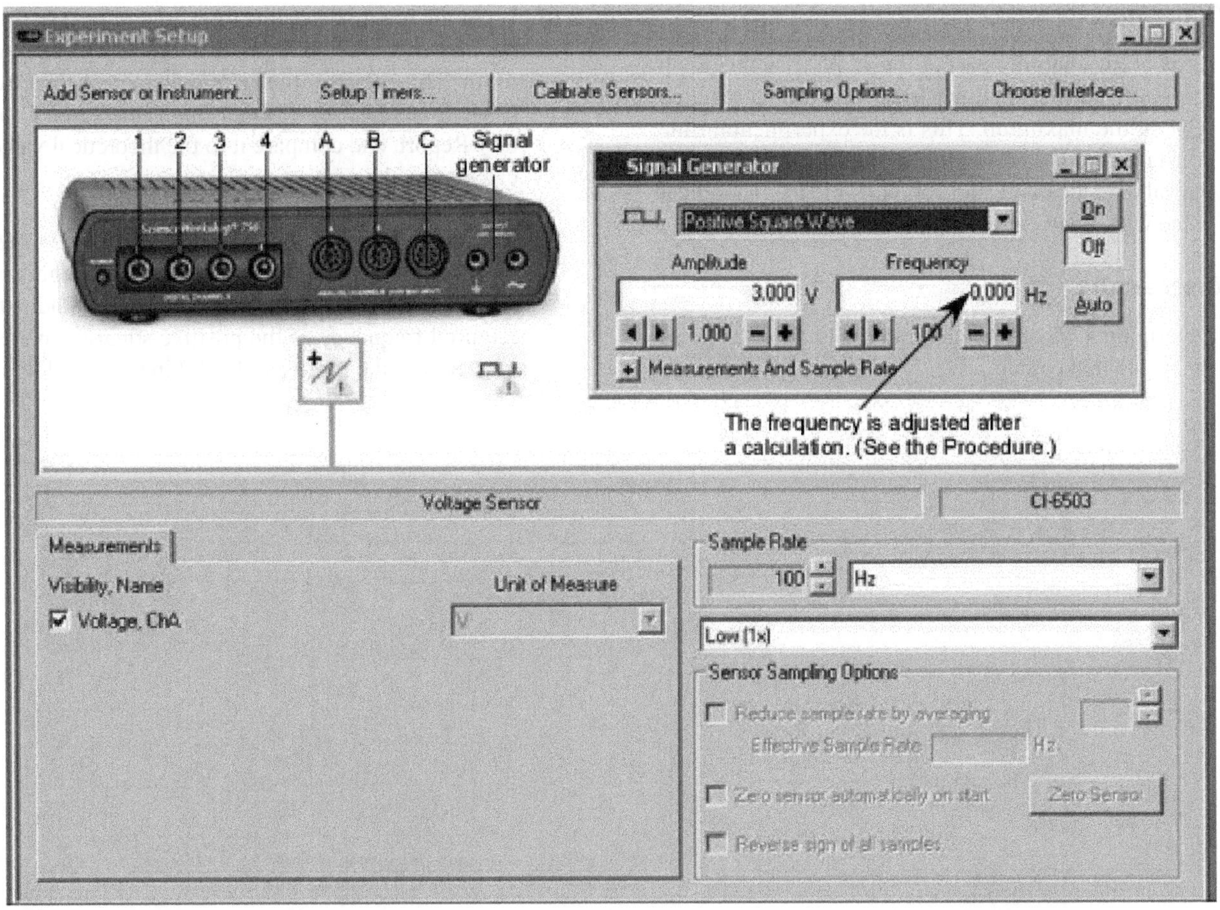

CI Figure 29.4 The Experiment Setup Window. The voltage sensor is connected to Channel A and works as a voltmeter. The signal generator of the Science Workshop interface is used as the voltage source that produces a positive square wave function. (Reprinted courtesy of PASCO Scientific.)

4. Calculate the approximate time needed to consider the capacitor fully charged. (See CI Eq. 29.4.) Enter the value in CI Data Table 1.

5. As explained in the CI Theory section, the frequency of the square wave needs adjusting so that the voltage source remains "ON" for enough time to charge the capacitor fully before it automatically turns "OFF" and discharges, as shown in CI Fig. 29.3. This is accomplished by following these steps:

a. The time to charge, calculated in step 4, is half the required period of the square wave. (See CI Fig. 29.3.) Calculate the required period, and enter it in CI Data Table 1.

b. Calculate the frequency, remembering that the frequency is the inverse of the period. Report the frequency in Data Table 1.

c. Enter the required frequency in the Signal Generator window, and set the generator to AUTO.

6. Set up the circuit shown in CI Fig. 29.2. The resistor, the capacitor, and the voltage source are connected in series. The voltage source is the output source of the 750 Interface, set to 3 V.*

7. Connect the voltage sensor across the capacitor, as shown in CI Fig. 29.2.

8. Press the START button. The capacitor will begin to charge and discharge. Press the STOP button after two cycles have been completed. Press the Scale-to-Fit button (leftmost button on the graph toolbar) to scale all data to fit on the screen.

9. Print the graph. If no printer is available, make a careful drawing of the graph. Paste the graph to the Laboratory Report.

10. Record the maximum voltage across the capacitor. Then calculate 63% of this value. Report these values in CI Data Table 1.

*The voltage value of 3 V is suggested for the values of *R* and *C* specified before because it produces an easy-to-read plot. The voltage sensor can measure a high range of voltages, and you may use a different value.

A. Charging

11. Look at the charging part of the graph. Use the graph tools to find the time at which the voltage reached 63% of the maximum. This is the experimental time constant of the circuit. [Refer to CI Eq. 29.5.] Enter the value in the table, and compare it to the theoretical value with a percent error.

B. Discharging

12. Determine 37% of the maximum voltage, and record this value in the table.

13. From the graph, determine how long *after the start of the discharge* the voltage was only 37% of the maximum. This is again the time constant of the circuit. [Refer to CI Eq. 29.7.] Enter this value in the Laboratory Report, and compare it to the theoretical value by calculating the percent error.

14. Repeat the experiment with a different value of resistance, keeping the capacitor and the voltage source constant. Do not forget to recalculate and adjust the required frequency of the positive square-wave function. Report the results as Trial 2 in CI Data Table 1.

Name ______________________ Section ______________ Date ______________

Lab Partner(s) ____

CI EXPERIMENT 29

The *RC* Time Constant

(Electronic Timing)

CI *Laboratory Report*

CI DATA TABLE 1

Purpose: To experimentally determine the time constant of the *RC* circuit.

			Trial 1	Trial 2
Theoretical Values	R			
	C			
	τ_{theo}			
	Time to fully charge $\approx 7\tau_{theo}$			
Output Signal	Period, T			
	Frequency, $f = \frac{1}{T}$			
Experimental Values	V_{max}			
	Charging	0.63 of V_{max}		
		τ_{exp}		
		Percent error		
	Discharging	0.37 of V_{max}		
		τ_{exp}		
		Percent error		

Don't forget units

(continued)

CI QUESTIONS

1. Show, by dimensional analysis, that the time constant $\tau = RC$ has units of time.

2. Compare the charging and discharging of the capacitors from Trial 1 and Trial 2. What things were similar and what things were different? Be specific.

3. Suppose that a particular *RC* series circuit has a time constant of 5.0 seconds. What does that mean in terms of the charging and discharging? How would this circuit compare to the ones you tried? Explain qualitatively and quantitatively.

4. What could be a practical application of an *RC* circuit?

Name ______________________ Section ______________ Date ______________

Lab Partner(s) ____

EXPERIMENT 30

Reflection and Refraction

TI Advance Study Assignment

Read the experiment and answer the following questions.

1. What is the law of reflection, and does it apply to all reflecting surfaces?

2. Distinguish between regular (specular) and irregular (diffuse) reflection. Give an example of each.

3. Why is light refracted when it passes from one medium into an optically different medium?

(continued)

4. Show by Snell's law that if the speed of light is less in a particular medium, then a light ray is bent toward the normal when entering that medium. What happens if the speed of light is greater in the medium?

5. What is the difference between the relative index of refraction and the absolute index of refraction? Explain why the absolute index of refraction can be determined experimentally fairly accurately using air as a medium.

PRE-LAB DEMONSTRATION | Instructor's Choice

Mirror Right–Left Reversal and Nonreversal [● Fig. D8]
Now You See It, Now You Don't

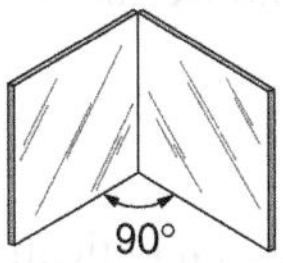

Figure D8 Right-left Reversal and Nonreversal

EXPERIMENT 30

Reflection and Refraction

INTRODUCTION AND OBJECTIVES

Reflection and refraction are two commonly observed properties of light. The reflection of light from smooth and polished surfaces, such as ponds of water and mirrors, enables us to view the images of objects, including ourselves. When light passes from one medium into another, it is bent, or refracted. As a result, a stick in a pond or a pencil in a glass or cup of water appears to be bent (● Fig. 30.1).

As part of geometrical optics, these phenomena are explained by the behavior of light rays. Through ray tracing, the physical laws of reflection and refraction can be conveniently investigated in the laboratory. In this experiment, a plane mirror and a glass plate will be employed to study these laws and the parameters used in describing the reflection and refraction of light.

After performing this experiment and analyzing the data, you should be able to:

1. Describe the law of reflection and explain how it can be verified experimentally.
2. Explain Snell's law and its application to transparent materials.
3. Explain what the index of refraction tells you about a transparent material and how it can be measured experimentally.

EQUIPMENT NEEDED

- Pins
- Pin board (cardboard or poster board suffices)
- Sheets of white paper ($8\frac{1}{2} \times 11$ in.)
- Ruler and protractor
- Short candle (less than 5 cm) or some similar light source
- Rectangular mirror (and holder if available)
- Thick glass plate (approximately 8×10 cm)

Note: Ray boxes may be used if available.

THEORY

A. Reflection

When light strikes the surface of a material, some light is usually reflected. The reflection of light rays from a plane surface such as a glass plate or a plane mirror is described by the **law of reflection**:

> *The angle of incidence* (θ_i) *is equal to the angle of reflection* (θ_r), *that is,* $\theta_i = \theta_r$

These angles are measured from a line perpendicular or *normal* to the reflecting surface at the point of incidence (● Fig. 30.2). Also, the incident and reflected rays and the normal lie in the same plane.

The rays from an object reflected by a smooth plane surface appear to come from an image behind the surface, as shown in the figure. From congruent triangles, it can be seen that the image distance d_i from the reflecting surface is the same as the object distance d_o. Such reflection is called **regular** or **specular reflection.**

The law of reflection applies to any reflecting surface. If the surface is relatively rough, like the paper of this page, the reflection becomes diffused or mixed, and no image of the source or object will be produced. This type of reflection is called **irregular** or **diffuse reflection.**

Figure 30.1 Refraction. Because of refraction, the pencil appears to be bent. (Charles D. Winters/Cengage Learning.)

B. Refraction

When light passes from one medium into an optically different medium at an angle other than normal to the surface, it is "bent," or undergoes a change in direction, as illustrated in ● Fig. 30.3 for two parallel rays in a beam of light. This is due to the different velocities of light in the two media. In the case of refraction, θ_1 is the angle of incidence and θ_2 is the angle of refraction.

From the geometry of Fig. 30.3, where d is the distance between the parallel rays at the boundary, we have

$$\sin\theta_1 = \frac{v_1 t}{d} \quad \text{and} \quad \sin\theta_2 = \frac{v_2 t}{d}$$

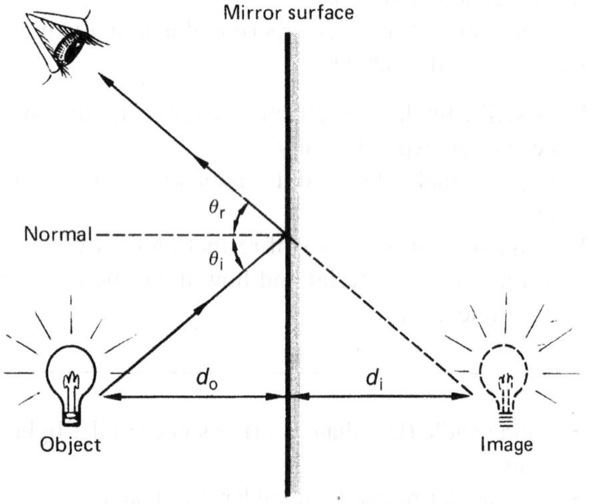

Figure 30.2 Law of reflection. The angle θ_i between the incident ray and the normal to the surface is equal to the angle θ_r between the reflected ray and the normal; that is, $\theta_i = \theta_r$. (Only a single ray is shown.) The object distance d_o is also equal to the image distance d_i for a plane mirror.

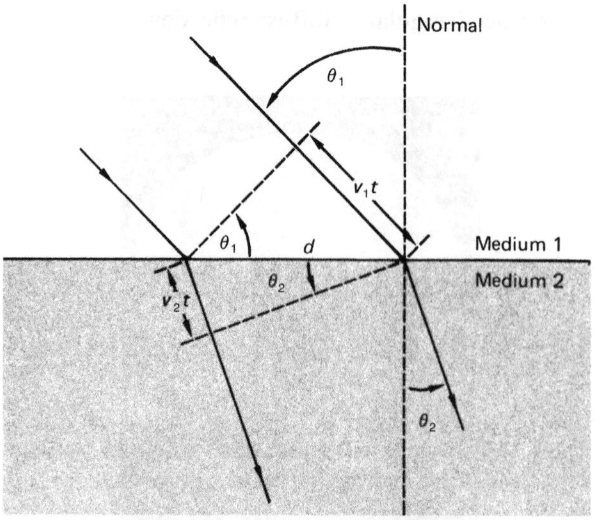

Figure 30.3 Refraction of two parallel rays. When medium 2 is more optically dense than medium 1, then $v_2 < v_1$ and the rays are bent toward the normal as shown here. If $v_2 > v_1$, the rays are bent away from the normal (as though the ray arrows were reversed in a reverse ray tracing here).

or

$$\boxed{\frac{\sin\theta_1}{\sin\theta_2} = \frac{v_1}{v_2} = n_{12}} \qquad \textbf{(30.1)}$$

where the ratio of the velocities n_{12} is called the **relative index of refraction**. Equation (30.1) is known as **Snell's law**.† If $v_2 < v_1$ (as in Fig. 30.3), the rays are bent toward the normal in the second medium. And if $v_2 > v_1$, the rays are bent away from the normal (for example, reversed rays in Fig. 30.2 with medium 2 taken as medium 1).

For light traveling initially in vacuum (or approximately for light traveling initially in air), the relative index of refraction is called the **absolute index of refraction** or simply the **index of refraction**, and

$$\boxed{n = \frac{c}{v}} \qquad \textbf{(30.2)}$$

where c is the speed of light in vacuum and v is the speed of light in the medium. Hence, the index of refraction of vacuum is $n = c/c = 1$, and for air $n \simeq c/c = 1$. For water, $n = 1.33$.

Snell's law can then be written

$$\frac{\sin\theta_1}{\sin\theta_2} = \frac{v_1}{v_2} = \frac{c/n_1}{c/n_2} = \frac{n_2}{n_1}$$

or

$$\boxed{n_1 \sin\theta_1 = n_2 \sin\theta_2} \qquad \textbf{(30.3)}$$

where n_1 and n_2 are in indices of refraction of the first and second media, respectively.

From Eq. (30.2), it can be seen that the index of refraction is a measure of the speed of light in a transparent material, or a measure of what is called the **optical density** of a material.* For example, the speed of light in water is less than that in air, so water is said to have a greater optical density than air. Thus the greater the index of refraction of a material, the greater its optical density and the lesser the speed of light in the material.

In terms of the indices of refraction and Snell's law [Eq. (30.3)], there are the following relationships for refraction:

- If the second medium is more optically dense than the first medium ($n_2 > n_1$), the refracted ray is bent *toward* the normal ($\theta_2 < \theta_1$), as in Fig. 30.3.
- If the second medium is less optically dense than the first medium ($n_2 < n_1$), the refracted ray is bent *away from* the normal ($\theta_2 > \theta_1$), as for the reverse ray tracing in Fig. 30.3.

EXPERIMENTAL PROCEDURE

A. Reflection

Glass Plate as a Mirror

1. Place a sheet of white paper on the table. As illustrated in ● Fig. 30.4, draw a line where the candle (or object)

†After a Dutch astronomer, Willebrord Snellius (1580–1626).
*Optical density does not correlate directly with mass density. In some instances, a material with a greater optical density than another may have a lower mass density.

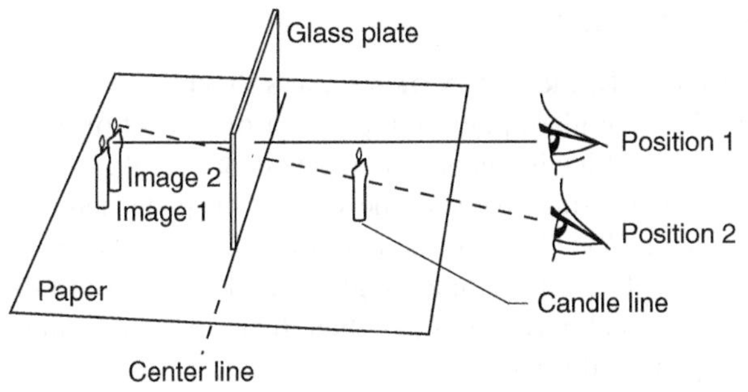

Figure 30.4 Glass plate as a mirror. The arrangement for the experimental procedure using a glass plate as a mirror. See text for description. (Images are displaced for illustration.)

will be placed. The line should be drawn parallel to the shorter edge of the page and about 3 cm to 4 cm from that edge. Make a mark near the center of the line, and place the candle on the mark.

Put the glass plate near the center of the paper, as shown in the figure. With the length of the plate parallel to the candle line, draw a line along the edge of the glass plate (side toward the candle). Light the candle.

Caution: *Take care not to burn yourself during the experimental procedure.*

Looking *directly over the candle* with your eye as in position 1 in Fig. 30.4, you will observe an image of the candle (image 1) in the glass plate. The glass plate reflects light and serves as a mirror. (Observation should be done with only one eye open.)

2. Observing the top of the flame from a side position (position 2 in Fig. 30.4), you will see a double image, one nearer than the other. Can you explain why?

   Place a pin in the pin board near the glass plate so that it is aligned (in the line of sight) with the front or nearer image of the candle (image 2 in Fig. 30.4; double image not shown in figure). Place another pin closer to you or to the edge of the paper so that both pins and the candle image are aligned. Mark the locations of the pins.

   Repeat this procedure, viewing from a position on the other side of the candle.

3. Remove the equipment from the paper. Draw straight lines through the pair of pin points extending from the candle line through the glass-plate line. (Extend the candle line, if necessary.) The lines will intersect on the opposite side of the plate line at the location of the candle image.

   Draw lines from the actual candle position or mark to the points of intersection of the previously drawn lines *and* the plate line. These lines from the candle (mark) to the glass-plate line and back to the observation positions are ray tracings of light rays.

4. Draw normal lines to the glass-plate line at the points of intersection of the ray lines. Label and measure the angles of incidence θ_i and reflection θ_r. Record the data in the Laboratory Report.

   Also, measure the perpendicular distances from the glass-plate line to the candle mark (the object distance d_o) and to the candle image position (the image distance d_i). Compute the percent differences of the quantities, as indicated in the Laboratory Report.

Plane Mirror

5. **(a)** Place the mirror near the center of a sheet of paper as with the glass plate used previously. (The mirror may be propped up by some means, or a holder may be used if available.) Draw a line along the silvered side of the mirror. Then lay an object pin about 10 cm in front of the mirror and parallel to its length (● Fig. 30.5).

   Mark the locations of the ends of the object pin on the paper with a pencil.

   (b) Stick a reference pin *R* in the board to one side of the object pin and near the edge of the paper, as illustrated in Fig. 30.5, and mark its location.

   (c) Place another pin nearer to the mirror so that it is visually aligned with the reference pin and the head of the object pin's image in the mirror. Mark the position of this pin, and label it with an *H*. Then move this pin over so that it aligns with the reference pin and the "tail" of the image pin. Mark this location, and label it with a *T*.

   (d) Repeat this procedure on the opposite side of the object pin with another reference pin.

6. Remove the equipment from the paper, and draw straight lines from the reference points through each of the H and T locations and the mirror line. The *H* lines and *T* lines will intersect and define the locations of the head and tail of the pin image, respectively.

   Draw a line between the line intersections (the length of the pin image). Measure the length of this

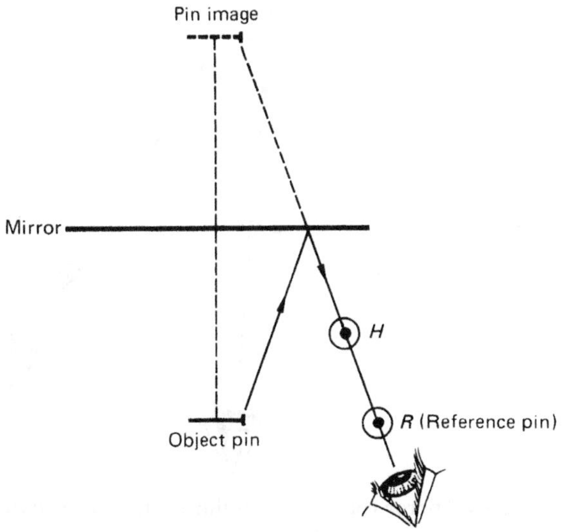

Figure 30.5 Plane mirror. The arrangement for the experimental procedure for a plane mirror. See text for description.

line and the length of the object pin, and record. Also, measure the object distance d_o and the image distance d_i from the mirror line, and record.

Compute the percent differences of the respective measured quantities.

Rotation of a Mirror

7. Place the mirror near the center of a sheet of paper (as described above), and draw a line along the length of the silvered side of the mirror. Measure so as to find the center of the line, and mark that location.

   Stick two pins (*A* and *B*) in the board to one side and in front of and in line with the center of the mirror, as in ● Fig. 30.6. Viewing the aligned images of these pins from the other side of the page, place two more pins (*C* and *D*) in alignment. Label the locations of the pins.

8. Leaving pins *A* and *B* in place, rotate the mirror a small but measurable angle θ (approximately 10° to 15°) about its center point, and draw a line along the silvered side of the mirror.

   Align two pins (*E* and *F*) with the aligned images of *A* and *B*, and mark and label the locations of *E* and *F*.

9. Remove the equipment from the paper, and draw the incident ray and the two reflected rays. Measure the angle of rotation θ of the mirror and the angle of deflection ϕ between the two reflected rays, and record in the Laboratory Report.

   Double θ, and compute the percent difference between 2θ and ϕ. Make a conclusion about the relationship between the angle of rotation of a mirror and the angle of deflection of a ray.

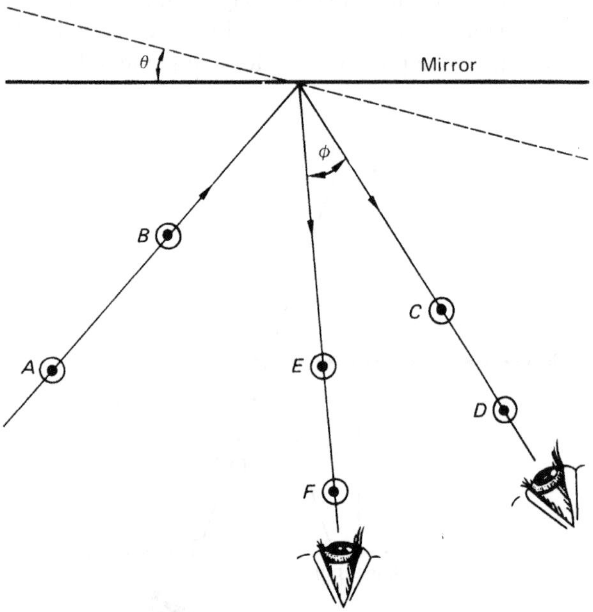

Figure 30.6 Mirror rotation. An illustration of the experimental arrangement and procedure for the rotation of a mirror. See text for description.

B. Refraction

Index of Refraction of a Glass Plate

10. Lay the glass plate in the center of a sheet of paper, and outline its shape with a pencil (● Fig. 30.7). Draw a line normal to one of the sides of the plate, and place a pin (*R*) at the intersection of this line and the face of the plate. Measure an angle θ_1 of 15° relative to this line, and place a pin (*A*) about 6 cm to 8 cm from the plate at this angle.

    Then, sighting through the edge of the plate from the eye position shown in Fig. 30.7, place a pin (A') adjacent to the face of the plate so that it is aligned with *R* and *A*. Mark and label the locations of the pins.

    Repeat with pins *B* and *C* at angles of 30° and 45°, respectively. For the 45°-angle case, align an additional pin (C'' Fig. 30.7).

11. Trace the various rays, and measure and record θ_1 and θ_2 for each case. Also measure and record the displacement *d* of ray $C'C''$ from the normal and the thickness of the plate. Using Eq. (30.3), compute the index of refraction of the glass.

    Compare the average experimental value of the index of refraction with the general range of the index of refraction of glass ($n = 1.5-1.7$, depending on type).

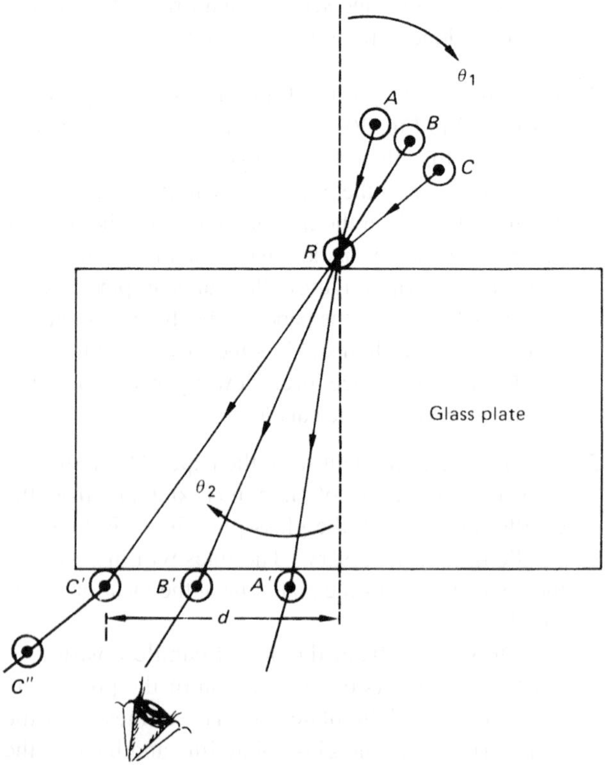

Figure 30.7 Index of refraction. An illustration (top view) of the experimental arrangement and procedure for determining the index of refraction of a glass plate. See text for description.

Name ______________________ Section ______________ Date ______________

Lab Partner(s) ____

EXPERIMENT 30

Reflection and Refraction

TI Laboratory Report

A. Reflection

Glass Plate as a Mirror

	θ_i	θ_r		
Ray 1	________	________	d_o	________
Ray 2	________	________	d_i	________

Percent differences between θ_i and θ_r

Ray 1 ________

Ray 2 ________

Percent differences between d_o and d_i ________

Plane Mirror

Length of pin ________ d_o ________

Length of image ________ d_i ________

Percent difference between pin length and image length ________

Percent difference between d_o and d_i ________

Rotation of a Mirror

Angle of rotation, θ ________ 2θ ________

Angle of deflection of ray, ϕ ________

Percent difference between ϕ and 2θ ________

Calculations
(show work)

Don't forget units

(continued)

B. Refraction

Index of Refraction of a Glass Plate

	θ_1	θ_2	Computed n
Ray ARA'	________	________	________
Ray BRB'	________	________	________
Ray CRC'	________	________	________

Average n ________

General range of the index of refraction of glass ________

Displacement d of ray $C'C''$ ________

Thickness of glass plate ________

Calculations
(show work)

TI QUESTIONS

1. (a) Why are two images seen in the glass plate when it is viewed from position 2 in Part A of the experiment? Why is only one image seen when it is viewed from position 1?

(b) Explain why reflection images are easily seen at night in a window pane from inside the house, whereas during the day they are not.

2. Judging on the basis of your experimental data, draw conclusions about (a) the relationship of the distance of the object in front of a plane mirror and the distance of its image "behind" the mirror; and (b) the image magnification (that is, how much bigger the image is than the object).

3. Explain the situation shown in ● Fig. 30.8. How can this be done without hurting one's hand? (*Hint*: The fearless author's hand extends inside the sliding glass-windowed door of a laboratory cabinet.)

Figure 30.8 See Question 3. (© Cengage Learning.)

(continued)

4. Prove mathematically that when a plane mirror is rotated an angle θ about an axis through its center (Part A of the experiment), the angle of deflection ϕ of a light ray is equal to 2θ. Draw a diagram and show the work involved in your proof. Attach an additional sheet, if necessary.

5. Referring to the situation in Fig. 30.7, show theoretically that ray $C'C''$ is parallel to ray CR. Compute the displacement d of the ray passing through the glass plate. Compare this with the measured experimental displacement.

6. Using the experimentally determined n for the glass plate, compute the speed of light in the glass plate.

Name ________ Section ________ Date ________
Lab Partner(s) ________

EXPERIMENT 31

Spherical Mirrors and Lenses

TI Advance Study Assignment

Read the experiment and answer the following questions.

1. Distinguish between concave and convex spherical mirrors.

2. What is the difference between a real image and a virtual image?

3. Distinguish between diverging and converging lenses.

4. What does the word *focal* mean with regard to the focal point of spherical mirrors and lenses?

(continued)

5. If an object is placed 15 cm in front of a concave mirror with a radius of curvature of 20 cm, what are the image characteristics? (Show your work.)

PRE-LAB DEMONSTRATION | Spherical Mirrors (right-side up and upside down)

When looking at one side of a shiny spoon, you'll see your image inverted. When looking at the other side, your image is right-side up. Why is this?

EXPERIMENT 31

Spherical Mirrors and Lenses

INTRODUCTION AND OBJECTIVES

Mirrors and lenses are familiar objects that are used daily. The most common mirror is a plane mirror, the type you look into in the morning to see your image. Spherical mirrors also have many common applications. For example, convex spherical mirrors are used in stores to monitor aisles and merchandise, and concave spherical mirrors are used as flashlight reflectors and as cosmetic mirrors that magnify.

Mirrors *reflect light*, whereas **lenses** *transmit light*. Spherical lenses are used to cause light rays to converge and hence focus them (convex spherical lenses) and to cause light rays to diverge (concave spherical lenses). Many of us wear lenses in the form of eyeglasses. Cameras and projectors use lens systems to form images. Cameras form reduced-size images on film or a chip (digital), and projectors form magnified images on a screen.

In this experiment, the fundamental properties of spherical mirrors and lenses will be investigated to learn the parameters that govern their use.

After performing this experiment and analyzing the data, you should be able to:

1. Distinguish among converging and diverging spherical mirrors and lenses.
2. Determine the image characteristics for spherical mirrors graphically using ray diagrams and analytically using the mirror equation and magnification factor.
3. Determine the image characteristics for spherical lenses graphically using ray diagrams and analytically using the thin-lens equation and magnification factor.

EQUIPMENT NEEDED

- Concave and convex spherical mirrors
- Convex lens (focal length 10 cm to 20 cm)
- Concave lens (focal length at least 5 cm longer than convex lens)
- Meterstick optical bench (or precision bench) with lens holder, screen, and screen holder (white cardboard can serve as the screen)
- Light source: candle and candle holder, or electric light source with object arrow

(Having cleaning tissue available might be helpful.)

THEORY

A. Spherical Mirrors

A **spherical mirror** is a section of a sphere and is characterized by a center of curvature *C* (● Fig. 31.1). The distance from the center of curvature to the vertex of the mirror along the optic axis is called the **radius of curvature** *R*. This also may be measured to any point on the surface of the mirror. (Why?)

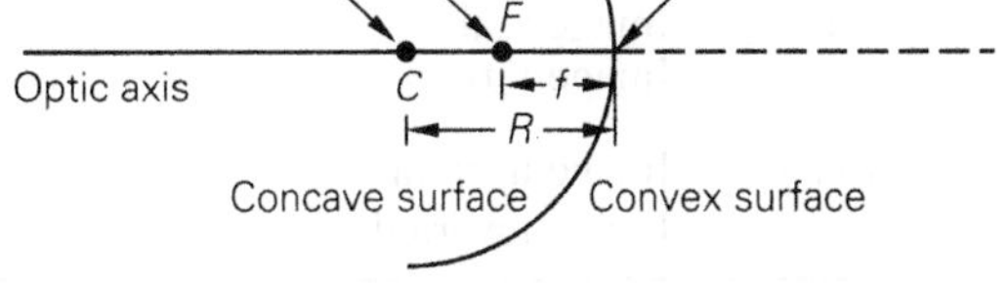

Figure 31.1 Spherical mirrors. The parameters used to describe spherical mirror surfaces. See text for description.

The focal point *F* is midway between *C* and the vertex, and the **focal length** *f* is one-half the radius of curvature:

$$f = \frac{R}{2} \tag{31.1}$$

If the reflecting surface is on the inside of the spherical section, the mirror is said to be **concave**. For a **convex** mirror, the reflecting surface is on the outside of the spherical section.*

The characteristics of the images formed by spherical mirrors can be determined either graphically or analytically. Examples of the graphical ray method are shown in the ray diagrams in ● Fig. 31.2.

As illustrated for a concave mirror (Fig. 31.2a):

1. A **chief ray** from the object goes through the center of curvature *C* and is reflected back through *C*.
2. A **parallel ray** from the object is parallel to the optic axis and is reflected through the focal point *F*.
3. A **focal ray** from the object passes through the focal point *F* and is reflected parallel to the optic axis.

* To help remember the difference, note that a con*cave* mirror is recessed, as though one were looking into a *cave*.

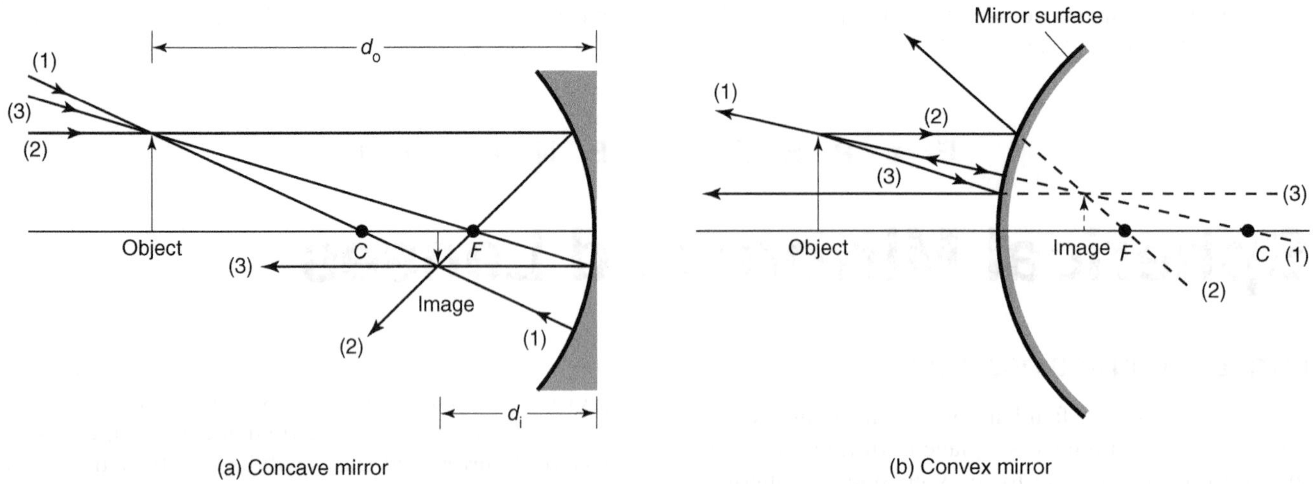

Figure 31.2 Mirror ray diagrams. Examples of the ray diagram method for determining the image characteristics for (a) a concave, or converging, spherical mirror and (b) a convex, or diverging, spherical mirror.

The intersection of these rays defines the location of the tip of the **image arrow**, which extends to the optic axis. The focal ray is a "mirror" image of the parallel ray and is not needed to locate the tip of the image. (The focal ray is often omitted, as the chief and parallel rays locate the image. However, the focal ray can be helpful when the object is inside the center of curvature.)

For a convex mirror, the chief and parallel rays appear to go through C and F, as illustrated in Fig. 31.2b.

A concave mirror is called a **converging mirror** because rays parallel to the optic axis converge at the focal point. Similarly, a convex mirror is called a **diverging mirror** because the rays parallel to the optic axis appear to diverge from the focal point.

If the image is formed on the same side of the mirror as the object, the image is said to be a **real image**. In this case, the light rays converge and are concentrated, and an image can be observed on a screen placed at the image distance. An image that is formed "behind" or "inside" the mirror is called a **virtual image**. Here, the rays appear to diverge from the image, and no image can be formed on a screen. Common plane mirrors form virtual images.

In general, an image is described in terms of whether it is

1. Real or virtual,
2. Upright (erect) or inverted (relative to the object orientation), and
3. Magnified or reduced (or smaller).

In Fig. 31.2a, the image is real, inverted, and reduced; in Fig. 31.2b, the image is virtual, upright, and reduced.

The distance from the object to the vertex along the optic axis is called the **object distance** d_o, and the distance from the vertex to the image is the **image distance** d_i. Knowing the focal length f of the mirror, the position of the image d_i can be found using the **spherical mirror equation**,

$$\frac{1}{d_o} + \frac{1}{d_i} = \frac{1}{f} \qquad \textbf{(31.2a)}$$

Another convenient form of this equation is

$$d_i = \frac{d_o f}{d_o - f} \qquad \textbf{(31.2b)}$$

In the case of a concave mirror, the focal length is taken to be positive (+); for a convex mirror, the focal length is taken to be negative (−). The object distance d_o is taken to be positive in either case. The resulting sign convention is as follows: If d_i is positive, the image is real, and if d_i is negative, the image is virtual. The **magnification factor** M is given by

$$M = -\frac{d_i}{d_o} \qquad \textbf{(31.3)}$$

If M is positive (with d_i negative), the image is upright; if M is negative (with d_i positive), the image is inverted. The sign convention is summarized in ● Table 31.1.

TABLE 31.1 Sign Convention for Spherical Mirrors and Lenses

Quantity	Conditions	Sign
Focal length f	Concave (converging) mirror	+
	Convex (diverging) mirror	−
	Convex (converging) lens	+
	Concave (diverging) lens	−
Object distance d_o	Usually* (always in this experiment)	+
Image distance d_i	Image real	+
	Image virtual	−
Magnification M	Image upright	+
	Image inverted	−

*In some cases of lens combinations, d_o may be negative when the image of one lens is used as the object for the next lens.

Example 31.1 An object is placed 45 cm in front of a concave mirror with a focal length of 15 cm (corresponding to the case in Fig. 31.2a). Determine the image characteristics analytically. (Neglect significant figures.)

Solution With $d_o = 45$ cm and $f = 15$ cm, Eq. 31.2a,

$$\frac{1}{45} + \frac{1}{d_i} = \frac{1}{15} = \frac{3}{45}$$

Then

$$\frac{1}{d_i} = \frac{3}{45} - \frac{1}{45} = \frac{2}{45} \quad \text{or} \quad d_i = \frac{45}{2} = 22.5 \text{ cm}$$

Then

$$M = -\frac{d_i}{d_o} = -\frac{22.5 \text{ cm}}{45 \text{ cm}} = -\frac{1}{2}$$

Thus, the image is real (positive d_i), inverted (negative M), and reduced by a factor of $\frac{1}{2}$ (that is, one-half as tall as the object).

B. Spherical Lenses

The shapes of biconvex and biconcave spherical lenses are illustrated in ● Fig. 31.3.* A radius of curvature is defined for each spherical surface, but only the focal points (one for each spherical surface) are needed for ray diagrams.

A convex lens is called a **converging lens** because rays parallel to the principal axis converge at the focal point. A concave lens is called a **diverging lens** because rays parallel to the principal axis appear to diverge from the focal point.

As with spherical mirrors, the characteristics of the images formed by spherical lenses can be determined graphically or analytically. The chief (1) and parallel (2) rays for the graphical method are illustrated in the ray diagrams in ● Fig. 31.4.

In the case of a convex lens (Fig. 31.4a), the chief ray (1) through the center of the lens passes straight through. A ray parallel (2) to the principal axis is refracted in such a way that it goes through the focal point on the far side of the lens. Also, a focal ray (3) through the near focal point is refracted by the lens, so it leaves parallel to the axis.

In the case of a concave lens (Fig. 31.4b), the chief ray (1) still goes straight through the center of the lens. The ray parallel (2) to the principal axis is refracted upward so that it appears to have passed through the focal point on the object side of the lens. The focal ray (3), which is headed for the focal point on the far side of the lens, is refracted so that it leaves parallel to the principal axis.

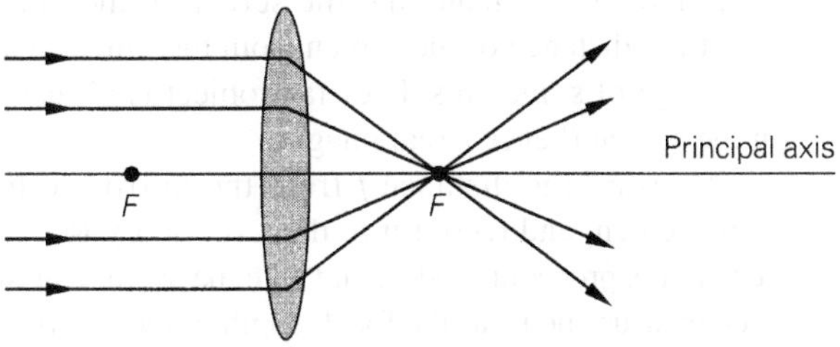

(a) Biconvex (Converging) Lens

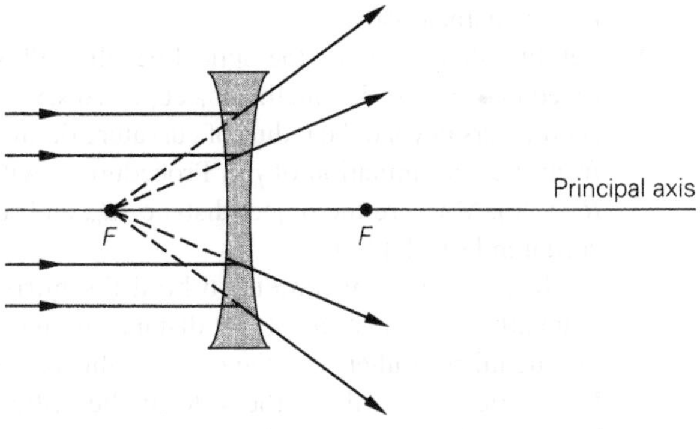

(b) Biconcave (Diverging) Lens

Figure 31.3 Spherical lenses. (a) A biconvex, or converging, lens and (b) a biconcave, or diverging, lens showing the refraction of parallel incident rays.

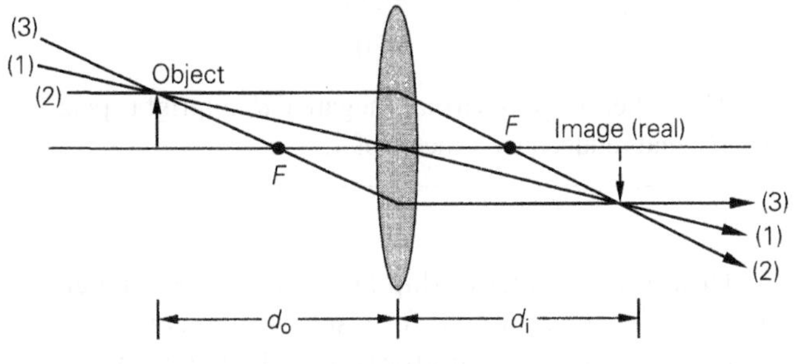

(a) Biconvex Lens

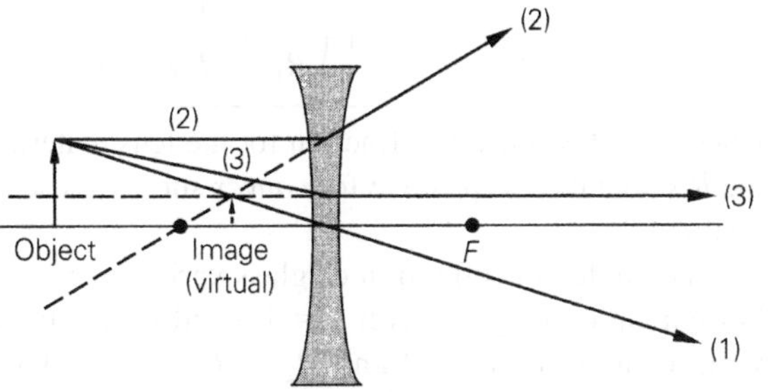

(b) Biconcave Lens

Figure 31.4 Lens ray diagrams. Examples of the ray diagram method for determining the image characteristics for (a) a biconvex, or converging, lens and (b) a biconcave, or diverging, lens.

*The bi- indicates two surfaces, for example, biconvex—two convex surfaces.

If the image is formed on the side of the lens opposite to the object, it is real and can be observed on a screen. However, if the image is on the same side of the lens as the object, it is virtual and cannot be seen on a screen.

The spherical **thin-lens equation** and **magnification factor** for analytically determining the image characteristics are identical to the equations for spherical mirrors (Eqs. 31.2 and 31.3). The sign convention is also similar (see Table 31.1). It should be noted that this lens equation applies only to *thin* lenses.

Example 31.2 An object is placed 30 cm from a biconcave lens with a focal length of 10 cm (corresponding to the case in Fig. 31.4b). Determine the image characteristics analytically.

Solution With $d_o = 30$ cm and $f = -10$ cm (negative by convention for a concave lens), using Eq. (31.2b) yields

$$d_i = \frac{d_o f}{d_o - f} = \frac{(30\text{ cm})(-10\text{ cm})}{30\text{ cm} - (-10\text{ cm})}$$

$$= \frac{-300\text{ cm}}{40\text{ cm}} = \frac{-30\text{ cm}}{4} = -7.5\text{ cm}$$

Then

$$M = -\frac{d_i}{d_o} = \frac{-(-30/4)}{30} = +\frac{1}{4}$$

Thus, the image is virtual (negative d_i), upright (positive M), and reduced by a factor of $\frac{1}{4}$.

However, the relationship between the focal length and the radius of curvature for a spherical lens is not as simple as for a spherical mirror (Eq. 31.1). For a lens, the focal length is given by what is known as the *lensmaker's equation:*

$$\frac{1}{f} = (n-1)\left(\frac{1}{R_1} + \frac{1}{R_2}\right) \qquad \textbf{(31.4)}$$

where n is the index of refraction for the lens material and the R's are taken as positive for *convex* surfaces. (See your textbook.)

The index of refraction of glass varies, $n = 1.5$–1.7. For example, for glass with $n = 1.5$ and symmetric converging lenses ($R_1 = R$ and $R_2 = R$), Eq. 31.4 yields $f = R$.* Keep in mind, however, that the focal length of a lens depends, in general, on the R values, which can be different, as well as on n. In computations, the experimentally determined value of f will be used.

*For f to be equal to $R/2$ for a symmetric lens, as may be for a spherical mirror, requires $n = 2$, which is greater than the index of refraction of glass.

EXPERIMENTAL PROCEDURE

Note: Handle optical pieces by the edges. Do not touch surfaces.

A. Spherical Mirrors

Concave Mirror

1. **(a)** Construct a ray diagram for a concave mirror with an object located at its focal point. (Drawing provided in the Laboratory Report.) It should be observed from the diagram that the reflected rays are parallel. In this case, we say that the rays "converge" at infinity or that the image is formed at infinity.

   Inversely, rays coming from an object at infinity converge to form an image at the focal point or in the focal plane (the plane perpendicular to the optic axis).

   (b) In the open area at the lower right corner of the Laboratory Report sheet, construct a ray diagram with several rays parallel to the optic axis to show that they converge at f.

   (c) Using the spherical-mirror equation, determine the image distance for an object at infinity (∞).

2. This focal property makes possible the experimental determination of the focal length of the mirror. An object, a great distance from the mirror, is essentially at infinity relative to the dimensions of the mirror.

   Take the mirror and screen to a window. Holding the mirror in one hand and the screen in the other, adjust the distance of the screen from the mirror until the image of some outside distant object is observed on the screen (hence a real image).†

   Measure the distance f from the mirror vertex to the screen, and record it in the Laboratory Report. Repeat this procedure twice, and take the average of the three measurements as the focal length of the mirror.

3. *Case 1: $d_o > R$.*

   (a) Sketch a ray diagram for an object at a distance slightly beyond R (that is, $d_o > R$), and note the image characteristics.

   (b) Set this situation up on the optical bench as illustrated in ● Fig. 31.5, with the object placed several centimeters beyond the radius of curvature (known from the determination of f in Procedure 2, with $R = 2f$). Measure the object distance d_o, and record it in Data Table 1.

   It is usually convenient to hold the mirror manually and adjust the object distance by moving the mirror rather than the object light source. Move the screen along the side of the optical bench until an image is observed on the screen.

†If a window is not available or it is a dark day, use Procedure 3 to determine f experimentally. In this case, show first that $d_i = d_o = R$ and $M = 1$. Then, d_i having been measured, the focal length is $f = d_i/2 = R/2$.

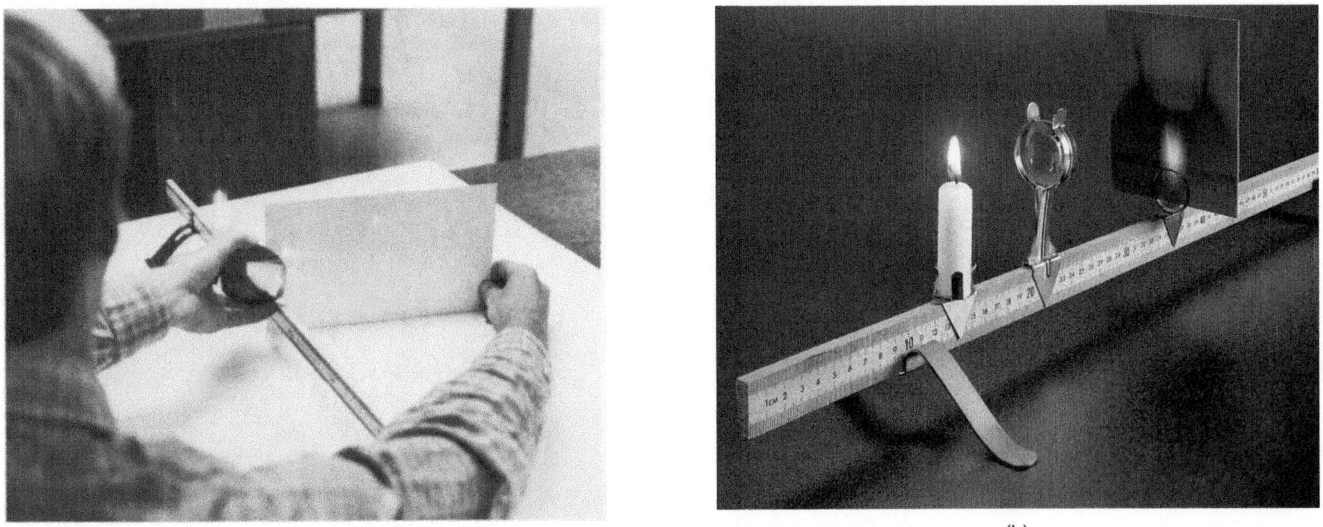

(a) (b)

Figure 31.5 Experimental arrangements. Arrangements for experimental procedures for (a) spherical mirrors and (b) spherical lenses. (© Cengage Learning.)

This is best observed in a darkened room. The mirror may have to be turned slightly to direct the rays toward the screen.

(c) Estimate the magnification factor M, and measure and record the image distance d_i.

(d) Using the mirror equation, compute the image distance and the magnification factor.

(e) Compare the computed value of d_i with the experimental value by computing the percent difference.

4. *Case 2:* $d_o = R$. Repeat Procedure 3 for this case.

5. *Case 3:* $f < d_o < R$. Repeat Procedure 3 for this case.

6. *Case 4:* $d_o < f$. Repeat Procedure 3 for this case.

Convex Mirror

7. Sketch ray diagrams for objects at (1) $d_o > R$, (2) $f < d_o < R$, and (3) $d_o < f$, and draw conclusions about the characteristics of the image of a convex mirror. Experimentally verify that the image of a convex mirror is virtual (that is, try to locate the image on the screen).

B. Spherical Lenses

Convex Lens

8. (a) Sketch a ray diagram for a convex lens with the object at its focal point. As with the concave mirror (Procedure 1), the image is formed at infinity.

(b) Using the lens equation, determine the image characteristics for an object at infinity.

(c) Experimentally determine the focal length of the lens by a procedure similar to that used for the concave mirror. (The lens may be placed in a lens holder and mounted on a meterstick.)*

*In general, for a lens, $f \neq \frac{R}{2}$. However, it can be shown for the case of $d_i = d_o$ that $d_o = 2f$. See Question 4 at the end of the experiment.

9. Repeat the four cases for the lens as was done for the concave mirror in Procedures 3 to 6, with R replaced by $2f$ (see Fig. 31.5). Record in Data Table 2. It is initially instructive to move the lens continuously toward the object light source (decreasing d_o) from a $d_o > 2f$ and to observe the image on the screen, which also must be moved continuously to obtain a sharp image. In particular, notice the change in the size of the image as d_o approaches f.

Concave Lens

10. Repeat the procedures carried out for the convex mirror in Procedure 7 for the concave lens, with R replaced by $2f$.

11. It is possible to determine the focal length of a concave lens experimentally by placing it in contact with a convex lens so as to form a lens combination. The combination forms a real image. If two lenses of focal lengths f_1 and f_2 are placed in contact, the lens combination has focal length f_c, given by

$$\boxed{\frac{1}{f_c} = \frac{1}{f_1} + \frac{1}{f_2}} \qquad \textbf{(31.5)}$$

Place the concave lens in contact with the convex lens (convex surface to concave surface) in a lens holder, and determine the focal length of the lens combination by finding the image of a distant object as in f_c, Procedure 8. Record in the Laboratory Report.

Using Eq. 31.5 with the focal length of the convex lens determined in Procedure 8, compute the focal length of the concave lens.

Name ______________________ Section ______________ Date ______________
Lab Partner(s) ____

EXPERIMENT 31

Spherical Mirrors and Lenses

TI *Laboratory Report*

A. Spherical Mirrors

Concave Mirror: Ray diagrams

$d_o = f$

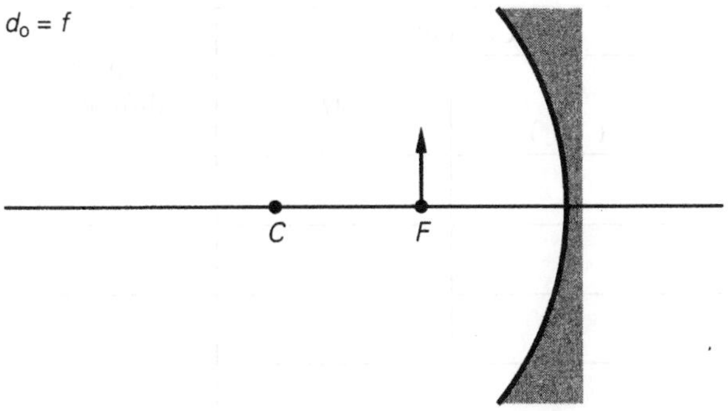

$f < d_o < R$

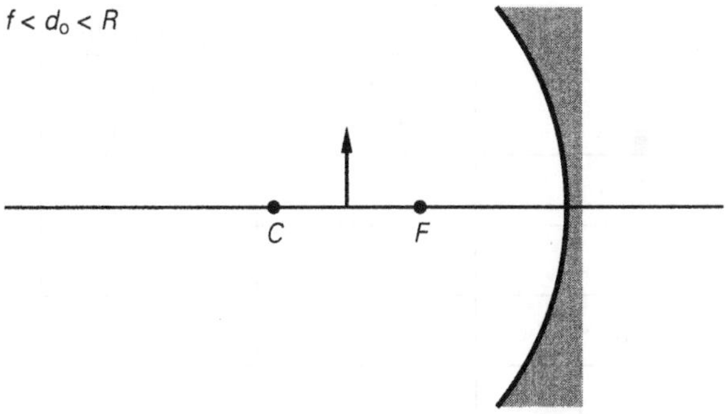

$d_o > R$

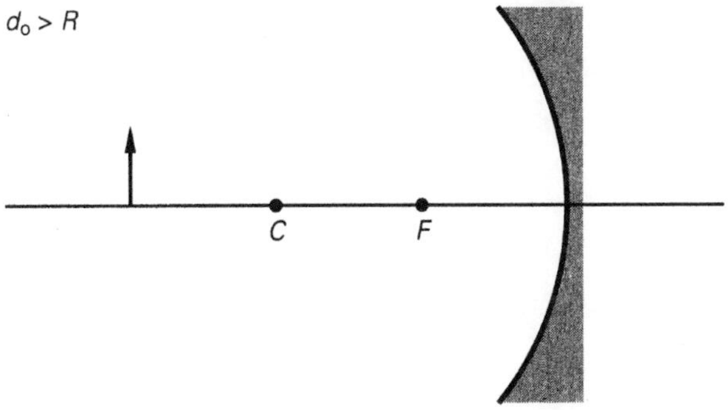

$d_o < f$

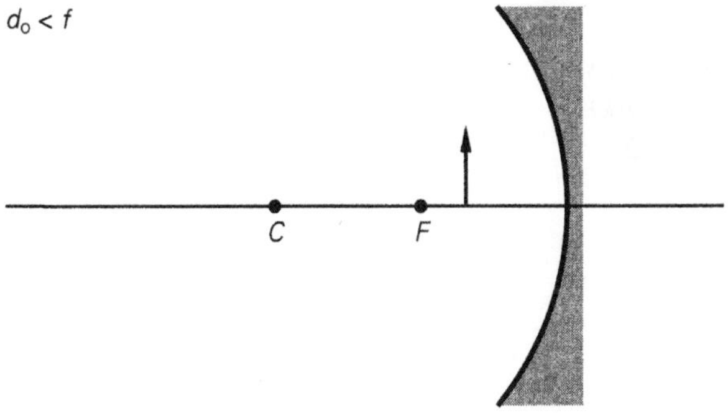

$d_o = R$

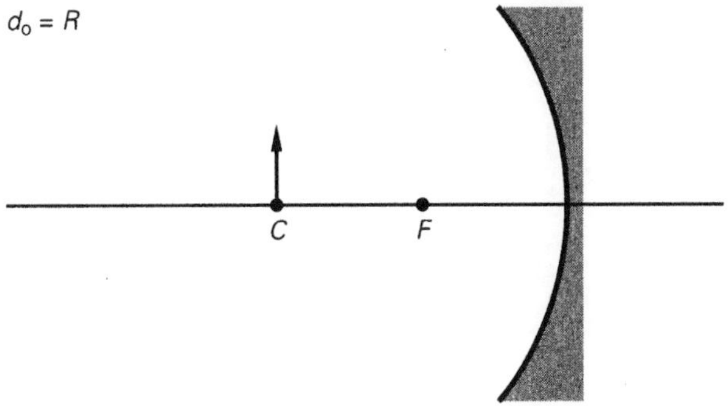

Don't forget units

(continued)

Calculation of d_i for object at ∞

Experimental focal length f ______________

____

____

Average ______________

DATA TABLE 1

Purpose: To determine the image distance and magnification (concave mirror).

	Experimental			Computed		d_i percent difference
	d_o ( )	d_i ( )	M factor (estimated)	d_i ( )	M	
$d_o > R$						
$d_o = R$						
$f < d_o < R$						
$d_o < f$						

Calculations
(show work)

Name ______________________ Section ______________ Date ______________

Lab Partner(s) ____

EXPERIMENT 31 **Spherical Mirrors and Lenses** ***Laboratory Report***

Convex mirror: Ray diagrams

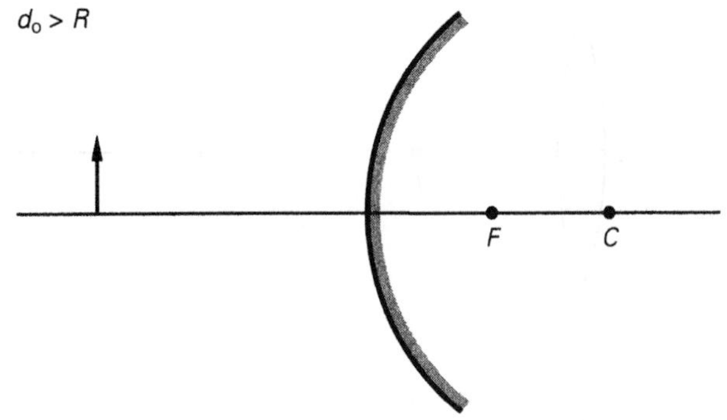

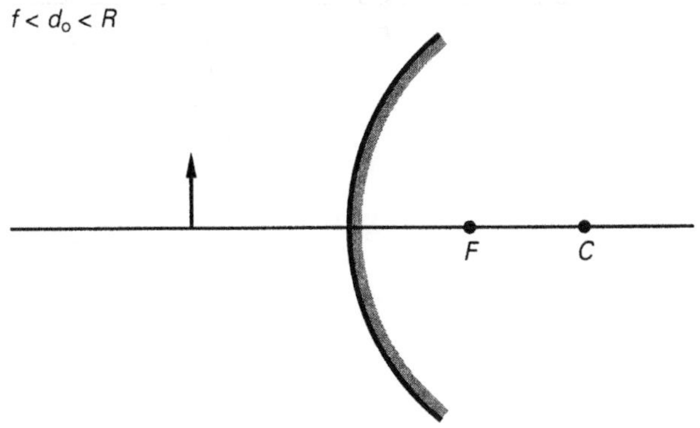

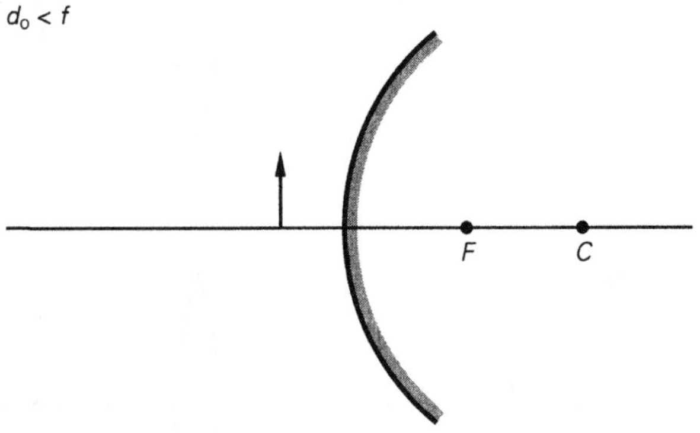

Conclusions

(continued)

B. Spherical Lenses

Convex lens: Ray diagrams

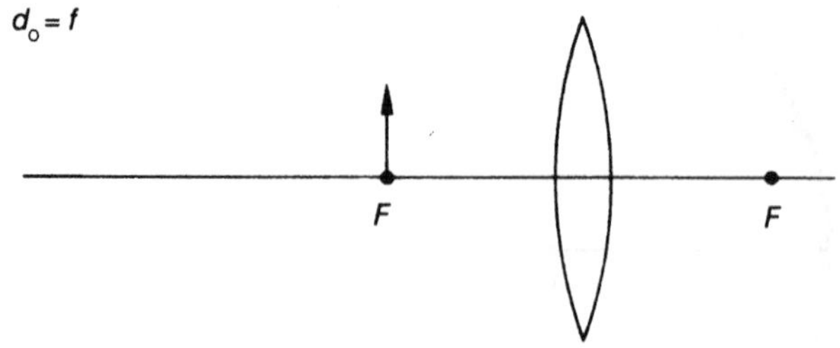

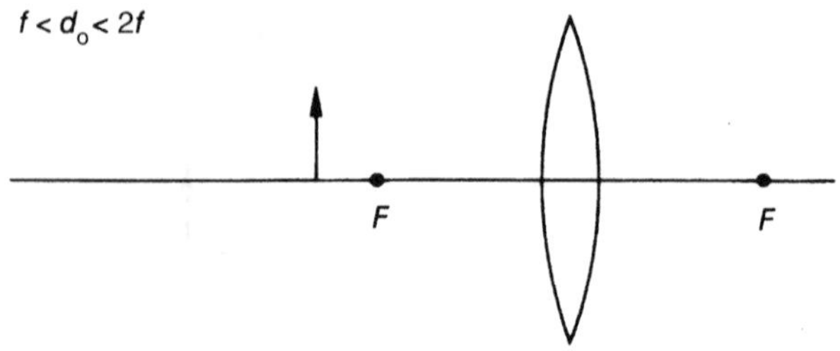

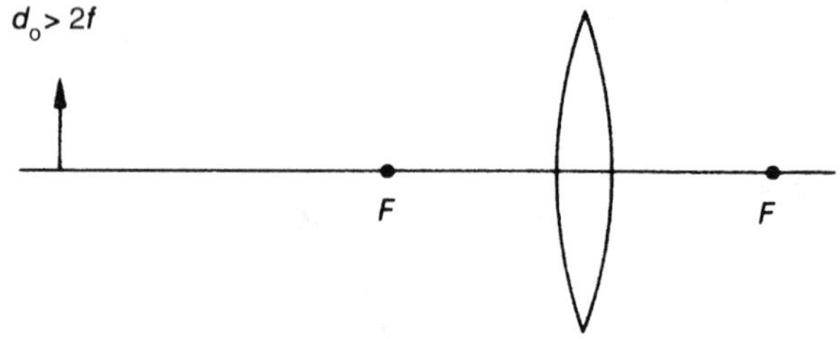

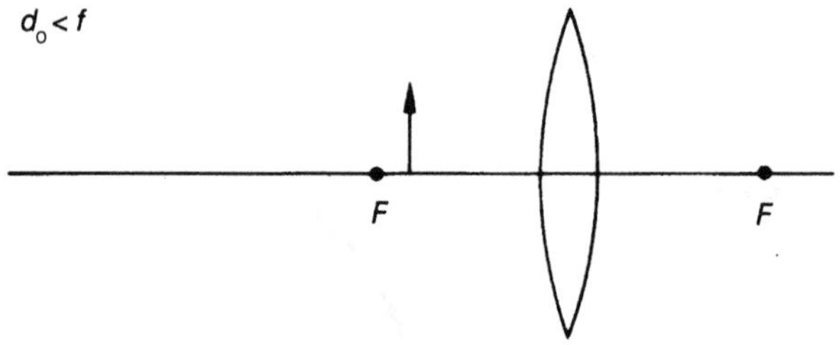

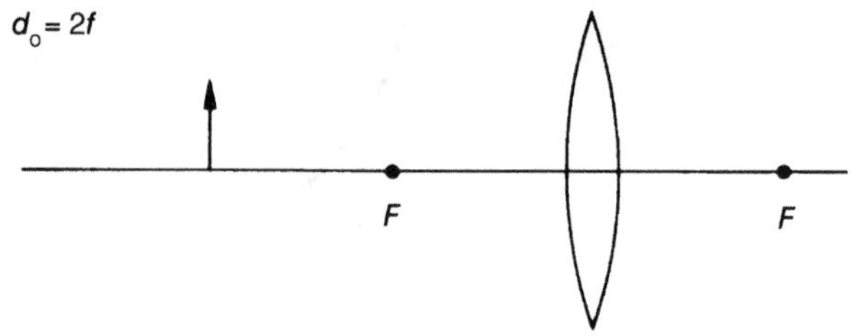

Calculation of d_i for object at ∞

Experimental focal length f ________________

____

____

Average ________________

Name ______________________ Section ______________ Date ______________

Lab Partner(s) ____

DATA TABLE 2

Purpose: To determine the image distance and magnification (convex lens).

	Experimental			Computed		d_i percent difference
	d_o ( )	d_i ( )	M factor (estimated)	d_i ( )	M	
$d_o > 2f$						
$d_o = 2f$						
$f < d_o < 2f$						
$d_o < f$						

Calculations
(show work)

(continued)

Concave lens: Ray diagrams

$d_o > 2f$

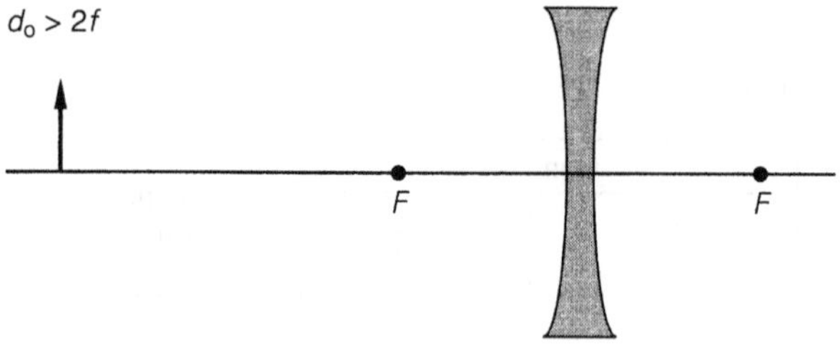

Conclusions

$f < d_o < 2f$

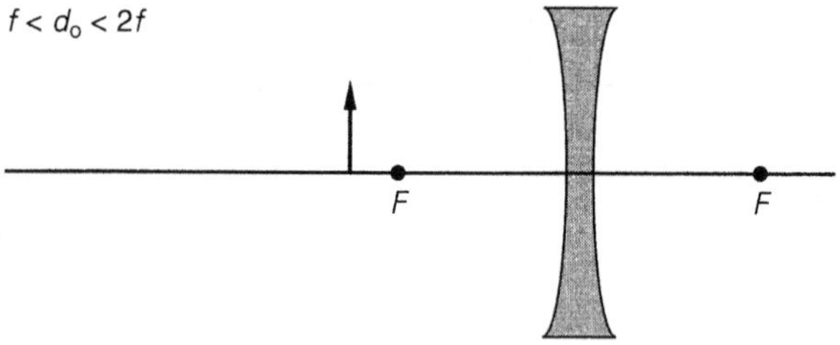

$d_o < f$

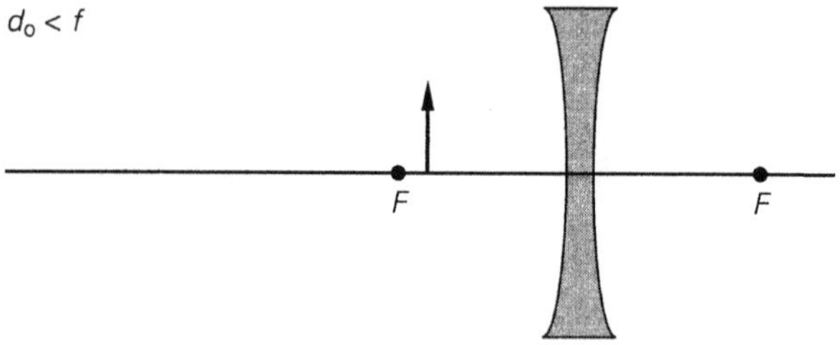

Focal length determination:

f_c, focal length of the combination ________________

f, focal length of convex lens ________________

f, focal length of concave lens ________________

TI QUESTIONS

1. A plane mirror essentially has a radius of curvature of infinity. Using the mirror equation, show that (a) the image of a plane mirror is always virtual, (b) the image is "behind" the mirror the same distance as the object is in front of the mirror, and (c) the image is always upright.

Name ______________________ Section ______________ Date ______________

Lab Partner(s) ____

2. Show that the magnification factor for a mirror or lens $M = d_i / d_o$ (sign convention omitted) is the lateral magnification, or the ratio of the height (lateral size) of the image to that of the object. (*Hint:* Draw a ray diagram.)

3. Explain what characteristics make convex spherical mirrors applicable for store monitoring and concave spherical mirrors applicable as flashlight reflectors.

4. Prove that for a converging lens, for the case $d_i = d_o$, that $d_i = d_o = 2f$.

(continued)

5. Using the thin-lens equation and the magnification factor, show that for a spherical diverging lens the image of a real object is always virtual, upright, and reduced. Does the same apply for a spherical diverging mirror?

6. (*Optional*) (a) Using the experimental value of f for the biconvex converging lens and $n = 1.5$, compute the radius of curvature of the lens's surfaces using the lensmaker's equation. (The radius of curvature for each surface is the same.)

(b) A student incorrectly assumes that $f = R/2$ for the lens and computes f using the value of R found in part (a). Compare this computed value of f with the experimental value.

(c) The index of refraction of the lens could have a different value (n of glass varies generally from 1.5 to 1.7). Would this make a difference? Explain.

Name ______________ Section ______________ Date ______________

Lab Partner(s) ______________

EXPERIMENT 32

TI Polarized Light

CI Malus's Law

TI *Advance Study Assignment*

Read the experiment and answer the following questions.

1. Is the plane of polarization of a polarizing polymer sheet in the same direction as the molecular chain orientation? Explain.

2. What is the condition for optimum polarization by reflection? Is the polarization angle the same for every material? Explain.

3. Describe how light can be polarized by refraction. Would two images be formed? Explain.

4. Why is sky light partially polarized, and why does it appear blue?

(continued)

5. What is meant by optical activity?

6. Describe the principle of optical stress analysis.

7. On a wristwatch or calculator, why are some portions of an LCD light and other portions dark?

CI Advance Study Assignment

Read the experiment and answer the following questions.

1. What is an analyzer? How is it different from and how is it similar to a polarizer?

2. What is the range of intensities for light passing through a polarizer and an analyzer? On what does this depend?

EXPERIMENT 32

TI Polarized Light
CI Malus's Law

OVERVIEW

Experiment 32 examines the polarization of light, but the TI and CI procedures differ in focus. The TI procedure examines the plane of polarization and illustrates some methods of polarization: reflection, refraction, and crystal double refraction, along with optical activity. The CI procedure focuses on Malus's law and examines the intensity of light transmitted through a polarizer and an analyzer.

INTRODUCTION AND OBJECTIVES

When speaking of polarized light, polarizing sunglasses usually come to mind, as this is one of the most common applications of polarization. However, few people understand how such sunglasses reduce "glare."

Since the unaided human eye cannot distinguish between polarized light and unpolarized light, we are not normally aware of the many instances of polarized light around us. Bees, on the other hand, with their many-faceted eyes, can detect polarized light and use scattered polarized sunlight in navigating.

Although the unaided human eye cannot detect polarized light, with a little help, polarization can be investigated experimentally. This is the purpose of the experiment.

TI OBJECTIVES

After performing this experiment and analyzing the data, you should be able to:

1. Explain what the polarization of light means.
2. Describe several means by which light can be polarized.
3. Explain some practical applications of polarized light.

CI OBJECTIVES

The purpose of the CI activity is to investigate the transmission of light through two polarizer filters (a polarizer and an analyzer) as a function of the angle θ between the planes of polarization. A light sensor is used to measure the intensity of the transmitted light. After performing this experiment and analyzing the data, you should be able to:

1. Explain the polarization of light using a polarizer and an analyzer.
2. Describe the intensity of light transmission through a polarizer and an analyzer from complete transmission to crossed polaroids.

TI EXPERIMENT 32

Polarized Light

TI EQUIPMENT NEEDED

- 3 polarizing sheets
- Polarizing sunglasses
- (*Optional*) Light meter
- Lamp (and converging lens for parallel beam if needed)
- Protractor
- 6–8 glass microscope slides
- Glass plate
- Tripod stand (open ring top)
- Calcite crystal
- Mica sheet
- Cellophane tape (*not* polymer tape)
- Lucite or other plastic pieces (for example, U-shaped hook or hollow triangle)
- LCD (as on a wristwatch or hand calculator)

TI THEORY

Light, like all electromagnetic radiation, is a **transverse wave**. That is, the directions of the vibrating electric and magnetic field vectors are at right angles to the direction of propagation, as illustrated schematically in ● TI Fig. 32.1. (If the vector vibrations were parallel to the direction of propagation, light would be a *longitudinal wave*.) The phenomenon of *polarization* is firm evidence that light is a transverse wave.

The term **polarization** refers to the orientation of the vibrating vectors of electromagnetic radiation. Light from an ordinary light source consists of a large number of waves emitted by the atoms or molecules of the source. Each atom produces a wave with its own orientation of the $\vec{\mathbf{E}}$ (and $\vec{\mathbf{B}}$) vibration corresponding to the direction of the atomic vibration.

However, with many atoms, all directions are possible. The result is that the emitted light is **unpolarized**. The vibration vectors are randomly oriented, with all directions equally probable. This is represented schematically in ● TI Fig. 32.2a, which views the $\vec{\mathbf{E}}$ vectors along the axis of propagation. ($\vec{\mathbf{B}}$ vectors are not represented.)

If for some reason the light vectors become preferentially oriented, the light is **partially polarized** (TI Fig. 32.2b). Should there be only one direction of vibration for the $\vec{\mathbf{E}}$ vectors (TI Fig. 32.2c), the light is then **linearly polarized**. This is sometimes called **plane polarized** or simply polarized light. The direction of vibration of the $\vec{\mathbf{E}}$ vector defines the plane, or direction, of polarization.

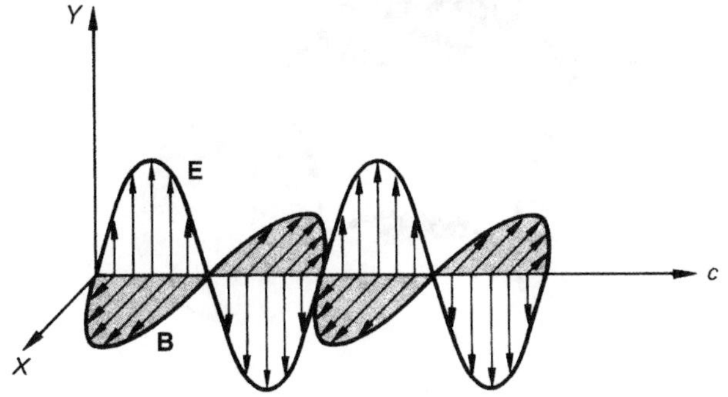

TI Figure 32.1 Electromagnetic wave. An illustration of an electromagnetic wave. The electric and magnetic field vectors ($\vec{\mathbf{E}}$ and $\vec{\mathbf{B}}$) vibrate at right angles to each other and perpendicularly to the direction of propagation (transverse).

The polarization of light may be effected by several means: (a) selective absorption, (b) reflection, (c) refraction, and (d) scattering. Let's take a look at these.

A. Polarization by Selective Absorption

Certain crystals are doubly refracting, or exhibit **birefringence**. That is, they have different indices of refraction in different directions, and light passing through the crystal is separated into two components or rays. The rays are also linearly polarized. (Birefringence is discussed in more detail in Section C.)

Some birefringent crystals, such as tourmaline, exhibit the interesting property of absorbing one of the polarized components more than the other—selective absorption, so to speak. This property is called **dichroism**.

Another dichroic crystal is quinine sulfide periodide (commonly called *herapathite*, after W. Herapath, an English physician who discovered its polarizing properties in 1852). This crystal was central in the development of modern polarizers. Around 1930, Edwin H. Land, an American scientist, found a way to align tiny dichroic crystals in sheets of transparent celluloid. The result was a thin sheet of polarizing material that was given the name *polaroid*.

Better polarizing films have been developed using polymer materials. During the manufacturing process, this kind of film is stretched in order to align the long molecular chains of the polymer.

With proper treatment, the outer (valence) molecular electrons can move along the oriented chains. As a result, the molecules readily absorb light with $\vec{\mathbf{E}}$ vectors parallel to the oriented chains and transmit light with the $\vec{\mathbf{E}}$ vectors perpendicular to the chains.*

*In a common analogy, plane polarization is likened to a picket fence. The slats of the fence represent the oriented chains, and the light $\vec{\mathbf{E}}$ vector passes through vertically, parallel to the pickets. In actuality, however, the $\vec{\mathbf{E}}$ vector passes perpendicular to the molecular chain "pickets."

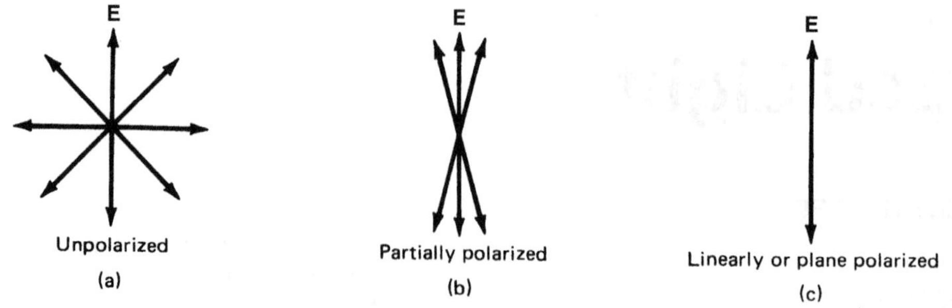

TI Figure 32.2 Polarization. An illustration of the polarization of light as denoted by the $\vec{E}$ vector viewed along the axis of propagation.

The direction perpendicular to the oriented molecular chains is commonly called the **transmission axis**, **plane of polarization**, or **polarization direction**. Hence, when unpolarized light falls on a polarizing sheet (polarizer), polarized light is transmitted. This is illustrated in ● TI Fig. 32.3.

The polarization of light may be analyzed (detected) by means of another polarizer, which acts as an analyzer (TI Fig. 32.3). The magnitude of the component of the $\vec{E}$ vector parallel to the transmission axis of the analyzer is $E_o \cos \theta$. Since the intensity varies as the square of the amplitude, the transmitted intensity of light through the analyzer is

$$\boxed{I = I_o \cos^2 \theta} \qquad \textbf{(TI 32.1)}$$

where I_o is the maximum intensity of light through the analyzer and θ is the angle between the transmission axes of the polarizer and analyzer. If $\theta = 90°$, we have a condition of cross-polarizers, and no light is transmitted through the analyzer.[†]

B. Polarization by Reflection

When light is incident on a material such as glass, some of the light is reflected and some is transmitted. The reflected light is usually partially polarized, and the degree of polarization depends on the angle of incidence. For angles of incidence of 0° and 90° (grazing and normal angles), the reflected light is unpolarized. However, for intermediate angles, the light is polarized to some extent.

Complete polarization occurs at an optimum angle called the **polarization angle** $\boldsymbol{\theta_p}$ (● TI Fig. 32.4). This occurs when the reflected and refracted beams are 90° apart, and θ_p is specific for a given material.

Referring to TI Fig. 32.4, since $\theta_1 = \theta_p$, then $\theta_1 + 90° + \theta_2 = 180°$, and $\theta_1 + \theta_2 = 90°$ or $\theta_2 = 90° - \theta_1$. By Snell's law (Experiment 30),

[†]The expression is known as *Malus's law*, after E. I. Malus (1775–1812), the French physicist who discovered it.

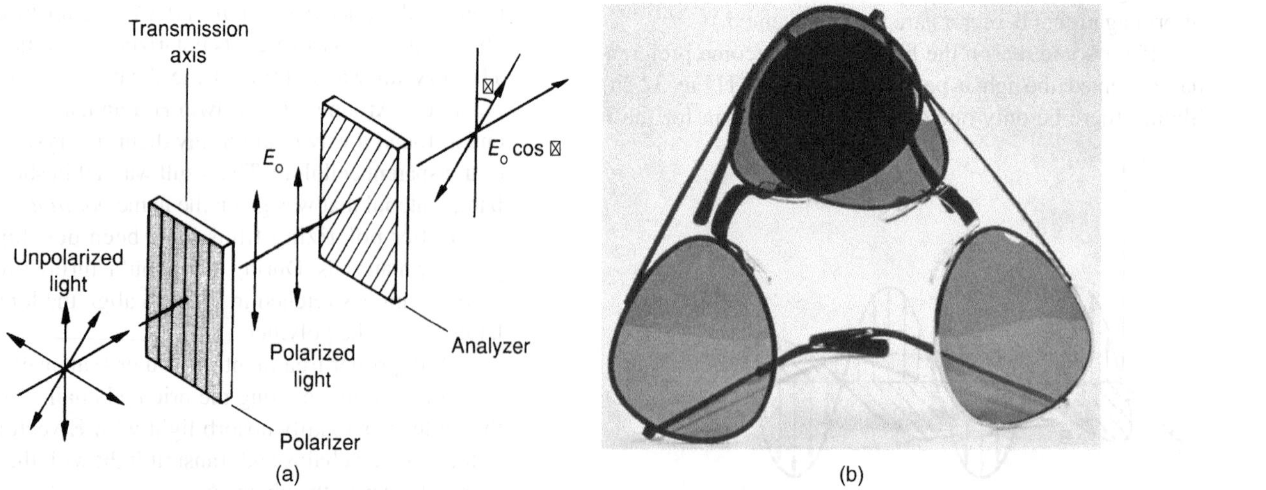

TI Figure 32.3 Polarizer and analyzer. (a) The transmission axis (or plane of polarization or polarization direction) is perpendicular to the oriented molecular chains. When the transmission axes of the polarizer and analyzer are not parallel, less light is transmitted. (b) For "cross polaroids" ($\theta = 90°$), little light (ideally no light) is transmitted, as shown in the "crossed" polarizing sunglasses lenses. (© Cengage Learning.)

$$\frac{\sin\theta_1}{\sin\theta_2} = n$$

where n is the index of refraction of the glass and $\sin\theta_2 = \sin(90° - \theta_1) = \cos\theta_1$. Thus

$$\frac{\sin\theta_1}{\sin\theta_2} = \frac{\sin\theta_1}{\cos\theta_1} = \tan\theta_1 = n$$

or

$$\boxed{\tan\theta_p = n} \qquad \textbf{(TI 32.2)}$$

This expression is sometimes called **Brewster's law,*** and θ_p is called **Brewster's angle**.

TI Example 32.1 A glass plate has an index of refraction of 1.48. What is the angle of polarization for the plate?

Solution With $n = 1.48$,

$$\theta_p = \tan^{-1}(1.48) = 56°$$

Notice from TI Fig. 32.4 that the reflected beam is horizontally polarized. Sunlight reflected from water, metallic surfaces (for example, from a car), and the like is partially polarized. If the surface is horizontal, the reflected light has a strong horizontal component. This fact is used in polarizing sunglasses. The transmission axis of the lenses is oriented vertically so as to absorb the reflected horizontal component and reduce the glare or intensity.

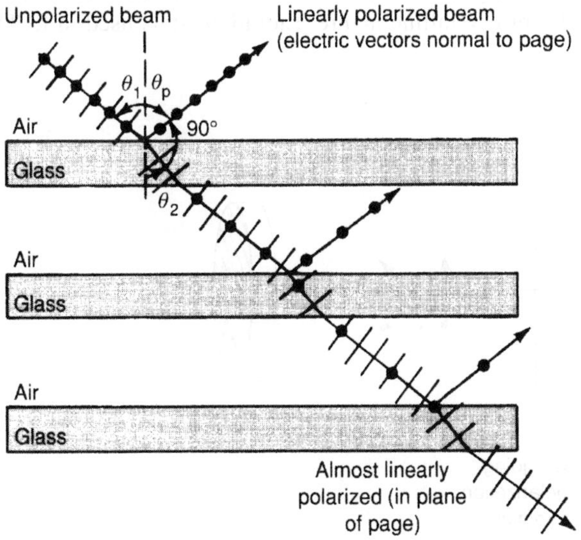

TI Figure 32.4 Polarization by reflection. Maximum polarization occurs for a particular polarization angle θ_p, which depends on the index of refraction of the material. Note that the transmitted beam is partially polarized.

*After its discoverer, David Brewster (1781–1868), a Scottish physicist.

Also notice that the refracted beam is partially polarized. If a stack of glass plates is used, the transmitted beam becomes more linearly polarized.

C. Polarization by Refraction

In a sense, the preceding case of a polarized transmitted beam from a stack of glass plates might be thought of as polarization by refraction since the refracted beam is polarized. However, polarization by refraction generally refers to the *double refraction* exhibited by some crystals.

In an optically isotropic medium, such as glass, light travels with the same speed in all directions. As a result, the material is characterized by a single index of refraction. In certain anisotropic crystals, however, the speed of light is not the same in all directions. The crystal calcite ($CaCO_3$, Iceland spar), for example, exhibits double refraction, or *birefringence*, and is characterized by two indices of refraction.

When an unpolarized beam of light enters a calcite crystal, it splits into two polarized rays with polarizations in mutually perpendicular directions (● TI Fig. 32.5). One beam is called the ordinary (o) ray, and the other the extraordinary (e) ray. Because of this property, when something (that is, a printed line) is viewed through a calcite crystal, a double image is seen, corresponding to the two emergent rays.

D. Polarization by Scattering

Scattering is the process of a medium absorbing light, then reradiating it. For example, if light is incident on a gas, the electrons in the gas atoms or molecules can absorb and reradiate (scatter) part of the light. In effect, the electrons absorb the light by responding to the electric field of the light wave. An electron can be thought of as a small antenna; it radiates light in all directions, *except* along its axis of vibration. Hence the scattered light is partially polarized.

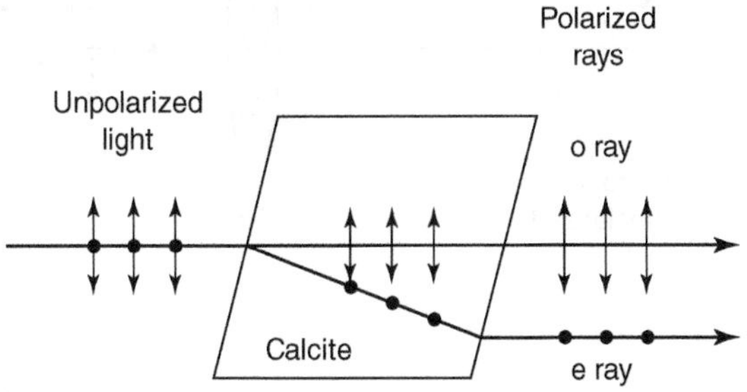

TI Figure 32.5 Polarization by double refraction, or birefringence. An unpolarized beam entering a crystal is split into two polarized beams. (Dots indicate electric field vectors oscillating normal to the page.)

Such scattering of sunlight occurs in the atmosphere and is known as **Rayleigh scattering**.* The condition for interaction and scattering is that the size d (diameter) of the molecules be much less than the wavelength λ of the light, $d \ll \lambda$. The intensity of the scattering then varies as $1/\lambda^4$. This condition is satisfied for O_2 (oxygen) and N_2 (nitrogen) molecules, the main gases in the atmosphere.

Sunlight incident on the atmosphere is white light with a spectrum of wavelength components (or colors). As a result of the $1/\lambda^4$ scattering relationship, the blue end (shorter wavelength) of the spectrum is scattered more than the red end (longer wavelength). The blue light is scattered and rescattered. When looking at the sky, we see this scattered light, and as a result, the sky appears to be blue. Hence the blue "sky light" we see is partially polarized, even though we can't visually detect it (without a little help).

E. Optical Activity

Certain substances have the property of being able to rotate the plane of polarization of a beam of polarized light. This rotation, called **optical activity**, is exhibited by crystalline mica, quartz, some sugars, and many long-chain molecular polymers. The principle is illustrated for an optically active crystal in ● TI Fig. 32.6.

Notice that the crystal essentially changes the direction of one of the vector components of the polarized light along one of its optical axes. The (vector) resultant is then rotated 90° to the polarization plane of the incoming light and transmitted through the "crossed" analyzer, which would not be the case without rotation.

*After Lord Rayleigh (1842–1919), the British physicist who described the effect.

In traveling through the crystal, the components travel at different speeds. Suppose the thickness of the crystal is such that the vertical component gains (or falls behind) by one-half wavelength compared with the horizontal component. The effect is a reversed vertical component and a 90° rotation of the plane of polarization, as shown in the figure.

Only for this particular wavelength of light and this particular crystal thickness (or an appropriate multiple thereof) will a 90° rotation occur, and only this wavelength (color) of light will be transmitted. As a result of nonuniform crystal thickness, a colored pattern is observed when the crystal is viewed through crossed polarizers.

F. Optical Stress Analysis

An important application of polarized light is optical stress analysis of transparent materials. Glasses and plastics are usually optically isotropic. If polarized light is transmitted through an isotropic material and viewed with a crossed polarizer, the transmitted light intensity is minimal.

However, when many materials are mechanically stressed, they become optically anisotropic (indices of refraction vary with direction), and the polarization of the transmitted polarized light is affected. Areas of strain may then be identified and studied through an analyzer (● TI Fig. 32.7). For example, improperly annealed glass may have internal stresses and strains that may later give rise to cracks.

G. Liquid Crystal Displays (LCDs)

LCDs, or liquid crystal displays, are now commonly used in wristwatches, hand calculators, and even gas pumps. A liquid crystal is a liquid in which the molecules have some order or crystalline nature. Liquid crystals used in LCDs

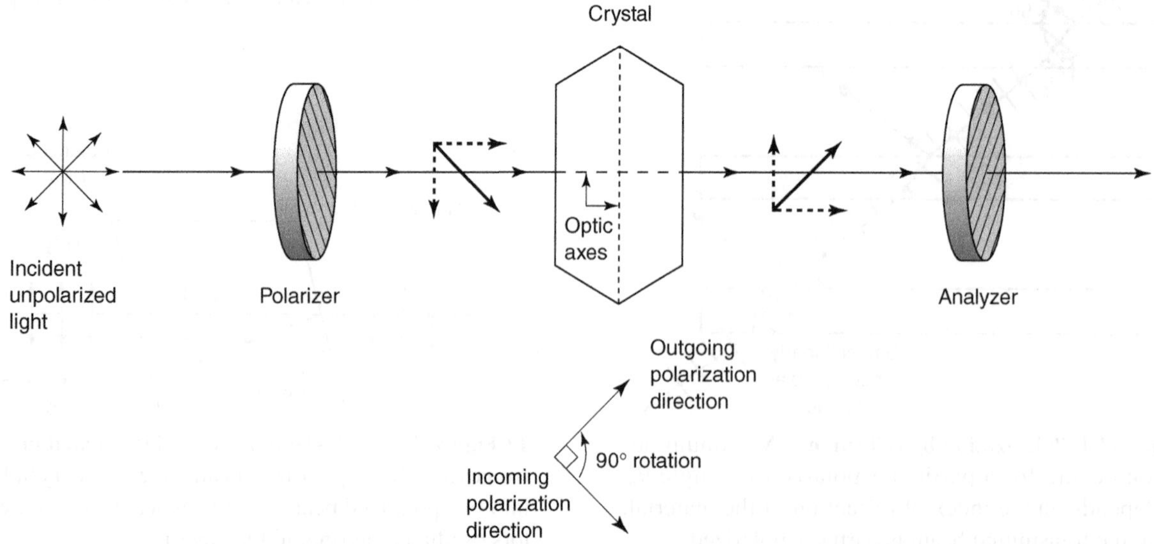

TI Figure 32.6 Optical activity. Certain substances have the property of being able to rotate the plane of polarization of a beam of linearly polarized light through 90°.

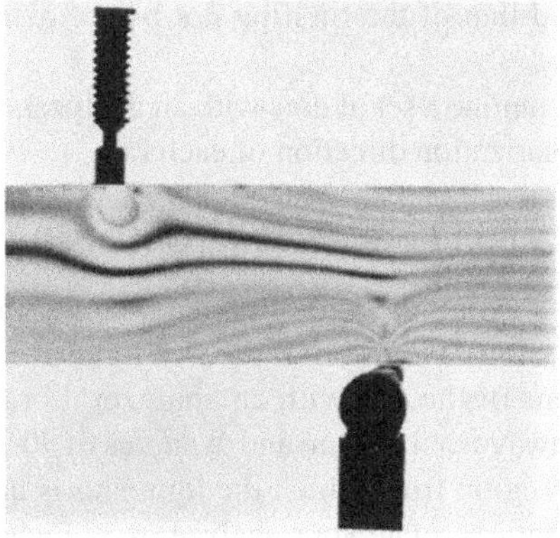

TI Figure 32.7 Optical stress analysis with polarized light. A transparent bar reveals stress concentrations near the support and loading point because of optical activity. (© Cengage Learning.)

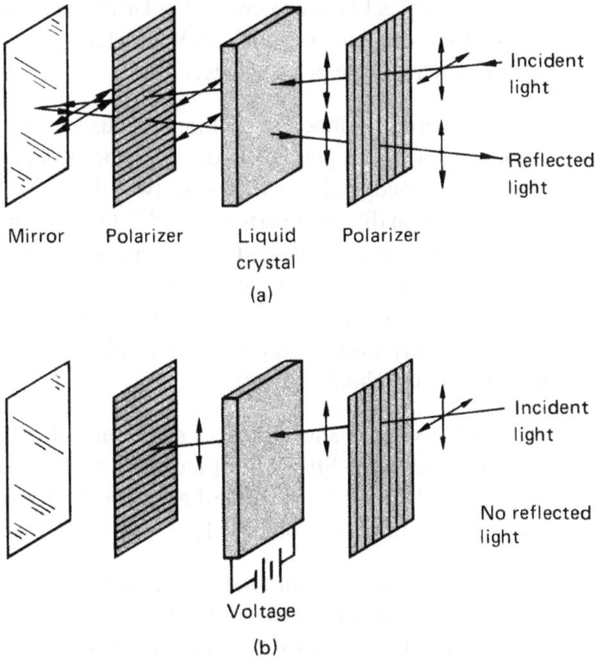

TI Figure 32.8 LCD (liquid crystal display). (a) A liquid crystal has the property of being able to "twist," or rotate, the plane of linearly polarized light by 90°. (b) When a voltage is applied to the crystal, this property is lost. With no light reemerging, the crystal appears dark.

have the ability to rotate, or "twist," the plane of polarization of polarized light. In a so-called *twisted nematic display*, a liquid crystal is sandwiched between crossed polarizing sheets and backed by a mirror (● TI Fig. 32.8).

Light falling on the surface of an LCD is polarized, twisted, reflected, and twisted again and then leaves the display. Hence, the display normally appears light because the doubly twisted, polarized light coming back from the surface is seen. However, when a voltage is applied to the crystal, the polarized light is not twisted the last time and hence is absorbed by the final polarizing sheet. Such dark regions of the crystal are used to form numbers and letters in the display.

TI EXPERIMENTAL PROCEDURE

A. Plane of Polarization (Transmission)

1. Inspect your polarizers to see whether the planes of polarization or transmission are indicated on them. If not, these planes need to be determined. (Polarizer and analyzer directions will be needed later.) One method is to use a pair of polarizing sunglasses that has a vertical plane of polarization.
   Determine the planes of polarization for your polarizers by observing the orientations of maximum and minimum transmissions through a polarizer and a polarizing sunglass lens. (Remember, sunglasses are vertically polarized.) Mark the planes of polarization of the polarizers by some means (for example, a wax pencil or pieces of tape).

2. Investigate the transmission through two polarizers as a function of the angle θ between the planes of polarization (TI Fig. 32.3). At what angles is the transmission estimated by brightness or intensity (*optional* lightmeter):
   (a) a maximum,
   (b) reduced by one-half, and
   (c) a minimum?
   Record these angles in the Laboratory Report, and show the theoretical prediction (using TI Eq. 32.1) of the angle for one-half transmission (that is, $I/I_o = 0.5$).

3. Orient two polarizers in a crossed position for minimum intensity. Place the third polarizer between these two. Viewing through the polarizer, rotate the middle one (keeping the outer two in a crossed orientation), and observe any intensity changes. As the middle polarizer is rotated, you should observe variations in the transmitted light intensity.
   Note the orientation of the plane of polarization of the middle polarizer (relative to the planes of the outer two) for maximum transmission. Make a sketch of the polarizers in the Laboratory Report, and indicate the planes of polarization for the polarizers for this condition.
   (*Hint:* Draw three polarizers of different sizes, and label the outer two as 1 and 2 and the middle polarizer as 3.) Is there more than one orientation of the middle polarizer for maximum transmission? (Rotate through 360°.) If so, are the angles between the polarization planes different? Explain in the Laboratory Report why this transmission through the outer crossed polarizers is observed.

B. Polarization by Reflection and Refraction

4. Using a general value of the index of refraction of glass as $n = 1.6$, compute in the Laboratory Report the reflection polarization angle θ_p.

5. (a) With a single glass microscope slide on the tripod stand positioned near the edge of the table, shine light on the slide at an angle of incidence equal to the computed θ_p. Observe the reflected light through an analyzer at the angle of reflection, and examine for polarization. Note the axis of polarization of the reflected light. Observe the reflected light through polarizing sunglasses, and comment.

Observe the light transmitted through the glass slide with an analyzer for any evidence of polarization.

(b) Observe the transmitted light for an increasing number of glass slides, and report and explain any observable differences.

C. Polarization by Crystal Double Refraction

6. Place the calcite crystal on some written or printed material and observe the double image. (The images may appear slightly fuzzy because of small defects in the crystal. Lay the crystal on the side that gives the clearest images.) Notice that when the crystal is rotated, the images move, one more than the other. Examine the images with an analyzer.

7. With a pencil or pen, make a linear series of small, heavy dots on a piece of paper. The line of dots should be long enough to extend beyond the edges of the crystal.

Placing the crystal on the line of dots, rotate the crystal. Notice that as the crystal is rotated, one of the dots of a double image remains relatively stationary and the second dot rotates about the first. The image of the nearly stationary dot is formed by the ordinary (o) ray and that of the rotating dot by the extraordinary (e) ray.

Examine a set of dots with an analyzer, and record the polarization direction of each ray.

D. Polarization by Scattering (Optional)

8. If it's a sunny day, go outside (with the instructor's permission) and observe the sky light from different portions of the sky with an analyzer. Look in directions away from the sun and at angles of 90°. Once you find a region from which the light shows appreciable polarization, rotate your analyzer to see whether there is any preferential direction of polarization. (Should it not be a sunny day, try this with your own polarizing sunglasses some fine day.)

E. Optical Activity

9. View a mica sheet between crossed polarizers. Rotate the analyzer, and note the change. What changes, the general pattern shape or its colors?

10. Form a pattern or symbol by sticking various layers of cellophane tape on a glass plate or slide. For example, try a "V" or wedge shape with one, two, three, etc., layers of tape in different parts of the "V." Observe the tape symbol between crossed polarizers. Rotate the analyzer. (You may wish to make letters or symbols, for example your school letters, with pieces of tape cut with a sharp knife or razor blade. Could you give the letters your school colors?)

11. Observe the various-shaped pieces of plastic between crossed polarizers. Stress the pieces by pulling or pushing on them (but not so hard as to break them). Can you explain what is observed?

12. View an LCD through an analyzer. Rotate the analyzer. Is the light coming from the lighted portion of the display polarized? (*Note:* You can do this at home using polarizing sunglasses.)

Name ______________________ Section ______________ Date ______________

Lab Partner(s) ____

TI EXPERIMENT 32

Polarized Light

TI Laboratory Report

A. Plane of Polarization (Transmission)

1. Transmission

Angle θ between polarizer planes

Maximum intensity ______________

One-half intensity ______________

Minimum intensity ______________

Calculation of angle for one-half intensity

2. Three polarizers

Explanation

Sketch

(continued)

B. Polarization by Reflection and Refraction

3. Polarization angle calculation

4. (a) Observations on reflected light

   (b) Observations on transmitted refracted light

C. Polarization by Crystal Double Refraction

Observations

Name ______________________ Section ______________ Date ______________

Lab Partner(s) ____

D. Optical Activity

Observations

TI QUESTIONS

1. In the polarization of light by reflection with a transparent material, would it be possible to have an optimum polarization angle θ_p less than 45°? Justify your answer. (*Hint:* Consider the definition of the index of refraction.)

2. At a sale of polarizing sunglasses, there is a half-price special on some glasses with lenses that have a horizontal polarization direction. Would you buy them? If not, why not?

(continued)

3. In the procedure using microscope slides, why was the polarization of the transmitted light more observable with an increasing number of slides?

4. The light coming from a liquid crystal display (LCD), as on a watch or calculator, is polarized. How could you conveniently show this to be the case using a commonly available item? (*Hint:* What is a common polarizing material or application of polarization?)

5. You hold two polarizing sheets in front of you and look through both of them. How many times would you see the sheets lighten and darken (a) if one of them were rotated through 360°, or one complete rotation, and (b) if both of them were rotated through one complete rotation in opposite directions at the same rate?

CI EXPERIMENT 32

Malus's Law

CI EQUIPMENT NEEDED

- Light sensor (PASCO CI-6504A)
- Rotary motion sensor (PASCO CI-6538)
- Polarization analyzer (PASCO OS-8533). (Polarizer with groove, plastic belt, and mounting screws are included in the kit.)
- Diode laser (PASCO OS-8525)
- Aperture bracket (PASCO OS-8534)
- Optics bench (PASCO OS-8515 or OS-8541)

CI THEORY

(See the methods of polarization in the TI Theory section.)

When unpolarized light is incident upon a polarizer, the transmitted light is reduced in intensity and linearly polarized, as shown in TI Fig. 32.3a. When this polarized light falls on a second polarizer (usually referred to as an analyzer), the transmitted intensity is given by Malus's law:

$$I = I_o \cos^2 \theta \qquad \textbf{(CI 32.1)}$$

Notice that the transmitted intensity can vary from complete transmission $(I = I_o)$ to no transmission $(I = 0)$ and can take on any intermediate value between the maximum and minimum, depending on the angle between the polarizing planes of the polarizer and analyzer. Here I_o is the maximum intensity of light through the analyzer when $\theta = 0°$.

In this experiment, we will investigate the relative orientations of polarizer and analyzer that produce maximum and minimum transmission. Light from a laser will be incident on a fixed polarizer. An analyzer placed in front of the polarizer will be rotated. Both the transmitted light intensity and the angular rotation of the analyzer will be measured simultaneously using sensors.

SETTING UP DATA STUDIO

1. Open Data Studio, and choose "Create Experiment."
2. The Experiment Setup window will open, and you will see a picture of the Science Workshop interface. There are seven channels to choose from. (Digital Channels 1, 2, 3, and 4 are the small buttons on the left; analog Channels A, B, and C are the larger buttons on the right, as shown in ● CI Fig. 32.1.)
3. Click on the Channel A button in the picture. A window with a list of sensors will open.
4. Choose the Light Sensor from the list, and press OK.
5. Connect the sensor to Channel A of the interface, as shown on the computer screen.
6. In the same window, set the Sample Rate to 20 Hz.
7. Now click on the Channel 1 button in the picture to access the list of sensors again.
8. Choose the Rotary Motion Sensor (RMS) from the list, and press OK.
9. Connect the RMS to Channels 1 and 2 of the interface, as shown on the computer screen.
10. On the same window, adjust the properties of the RMS as follows:

    First Measurements tab: select Angular Position, Ch 1 and 2 and select the unit of measure to be degrees. Deselect all others.

    Rotary Motion Sensor tab: set the Resolution to low (360 divisions/rotations), and set the Linear Scale to Large Pulley (Groove).

    Set the Sample Rate to 20 Hz.

    The Data list on the left of the screen should now have three icons: one for voltage, one for light intensity, and one for the angular position data.
11. Create a graph by dragging the "Voltage(ChA)" icon from the Data list and dropping it on the "Graph" icon on the displays list. A graph of voltage versus time will open in a window called Graph 1. This graph will be used later to adjust the light sensor gain.
12. Create a second graph by dragging the "Light Intensity(ChA)" icon and dropping it on the "Graph" icon on the displays list. A graph of light intensity versus time will open in a window called Graph 2.
13. Drag the "Angular Position" icon from the Data list, and drop it on top of the time axis of Graph 2. The time axis will change into an angular position axis. Graph 2 should now be a plot of light intensity versus angular position. ● CI Fig. 32.2 shows how the screen should appear after the setup is complete.

CI EXPERIMENTAL PROCEDURE

1. CI Figs. 32.3 through 32.6 show the equipment setup.
   (a) ● CI Fig. 32.3: The aperture disk is mounted on the aperture bracket holder. The light sensor is then mounted on the aperture bracket, behind the aperture disk, and connected to the interface.

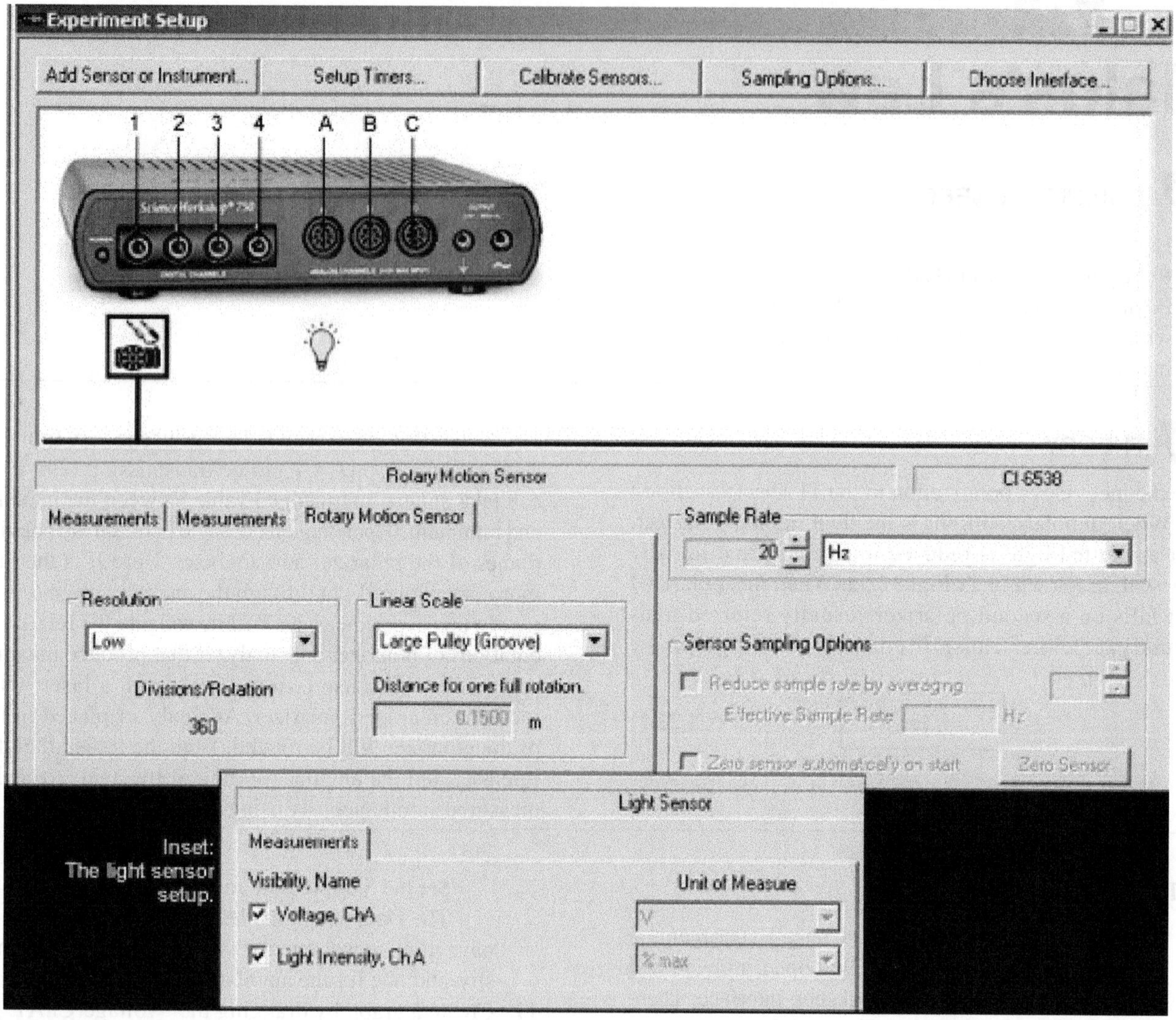

CI Figure 32.1 The Experiment Setup window. A light sensor will measure the intensity of the light that crosses the analyzer. The rotary motion sensor will measure the angular rotation of the analyzer with respect to the polarizer. (Reprinted courtesy of PASCO Scientific.)

(b) ● CI Fig. 32.4: Rotate the aperture disk so that the translucent mask covers the opening of the light sensor.

(c) ● CI Fig. 32.5a: Mount the polarizers into the holders. Mount the RMS bracket on the holder that has the polarizer with groove.

(d) ● CI Fig. 32.5b: The rotary motion sensor is mounted on the polarizer bracket, and the plastic belt is used to connect the large pulley of the RMS with the polarizer groove. The polarizer with the RMS will be the analyzer in this experiment.

(e) ● CI Fig. 32.6: The components are placed on the optics track in the order shown in this figure. The light from the laser will pass a polarizer first, then pass the analyzer (with the RMS), and finally make it into the light sensor.

2. Setting the correct light sensor gain.

(a) Bring Graph 1 (voltage versus time) to the front on the screen. Increase its size if needed to see it well.

(b) Remove the holder that contains the RMS from the track, and set it aside.

(c) Slide all the other components close to each other, and dim the room lights.

(d) Turn on the laser. Use the horizontal and vertical adjust (on the back of the laser) if needed so that the light shines centered on the light sensor opening.

(e) Press the START button, and rotate the polarizer until the voltage on the graph reaches a maximum. You may have to press the Scale-to-Fit button (the leftmost button on the graph toolbar) to expand the graph scale while data are collected. Press STOP once the maximum is reached.

(f) Now put the analyzer (with RMS) back on the track. Press the START button, and rotate the analyzer until the voltage on the graph reaches a maximum. If this maximum exceeds 4.5 V, decrease the gain on the light sensor. (This is a

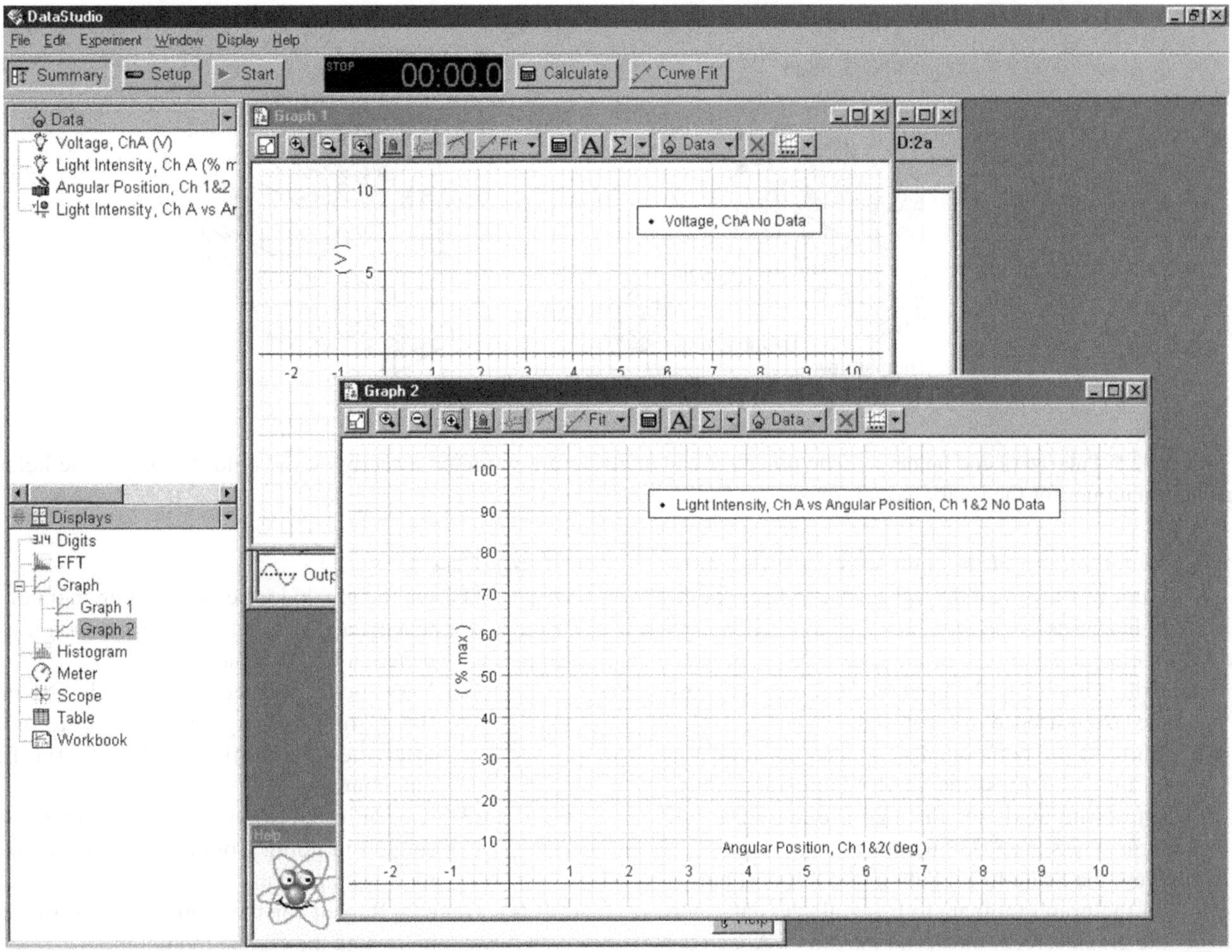

CI Figure 32.2 Data Studio setup. A graph of light intensity versus angular position will show the variations in light intensity as the analyzer is rotated with respect to the polarizer. The voltage-versus-time graph (background) is used to calibrate the sensor. (Reprinted courtesy of PASCO Scientific.)

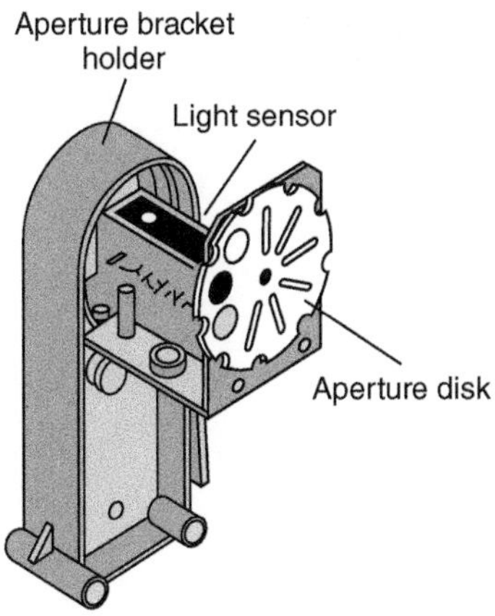

CI Figure 32.3 The aperture bracket. The aperture disk is mounted in front of the light sensor. The sensor and the apertures sit on the holder and install on the optics bench.

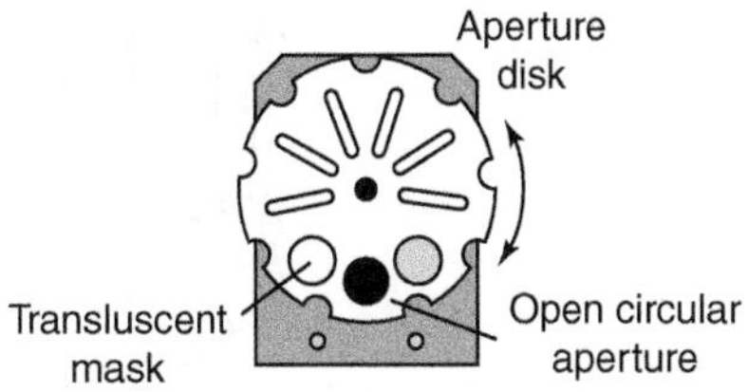

CI Figure 32.4 Details of the aperture bracket. Rotate the disk until the translucent mask covers the opening into the light sensor.

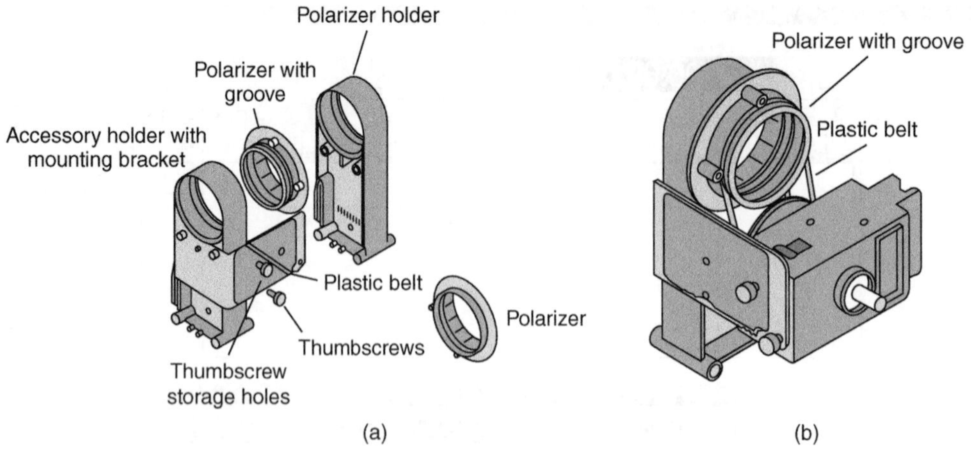

CI Figure 32.5 Polarizers and holders. The polarizer and analyzer are mounted directly on the holders. One of the holders must have a mount for the rotary motion sensor.

switch on top of the light sensor.) If the maximum was less than 0.5 V, increase the gain of the light sensor.

3. Data Collection

(a) Bring Graph 2 to the front on the screen. Increase the size if needed to see it well.

(b) With all the components in place, as shown in CI Fig. 32.6, press the START button and *slowly* rotate the analyzer through a one-and-a-half turns. Press the STOP button.

(c) Press the Scale-to-Fit button (the leftmost button on the graph toolbar) to bring all the data onto the graph screen; then print the graph. Attach the graph to the Laboratory Report.

(d) Case 1:

(1) On the graph, locate the first minimum of intensity, and record the angle at which it happened in CI Data Table 1.

(2) Now locate and record the angle at which the light intensity reached a maximum after the first minimum (that is, the first maximum).

(3) Determine by what angle the analyzer was rotated in going from the first minimum to the first maximum.

(e) Case 2: Repeat for the angle between the first maximum and the second minimum.

(f) Case 3: Repeat for the angle between the second minimum and the second maximum.

(g) Case 4: Repeat for the angle between the second maximum and the third minimum.

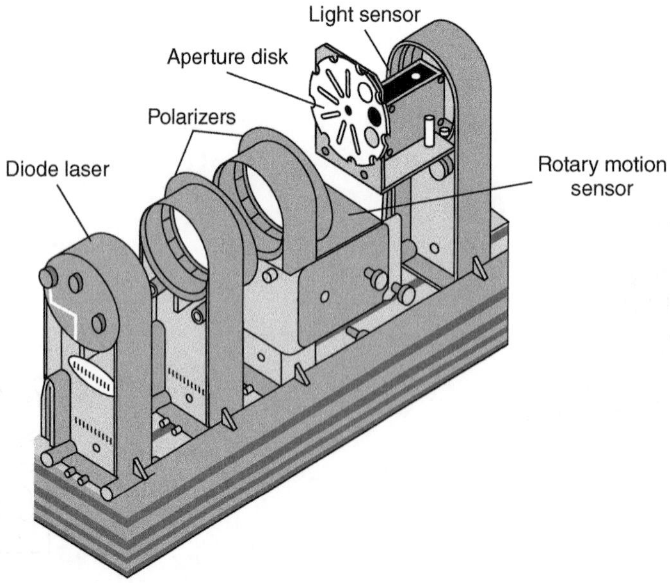

CI Figure 32.6 Order of components on the bench. Light from the laser will pass a polarizer and then the analyzer before entering the light sensor.

Name ______________________ Section ______________ Date ______________

Lab Partner(s) ____

CI EXPERIMENT 32

Malus's Law

CI *Laboratory Report*

CI DATA TABLE 1

Purpose: To investigate the relative orientations of polarizer and analyzer that produce minimum and maximum transmitted intensity.

	Successive minima and maxima of intensity	Total analyzer rotation between minimum and maximum $\lvert\theta_{min} - \theta_{max}\rvert$
Case 1	θ_{min}	
	θ_{max}	
Case 2	θ_{min}	
	θ_{max}	
Case 3	θ_{min}	
	θ_{max}	
Case 4	θ_{min}	
	θ_{max}	
	Average =	

Attach the graph to the Laboratory Report.

Don't forget units

(continued)

CI QUESTIONS

1. Discuss how well the experimental results match the theory.

2. What are sources of experimental error in this activity?

Name ______________________ Section ______________ Date ______________

Lab Partner(s) ____

EXPERIMENT 33

The Prism Spectrometer: Dispersion and the Index of Refraction

TI *Advance Study Assignment*

Read the experiment and answer the following questions.

1. What is meant by *dispersion?*

2. Is the index of refraction of a dispersive medium the same for all wavelengths? Explain this in terms of the speed of light in the medium.

3. What is meant by the *angle of deviation* for a prism?

(continued)

4. What is the condition for the angle of minimum deviation? At this angle, what is the relation of the transmitted ray to the base of the prism?

5. What are the four major components of the prism spectrometer and their functions?

6. Which color of visible light has the greatest prism dispersion?

EXPERIMENT 33

The Prism Spectrometer: Dispersion and the Index of Refraction

INTRODUCTION AND OBJECTIVES

In vacuum, the speed of light, $c = 3.0 \times 10^8$ m/s, is the same for all wavelengths or colors of light. However, when a beam of white light falls obliquely on the surface of a glass prism and passes through it, the light is spread out, or *dispersed*, into a spectrum of colors. This phenomenon led Newton to believe that white light is a mixture of component colors. The dispersion arises in the prism because the wave velocity is slightly different for different wavelengths.

A **spectrometer** is an optical device used to observe and measure the angular deviations of the components of incident light due to refraction and dispersion. Using Snell's law, the index of refraction of the prism glass for a specific wavelength or color can easily be determined.

After performing this experiment and analyzing the data, you should be able to:

1. Explain the dispersion of light in a dispersive medium.
2. Describe the operation of a prism spectrometer.
3. Tell how the index of refraction of a prism can be measured.

EQUIPMENT NEEDED

- Prism spectrometer
- Incandescent light source and support stand

THEORY

A monochromatic (single color or wavelength) light beam in air, obliquely incident on the surface of a transparent medium, and transmitted through the medium, is refracted and deviated from its original direction in accordance with Snell's law (see Experiment 30):

$$n = \frac{c}{c_m} = \frac{\sin\theta_1}{\sin\theta_2} \qquad \textbf{(33.1)}$$

where n is the index of refraction, c is the speed of light in vacuum (air), c_m is the speed of light in the medium, and θ_1 and θ_2 are the angles of incidence and refraction, respectively.

If the incident light beam is not monochromatic, each component wavelength (color) is refracted differently. This is why white light incident on a glass prism forms a spectrum (● Fig. 33.1). The material is said to exhibit **dispersion**.

The explanation of this effect has to do with the speed of light. In vacuum, the speed of light is the same for all wavelengths of light, but in a dispersive medium, the speed of light is slightly different for different wavelengths. (The frequencies of the light components are unchanged.) Since the index of refraction n of a medium is a function of the speed of light ($n = c/v = c/\lambda_m f$, where the wave speed in the medium is $v = \lambda_m f$), the index of refraction will then be different for different wavelengths. It follows from Snell's law (Eq. 33.1) that different wavelengths of light will be refracted at different angles.

The dispersion of a beam of white light spreads the transmitted emergent beam into a spectrum of colors, red through violet (see Fig. 33.1). The red component has the longest wavelength, so it is deviated least. The angle

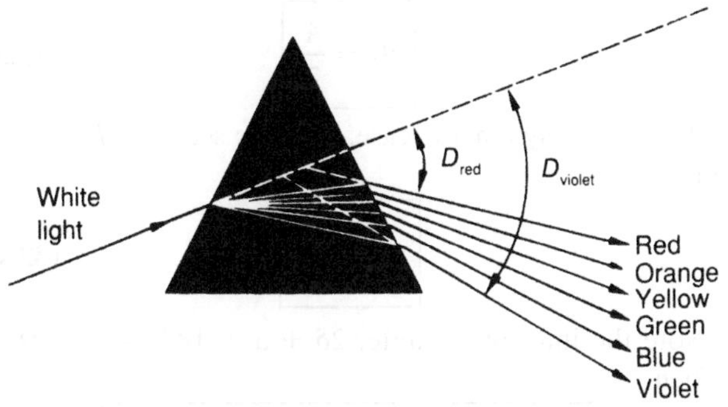

Figure 33.1 Dispersion. The dispersion of light by a glass prism causes white light to be spread out into a spectrum of colors. The angle between the original direction of the beam and the emergent component is called the *angle of deviation D* for that particular component.

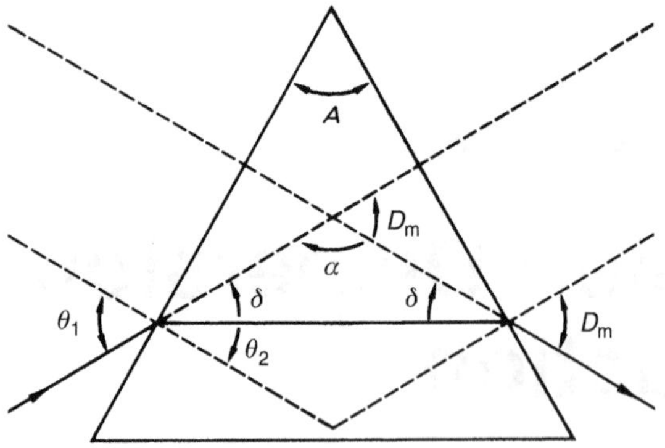

Figure 33.2 Minimum angle of deviation. The geometry for determining the minimum angle of deviation D_m for a light ray. See text for description.

between the original direction of the beam and an emergent component of the beam is called the **angle of deviation** *D*; it is different for each color or wavelength.

As the angle of incidence is decreased from a large value, the angle of deviation of the component colors decreases, then increases, and hence goes through an angle of minimum deviation, D_m. The angle of minimum deviation occurs for a particular component when the component ray passes through the prism symmetrically, that is, parallel to the base of the prism if the prism is isosceles (● Fig. 33.2).

The angle of minimum deviation and the prism angle *A* are related to the index of refraction of the prism glass (for a particular color component) through Snell's law by the relationship

$$n = \frac{\sin[(A + D_m)/2]}{\sin(A/2)} \tag{33.2}$$

The derivation of this equation can be seen from the geometry of Fig. 33.2. Note from the top triangle that

$$2(90° - \theta_2) + A = 180°$$

and therefore

$$\theta_2 = \frac{A}{2} \tag{33.3}$$

Also, for the symmetric case, it can be seen that $D_m = 2\delta$, or

$$\delta = \frac{D_m}{2} \tag{33.4}$$

(Note the interior triangle, $2\delta + \alpha = 180° = \alpha + D_m$.) Then

$$\theta_1 = \theta_2 + \delta = \frac{A}{2} + \frac{D_m}{2} = \frac{A + D_m}{2} \tag{33.5}$$

Substituting Eqs. 33.3 and 33.5 into Snell's law (Eq. 33.1) yields Eq. 33.2.

EXPERIMENTAL PROCEDURE

1. Two types of prism spectrometers are shown in ● Fig. 33.3, one of which is an adapted force table (see Experiment 6).* The four basic parts of a spectrometer are the (a) collimator and slit assembly, (b) prism, (c) telescope, and (d) divided circle.

   The **collimator** is a tube with a slit of adjustable width at one end and a converging lens at the other. Light from a light source enters the collimator. The length of the collimator tube is made equal to the focal length of the lens so as to make the rays of the emerging light beam parallel.

   The **prism** deviates and disperses the beam into a spectrum. The objective lens of the **telescope** converges the beam and produces an image of the slit, which is viewed through the telescope eyepiece. The eyepiece is fitted with cross hairs, which may be fixed on a

(a)

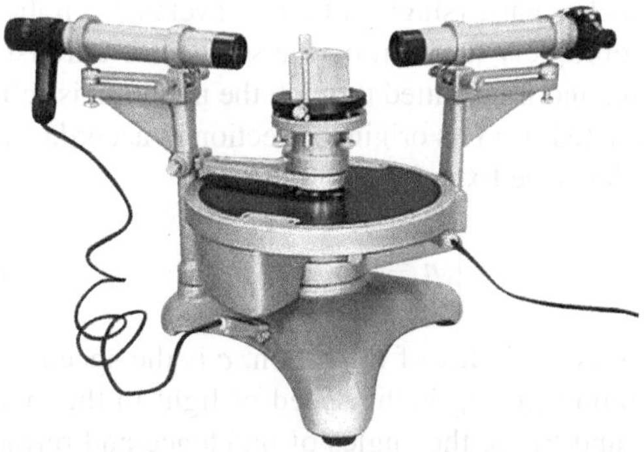

(b)

Figure 33.3 Prism spectrometer. (a) Simple spectrometer. The prism rests on a graduated (divided) circle used for angle measurements. Light directed into the collimator tube on the left is refracted by the prism to the adjustable telescope on the right. (b) Advanced spectrometer. The collimator has a built-in light source, and the angular measurement scale has a vernier scale that can be read to 1 minute of arc. (Fisher Scientific Company, LLC.)

* Some procedures may not apply to the force table apparatus.

particular spectral color. The **divided circle** makes it possible to measure the angle(s) of deviation.

2. After being given instructions by the instructor, study the various clamps and adjustment screws of your spectrometer. In particular, study the divided circle scale. Some spectrometers are equipped with vernier scales that permit readings to 1 min of arc. Be careful, because the adjustments and alignments of the spectrometer are critical, and it can be time-consuming to restore proper adjustment.

3. *For spectrometer with telescope.* With the prism not on the spectrometer table, sight the telescope on some distant object and adjust the eyepiece until the cross hairs are in good focus. Then mount the lamp near the collimator slit (adjusted to a small slit width). Move the telescope into the line of sight, and adjust so that a sharp image of the illuminated slit is seen focused on the cross hairs.

4. *Measurement of the prism angle A.* Mount the prism in the center of the spectrometer table, and orient it as shown in ● Fig. 33.4. With the unaided eye, locate the white image of the slit reflected from a face of the prism on either side of the prism angle A. The prism may have to be adjusted slightly. (You may also note the color spectrum in the prism face opposite A.)

   Move the telescope in front of the eye, and adjust the cross hairs on the center of the slit image (with the fine-adjustment screw, if available). Make the slit as narrow as possible so that the best setting can be made. On a force table apparatus, adjustment is not required. Read the angle from the divided circle, and record it in the Laboratory Report.

   Repeat this procedure for the other face of the prism. As shown in Fig. 33.4, the angle between the positions is equal to $2A$. Compute the angle A from the circle readings.

5. *Measurement of the angle of minimum deviation.* Remove the prism and move the telescope into the line of sight of the slit. (It is convenient, but not necessary, to adjust the setup so the telescope has a zero reading on the divided circle. This makes finding the deviation angles easy by reading directly.) Adjust the telescope so that a sharp image of the illuminated slit is seen on the cross hairs. Note and record the reading of the divided circle.

   Replace and rotate the prism to a position as shown in ● Fig. 33.5, and with the unaided eye, locate the emergent spectrum of colors. Move the telescope in front of the eye and examine the spectrum. (Change the slit width, if applicable, and note any difference.) List the sequence of colors, beginning with red, in the Laboratory Report.

6. With the slit set as narrow as possible, rotate the prism back and forth slightly, and note the reversal of the direction of motion of the spectrum when the prism is rotated in one direction.

   Stop rotating the prism at the position of the reversal of motion of the yellow component of the

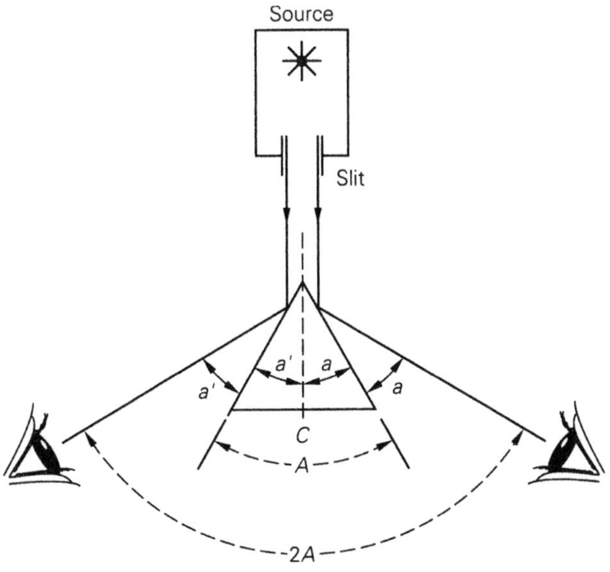

Figure 33.4 Determination of the prism angle. An illustration of the prism orientation for the experimental procedure to determine the prism angle A.

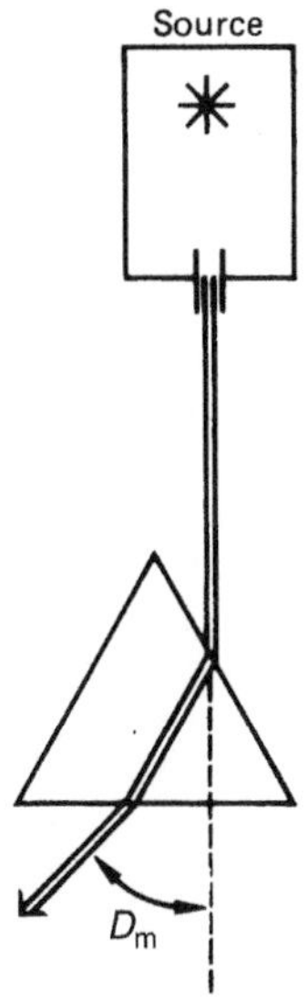

Figure 33.5 Determination of the angle of minimum deviation. An illustration of the prism orientation for the experimental procedure to determine the angle of minimum deviation.

spectrum. This is the position for minimum deviation of this component.

7. Being careful not to disturb the prism, center the telescope cross hairs on the middle of the yellow color band, and record the divided circle reading. Also measure the angle readings for each end of the visible spectrum [that is, the red and blue (violet) ends]. Do this by setting the cross hairs of the telescope at the locations where the spectrum ends are just visible (not at the center of the extreme bands).

8. Compute the index of refraction for yellow light using Eq. 33.2.

Name ______________________ Section ______________ Date ______________

Lab Partner(s) ____

EXPERIMENT 33

The Prism Spectrometer: Dispersion and the Index of Refraction

TI *Laboratory Report*

Measurement of Prism Angle A

Calculations
(show work)

Circle readings for reflected images ______________

Computation of $2A$ ______________

Prism angle A ______________

Measurement of Angle of Minimum Deviation

Spectrum (sequence) of colors			*Circle reading*	*Minimum deviation*
red ______	______	Line of sight	______	______
______	______	Yellow	______	______
______	______	Red end	______	______
______	______	Blue end	______	______

Calculations
(show work)

Index of Refraction for Yellow Light

$n =$ ______________

Don't forget units

(continued)

TI QUESTIONS

1. Show that the angle between the two telescope settings of the reflected slit images is equal to $2A$ (Procedure 4).

2. Judging on the basis of your experimental results, what is the speed of yellow light in the prism?

3. What is the range of indices of refraction of the prism for the wavelengths of visible light?

Name ______ Section ______ Date ______
Lab Partner(s) ______

EXPERIMENT 34

Line Spectra and the Rydberg Constant

TI Advance Study Assignment

Read the experiment and answer the following questions.

1. Distinguish between continuous spectra and line spectra, and describe their causes.

2. Why does a gas discharge tube (for example, a neon light) have a certain color?

3. What are (a) the Balmer series and (b) the Rydberg constant?

(continued)

4. Explain briefly how the prism spectrometer is calibrated.

5. Explain briefly how the Rydberg constant is determined experimentally.

EXPERIMENT 34

Line Spectra and the Rydberg Constant

INTRODUCTION AND OBJECTIVES

In spectroscopic analysis, two types of spectra are observed: continuous spectra and line or discrete spectra. The spectrum of visible light from an incandescent source is found to consist of a **continuous spectrum**, or band of merging colors, and contains all the wavelengths of the visible spectrum.

However, when the light from a gas discharge tube (for example, mercury or helium) is observed through a spectroscope, only a few colors, or wavelengths, are observed. The colored images of the spectroscope slit appear as bright lines separated by dark regions, hence the name **line** or **discrete spectra**.

Each gas emits a particular set of spectral lines and hence has a characteristic spectrum. Thus, spectroscopy (the study of spectra) provides a method of identifying elements. The discrete lines of a given spectrum depend on the atomic structure of the atoms and are due to electron transitions.

The line spectrum of hydrogen was explained by Bohr's theory of the hydrogen atom. However, before this, the line spectrum of hydrogen was described by an empirical relationship involving the Rydberg constant. In this experiment, line spectra will be observed and the relationship of the Rydberg constant to the theoretical quantities of the Bohr theory will be investigated.

After performing this experiment and analyzing the data, you should be able to:

1. Clearly distinguish between continuous and line (discrete) spectra.
2. Explain why gas discharge tubes emit line spectra.
3. Tell what is meant by the Balmer series and the Rydberg constant.

EQUIPMENT NEEDED

- Prism spectrometer
- Incandescent light source
- Mercury or helium discharge tube
- Hydrogen discharge tube
- Discharge-tube power supply
- 2 sheets of Cartesian graph paper

THEORY

The electrons in an incandescent light source undergo thermal agitation and emit electromagnetic radiation (light) of many different wavelengths, producing a continuous spectrum. However, when light emitted from excited gases or vaporized liquids or solids is analyzed, line spectra such as those illustrated in ● Fig. 34.1 are observed.

Modern theory explains spectra in terms of photons of light of discrete wavelengths being emitted as the result of electron transitions between atomic energy levels. Different substances have characteristic spectra, that is, they have a characteristic set of lines at specific wavelengths. In a manner of speaking, the spectrum of a substance acts as a "fingerprint" by which the substance can be identified.

The characteristic color of light from a gas discharge tube is often indicative of the most intense spectral line(s) in the visible region. For example, light from a hydrogen discharge tube has a characteristic red glow resulting from an intense emission line with a wavelength of 656.1 nm. Similarly, when table salt is vaporized in a flame, yellow light is observed because of the intense yellow discharge line in the spectrum of sodium. It is the presence of sodium that gives candles and wood fires their yellow glow. You may have noticed that many highway and parking lot lights are bright yellow. They are sodium lights, used because

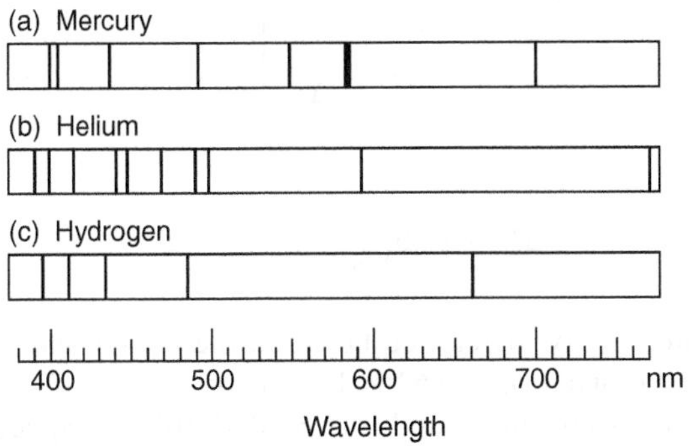

Figure 34.1 Line spectra. Illustrations of visible line spectra for (a) mercury, (b) helium, and (c) hydrogen. From Wilson/Buffa, *College Physics*, 5th ed. Copyright © 2010. Reprinted by permission of Pearson Education.

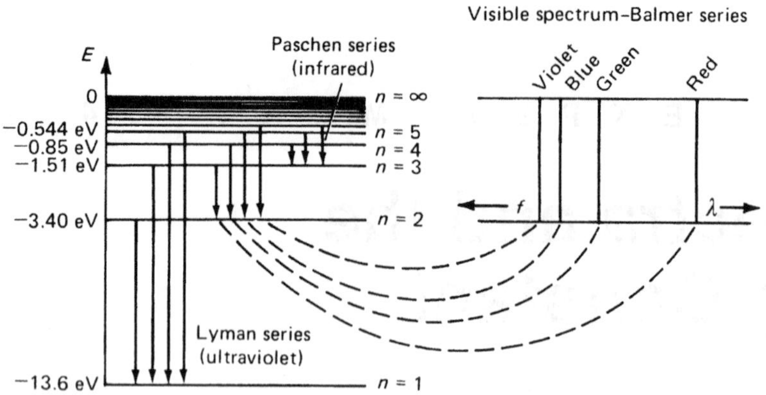

Figure 34.2 Energy-level transitions. The energy-level transmissions for the hydrogen atom. The Balmer series, $n_f = 2$, produces a line spectrum in the visible region.

sodium discharge is a very efficient way to produce light. Wavelengths are commonly measured in nanometers (nm):

$$1 \text{ nm} = 10^{-9} \text{ m} = 10^{-7} \text{ cm}$$

The systematic spacing of the spectral lines in the hydrogen spectrum was empirically described by spectroscopists in the late 1800s. For example, the wavelengths of spectral lines in the visible region, called the Balmer series, were found to fit the formula

$$\boxed{\frac{1}{\lambda} = R\left(\frac{1}{2^2} - \frac{1}{n^2}\right) \quad n = 3,4,5, \ldots} \quad \textbf{(34.1)}$$

where R is the Rydberg constant,* with a value of $1.097 \times 10^{-2} \text{ nm}^{-1}$.

The hydrogen spectrum is of particular theoretical interest because hydrogen, having only one proton and one electron, is the simplest of all atoms. Niels Bohr (1885–1962), a Danish physicist, developed a theory for the hydrogen atom that explains the spectral lines as resulting from electron transitions between energy levels, or discrete electron orbits (● Fig. 34.2), with the wavelengths of the spectral lines being given by the theoretical equation

$$\boxed{\lambda = \frac{hc}{\Delta E}} \quad \textbf{(34.2)}$$

where

$$\Delta E = 13.6\left(\frac{1}{n_f^2} - \frac{1}{n_i^2}\right) \quad \text{eV}$$

is the energy difference between the initial and final states, n_i and n_f, $h = 6.63 \times 10^{-34}$ J-s $= 4.14 \times 10^{-15}$ eV-s (Planck's constant), and $c = 3.00 \times 10^8$ m/s (speed of light in vacuum). The values $n = 1, 2, 3, 4, \ldots$ are called the **principal quantum numbers**. Different final states account for the different series.†

For spectral lines in the visible region, the final state is $n_f = 2$, and

$$\lambda = \frac{hc}{\Delta E} = \frac{hc}{13.6[(1/2^2) - 1/n^2]} \qquad n = 3, 4, 5, \ldots$$

or

$$\boxed{\frac{1}{\lambda} = \frac{13.6}{hc}\left(\frac{1}{2^2} - \frac{1}{n_i^2}\right) \quad n = 3, 4, 5, \ldots} \quad \textbf{(34.3)}$$

Comparing this theoretical equation with the empirical equation, Eq. 34.1 reveals that the forms are identical, with the prediction that $R = (13.6 \text{ eV})/hc$.

EXPERIMENTAL PROCEDURE

1. A prism spectrometer will be used to analyze and study spectra in this experiment. The prism spectrometer is illustrated and its use described in Experiment 33. Review the operation of this instrument. Place the incandescent source in front of the collimator slit, and observe the continuous spectrum that results from the prism dispersion (see Experiment 33). List the colors of the spectrum in the Laboratory Report, beginning with red.

2. A convenient type of discharge tube and power supply is shown in ● Fig. 34.3.
Caution: *Great care should be taken, because the discharge tube operates at high voltage and you could receive an electrical shock. When mounting a tube, have the power supply unplugged for safety.*

*After J. R. Rydberg (1854–1919), the Swedish physicist who developed the series relationship.

† The three such spectral series (shown in Fig. 34.2) are named after 19th-century scientists: the Swiss mathematician Johann Balmer and the German physicists Theodore Lyman and Friedrich Paschen.

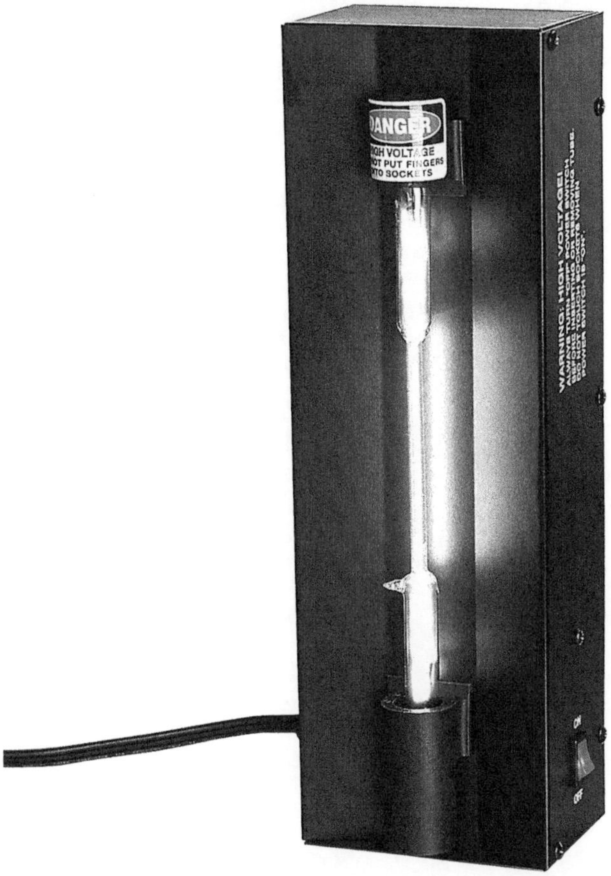

Figure 34.3 Experimental apparatus. A gas discharge tube and power supply. (Photo Courtesy of Sargent-Welch.)

Mount a mercury (or helium) discharge tube in the power supply holder, and place it in front of the collimator slit.

Caution: *If a larger mercury source is used, it should be properly shielded because of the ultraviolet radiation that may be emitted. Consult your instructor.*

Turn on the power supply, observe the mercury (or helium) spectrum through the telescope, and note its line nature.

With the slit as narrow as possible, rotate the prism slightly back and forth, and notice the reversal of direction of the motion of the spectrum when the prism is rotated in one direction. Focusing on the yellow line (for mercury, the brighter yellow line), stop rotating the prism at the position of the reversal of motion of this line.

This sets the prism for minimum deviation for the yellow line (see Experiment 33), which will be taken as an average for the spectrum. (The other lines have slightly different minimum deviations.)

3. **(a)** Without disturbing the prism, starting at the red end of the spectrum, set the crosshairs of the telescope on the extreme red line, and record the color and divided circle reading in the Laboratory Report. Repeat this procedure for each spectral line in order. (Turn off the discharge tube as soon as possible to conserve the life of the tube.)

(b) Find the wavelengths of the spectral lines for the discharge tube gas in Appendix A, Table A8, and match them to the line readings.

(c) Using these data, plot the wavelength λ versus the divided circle reading θ. This calibrates the spectrometer, and unknown wavelengths can be determined from divided circle readings from the calibration curve.

4. With the discharge tube power supply off and unplugged, replace the mercury (or helium) discharge tube with a hydrogen discharge tube. Turn on the power supply, and starting with the red line of the hydrogen spectrum, determine the divided circle reading for each spectral line with the crosshairs of the telescope positioned on the center of the line. Record in the Laboratory Report.

The red line is referred to as H_α in spectroscopic notation. The other sequential lines are referred to as H_β, etc., with subscripts in Greek alphabetical order.

5. Determine the wavelengths of the hydrogen lines from the calibration curve, and plot the reciprocal of the wavelength $1/\lambda$ versus $1/n^2$. (Begin the abscissa scale with zero.) Draw the best straight line that fits the data points and determine the slope of the line.

Note that Eq. 34.1:

$$\frac{1}{\lambda} = R\left(\frac{1}{2^2} - \frac{1}{n^2}\right) = \frac{R}{4} - \frac{R}{n^2}$$

has the form of a straight line, $y = mx + b$, with the negative slope equal to the Rydberg constant. Compare the slope of the line with the accepted value of the Rydberg constant by computing the percent error. Compare the intercept of this line with $R/4$.

Name ______________________ Section ______________ Date ______________

Lab Partner(s) ______________________

EXPERIMENT 34

Line Spectra and the Rydberg Constant

TI Laboratory Report

Colors of the Continuous Spectrum

DATA TABLE 1 Mercury (or Helium) Spectrum

Color	Divided circle reading	Wavelength ( ) (from Table A8)

Sequence of colors

red

______

______

______

______

______

______

______

______

______

Don't forget units

(continued)

DATA TABLE 2 Hydrogen Spectrum

Line	Color	Divided circle reading	Wavelength ( )	$1/\lambda$ ( )	$1/n^2$
H_α, $n = 3$					
H_β, $n = 4$					
H_γ, $n = 5$					
H_δ, $n = 6$					

Calculations (Slope of Graph)
(show work)

R (experimental) ____________

Accepted value ____________

Percent error ____________

R/4 from graph ____________

Accepted value ____________

Percent error ____________

TI QUESTIONS

1. Compute the value of the Rydberg constant from the Bohr theory, and compare it with the accepted empirical value.

Name ______________________ Section ______________ Date ______________

Lab Partner(s) ____

EXPERIMENT 34 Line Spectra and the Rydberg Constant *Laboratory Report*

2. Why are only four lines seen in the Balmer series? (Transitions for $n_i > 6$ also exist.) Justify your answer mathematically.

3. As n becomes very large, the wavelengths of the Balmer (and other) series approach a minimum wavelength, or series limit (Eq. 34.1). What is the wavelength of the series limit for the Balmer series?

Name ______________________ Section ______________ Date ______________

Lab Partner(s) ____

EXPERIMENT 35

TI The Transmission Diffraction Grating: Measuring the Wavelengths of Light

CI Single-Slit and Double-Slit Diffraction

TI *Advance Study Assignment*

Read the experiment and answer the following questions.

1. What is a diffraction grating? Distinguish between the two types of gratings.

2. What is the grating constant? What would be the grating constant for a grating with 300 lines/mm? (Express the constant in nanometers.)

(continued)

3. Explain why there is a spectrum for each diffraction order when multicolored light is analyzed.

4. Will the red or the violet end of the first-order spectrum be nearer the central maximum? Justify your answer.

5. It will be observed that the second-order spectrum is "spread out" more than the first-order spectrum. Why?

CI Advance Study Assignment

Read the experiment and answer the following questions.

1. What is diffraction?

Name ______________________ Section ______________ Date ______________

Lab Partner(s) ____

EXPERIMENT 35 *Advance Study Assignment*

2. What type of pattern is produced by a double slit? By a single slit?

3. What causes dark and bright fringes?

EXPERIMENT 35

TI The Transmission Diffraction Grating: Measuring the Wavelengths of Light

CI Single-Slit and Double-Slit Diffraction

OVERVIEW

Experiment 35 examines diffraction, but the TI and CI procedures differ in focus. The TI procedure uses a transmission diffraction grating to measure the wavelengths of light from an incandescent source and the mercury (Hg) line spectrum. The CI procedure complements this by investigating single-slit and double-slit diffraction. By using the diffraction patterns formed by laser light, it examines the conditions for single-slit dark fringes and double-slit bright fringes.

INTRODUCTION AND OBJECTIVES

In Experiment 33, the prism spectrometer had to be calibrated in terms of known wavelengths before being able to determine unknown wavelengths of light experimentally. How, then, are the wavelengths of spectral lines or colors initially determined? This is most commonly done with a diffraction grating, a simple device that allows for the study of spectra and the measurement of wavelengths.

By replacing the prism with a diffraction grating, a prism spectrometer (Experiments 33 and 34) becomes a grating spectrometer. When a diffraction grating is used, the angle(s) at which the incident beam is diffracted relate simply to the wavelength(s) of the light. In this experiment, the properties of a transmission grating will be investigated, and the wavelengths of several spectral lines will be determined.

TI OBJECTIVES

After performing this experiment and analyzing the data, you should be able to:

1. Describe the principle of a diffraction grating.
2. Explain the operation of a grating spectrometer.
3. Tell how the wavelength of light can be measured with a grating spectrometer.

CI OBJECTIVES

A diffraction grating has many slits. But what about using a single slit or a double slit? When illuminated with monochromatic light, these slits produce interference and diffraction patterns with bright and dark fringes. Geometric analysis gives equations for the positions of single-slit dark fringes and double-slit bright fringes. The CI portion of this experiment investigates these fringe relationships.

After performing this experiment and analyzing the data, you should be able to:

1. Verify that the positions of the minima in a diffraction pattern match the positions predicted by theory.
2. Use a diffraction and interference pattern to determine the wavelength of light.
3. Compare the patterns formed by single slits to those formed by double slits.
4. Investigate the effects of changing slit width and slit separation.

TI EXPERIMENT 35

The Transmission Diffraction Grating: Measuring the Wavelengths of Light

TI EQUIPMENT NEEDED

- Spectrometer*
- Diffraction grating and holder
- Mercury discharge tube
- Power supply for discharge tube
- Incandescent light source

**Instructor's note:* If a spectrometer is not available, an alternative, inexpensive method is described in the Instructor's Manual.

TI THEORY

A **diffraction grating** consists of a piece of metal or glass with a very large number of evenly spaced parallel lines or grooves. This gives two types of gratings: *reflection gratings* and *transmission gratings*.

Reflection gratings are ruled on polished metal surfaces; light is reflected from the unruled areas, which act as a row of "slits." **Transmission gratings** are ruled on glass, and the unruled slit areas transmit incident light.

The transmission type grating is used in this experiment. Common laboratory gratings have 300 grooves per mm and 600 grooves per mm (about 7500 grooves per in. and 15,000 grooves per in.) and are pressed plastic *replicas* mounted on glass. Glass originals are very expensive.

Diffraction consists of the "bending," or deviation, of waves around sharp edges or corners. The slits of a grating give rise to diffraction, and the diffracted light interferes so as to set up interference patterns (● TI Fig. 35.1).

Complete constructive interference of the waves occurs when the phase or path difference is equal to one wavelength, and the first-order maximum occurs for

$$d \sin\theta_1 = \lambda \qquad \textbf{(TI 35.1)}$$

where d is the grating constant, or distance between the grating lines, θ_1 is the angle the rays are diffracted from the incident direction, and $d \sin\theta_1$ is the path difference between adjacent rays. The **grating constant** is given by

$$d = 1/N \qquad \textbf{(TI 35.2)}$$

where N is the number of lines or grooves per unit length (usually per millimeter or per inch) of the grating.

A second-order maximum occurs for $d \sin\theta_2 = 2\lambda$, and so on, so that in general,

$$d \sin\theta_n = n\lambda \quad n = 1, 2, 3, \ldots \qquad \textbf{(TI 35.3)}$$

where n is the order of the image maximum. The interference is symmetric on either side of an undeviated and undiffracted central maximum of the slit image, so the angle between symmetric image orders is $2\theta_n$ (● TI Fig. 35.2).

In practice, only the first few orders are easily observed, with the number of orders depending on the grating constant. If the incident light is other than monochromatic, each order corresponds to a spectrum. (That is, the grating spreads the light out into a spectrum.)

As can be seen from TI Eq. 35.1, since d is constant, each wavelength (color) deviates by a slightly different angle so that the component wavelengths are separated into a spectrum. Each diffraction order in this case corresponds to a spectrum order. [The colorful displays seen on compact disks (CDs) result from diffraction.]

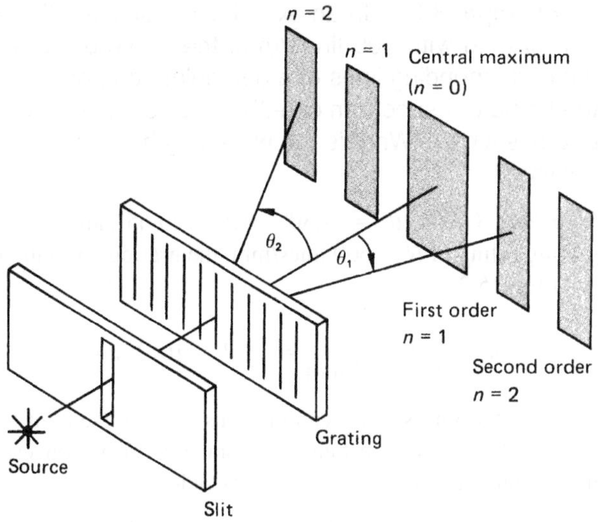

TI Figure 35.1 Diffraction pattern. A simplistic view of the diffraction pattern (two orders) produced by a diffraction grating. (Pattern and angles exaggerated illustration.)

(a)

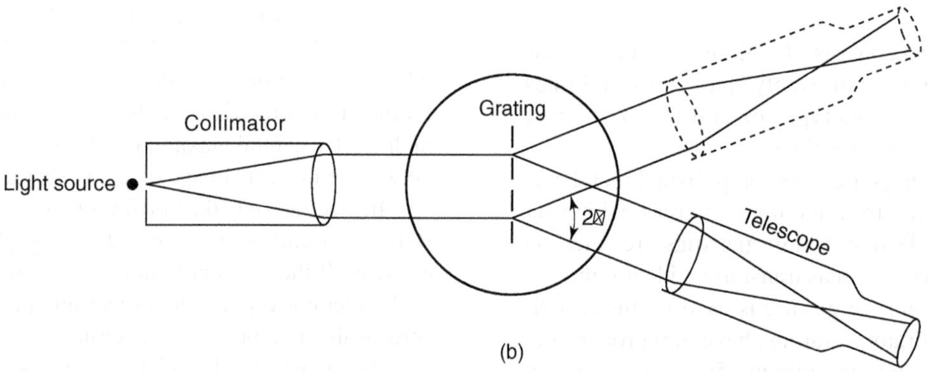

(b)

TI Figure 35.2 Grating spectrometer. (a) A student views a diffraction pattern through the telescope of a grating spectrometer. The light source and collimator are on the right. (b) A diagram of a top view of a grating spectrometer. When the symmetric images of a particular order n are viewed from both sides of the central maximum, the angle between the two viewing positions is $2\theta_n$. (Photo Courtesy of Sargent-Welch.)

TI Example 35.1 In an experiment using a diffraction grating with 600 lines/mm, the angle between the corresponding lines of a particular component of the first-order spectrum on either side of the incident beam is 41.30°. What is the wavelength of the spectral line?

Solution Given $2\theta_1 = 41.30°$, or $\theta_1 = 20.65°$, and with a grating ruling of $N = 600$ lines/mm, the grating constant d is TI Eq. 35.2

$$d = \frac{1}{N} = \frac{1}{600/\text{mm}} = 1.67 \times 10^{-3}\,\text{mm}$$

When doing several calculations, it is convenient to express the grating constant in nanometers (nm). Converting to nanometers (1 mm = 10^6 nm) yields

$$d = 1.67 \times 10^{-3}\,\text{mm}\,(10^6\,\text{nm/mm}) = 1.67 \times 10^3\,\text{nm}$$

Then for first-order ($n = 1$) interference, by TI Eq. 35.3,

$$d \sin\theta_1 = \lambda$$

or

$$\lambda = d \sin\theta_1 = (1.67 \times 10^{-3}\,\text{nm})(\sin 20.65°) = 589\,\text{nm}$$

TI EXPERIMENTAL PROCEDURE

1. Review the general operation of a spectrometer if necessary (Experiment 33). Record the number of lines per mm of your diffraction grating in the Laboratory Report. Mount the grating on the spectrometer table with the grating ruling parallel to the collimator slit and the plane of the grating perpendicular to the collimator axis.

Determination of the Wavelength Range of the Visible Spectrum

2. Mount an incandescent light source in front of the collimator slit. Move the spectrometer telescope into the line of the slit of the collimator, and focus the crosshairs on the central slit image.

   Notice that this central maximum, or "zeroth"-order image, does not depend on the wavelength of light, so a white image is observed. Then move the telescope to either side of the incident beam, and observe the first- and second-order spectra. Note which is spread out more.

3. **(a)** Focus the crosshairs on the blue (violet) end of the first-order spectrum at the position where you judge the spectrum just becomes visible. Record the divided circle reading (to the nearest minute of arc) in TI Data Table 1.

   (b) Move the telescope to the other (red) end of the spectrum, and record the divided circle reading of its visible limit.

   (c) Repeat this procedure for the first-order spectrum on the opposite side of the central maximum. The angular difference between the respective readings corresponds to an angle of 2θ (TI Fig. 35.2b).

4. Compute the grating constant d in millimeters, and with the experimentally measured θ's, compute the range of the wavelengths of the visible spectrum in nanometers.

Determination of the Wavelengths of Spectral Lines

5. Mount the mercury discharge tube in its power supply holder, and place in front of the collimator slit.

   Caution: *Work very carefully, as the discharge tube operates at high voltage and you could receive an electrical shock. Make certain the power supply is turned off and unplugged before inserting the tube. If a large mercury source is used, it should be properly shielded because of the ultraviolet radiation that may be emitted. Consult with your instructor.*

   Turn on the power supply, and observe the first- and second-order mercury line spectra on both sides of the central image.

6. Because some of the lines are brighter than others and the weaker lines are difficult to observe in the second-order spectra, the wavelengths of only the brightest lines will be determined. Find the listing of the mercury spectral lines in Appendix A, Table A8, and record the color and wavelength in TI Data Table 2.

   Then, beginning with either first-order spectra, set the telescope crosshairs on each of the four brightest spectral lines, and record the divided circle readings (read to the nearest minute of arc). Repeat the readings for the first-order spectrum on the opposite side of the central image.

7. Repeat the measurement procedure for the four lines in the second-order spectra and, using TI Eq. 35.2, compute the wavelength of each of the lines for both orders of spectra. Compare with the accepted values by computing the percent error of your measurements in each case.

   Note: In the second-order spectra, two yellow lines—a *doublet*—may be observed. Make certain that you choose the appropriate line. (*Hint:* See the wavelengths of the yellow lines in Appendix A, Table A8. Which is closer to the red end of the spectrum?)

Name ______________________ Section ______________ Date ______________

Lab Partner(s) ____

TI EXPERIMENT 35

TI The Transmission Diffraction Grating: Measuring the Wavelengths of Light

TI *Laboratory Report*

Number of lines per millimeter on grating ______________ Grating constant d ______________ (mm)

TI DATA TABLE 1

Purpose: To determine the wavelength range of the visible spectrum.

Spectrum limit	Divided circle reading		2θ	θ	$\sin\theta$	Computed wavelength
	Right	Left				( )
Violet end						
Red end						

Calculations
(show work)

Don't forget units

(continued)

TI **DATA TABLE 2**

Purpose: To determine the wavelengths of spectral lines.

Mercury Lines		Divided circle reading		2θ	θ	$\sin\theta$	Computed λ ( )	Percent error
Color	Wavelength	Right	Left					
First-order spectrum								
Second-order spectrum								

Calculations
(show work)

Name ______________________ Section ______________ Date ______________

Lab Partner(s) ____

TI QUESTIONS

1. If a grating with more lines per unit length were used, how would the observed angles or spread of the spectra be affected?

2. Was there any difference in the accuracy of the determination of the wavelengths of the mercury lines for the different-order spectra? If so, give an explanation.

3. Is it possible for the first-order spectrum to overlap the second-order spectrum? Explain, assuming a continuous spectrum.

4. Is there a theoretical limit to the order of the spectrum you would be able to observe with your diffraction grating? Justify your answer mathematically.

(continued)

5. (Reminder about rounding errors.) In Example 35.1, values were rounded to the proper number of significant figures in each step. It is sometimes suggested that one or two insignificant (extra) figures be carried along and stated that if a calculator is used, rounding off may be done only on the final result of multiple calculations.

   If you applied this procedure to the calculations in Example 35.1, would you get 589 nm? (Justify your answer.)

CI EXPERIMENT 35

Single-Slit and Double-Slit Diffraction

CI EQUIPMENT NEEDED

- Optics bench from the Basic Optics System (PASCO OS-8515)
- Diode laser (OS-8525)
- Linear translator (OS-8535)
- Aperture bracket (OS-8534)
- Rotary motion sensor (CI-6538)
- Light sensor (CI-6504A)
- Single-slit disk from Slit Accessory (OS-8523)
- Multiple-slit disk from Slit Accessory (OS-8523)

Note: The light sensor needs to be calibrated before use. Refer to the owner's manual for instructions on how to calibrate the sensor.

CI THEORY

A. Single-Slit Diffraction

Diffraction is the bending of a wave by means other than reflection or refraction. It occurs when a wave encounters an obstacle and bends around it, reaching places that would otherwise be shadowed. The amount of bending depends on the wavelength of the wave relative to the size of the obstacle. For waves of visible light that have wavelengths in the nanometer range (10^{-9} m), some obstacles that will produce diffraction are sharp edges, point objects, and thin slits.

Let's consider monochromatic light that passes through a single thin slit. The light "flares out" as it goes through, producing, on a screen a distance L away, what is called a single-slit diffraction pattern. A sketch of such a pattern is shown in ● CI Fig. 35.1. The diffraction pattern has a bright central region. Other, less intense regions are symmetrically distributed around the central region. These bright regions, or bright bands, are called maxima and are regions of constructive interference. The dark regions in between are called minima and are regions of destructive interference.

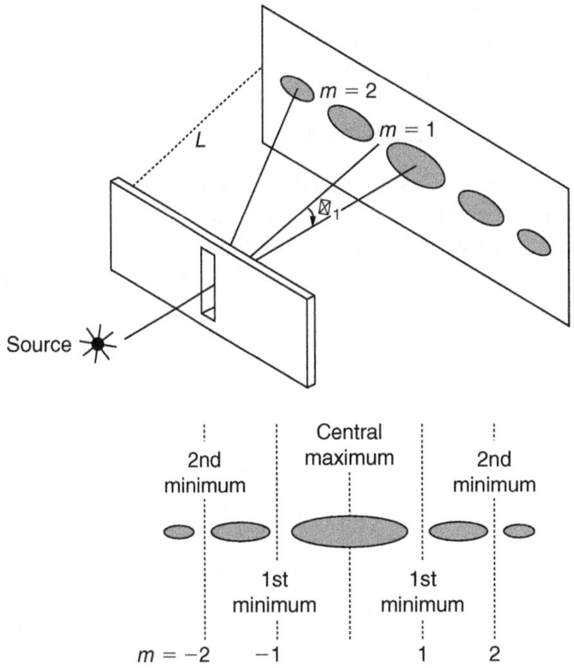

CI Figure 35.1 Single-slit diffraction. As the diagram shows, the minima are distributed symmetrically on both sides of the central maximum. The first minimum is designated $m = 1$ and occurs between the central maximum and the next bright band. After that, the centers of other dark bands are designated $m = 2$, $m = 3$, and so on. Traditionally, positive and negative numbers are used to distinguish one side from the other.

From an analysis of the geometry, it can be shown that the condition for dark bands, or minima, is given by

$$\boxed{\omega \sin\theta = m\lambda \quad m = 1, 2, 3, \ldots} \qquad \textbf{(CI 35.1)}$$

(condition for dark fringes)

where ω is the width of the slit, λ is the wavelength of the light, and θ is the angle to the center of a particular band minimum designated by $m = 1, 2, 3, \ldots$. The m number is called the order number, and the bands are referred to as first-order, second-order, third-order, and so on.*

Note the geometry in ● CI Fig. 35.2, where $\tan\theta_1 = y_1/L$. Experimentally, $y \ll L$, and using the small-angle approximation $\tan\theta_1 \approx \sin\theta_1 \approx y_1/L$ for the first-order minimum, we can in general write CI Eq. 35.1 as

$$\boxed{y_m \approx \frac{mL\lambda}{\omega} \quad m = 1, 2, 3, \ldots} \qquad \textbf{(CI 35.2)}$$

(small angles only)

*CI Eq. 35.1 is sometimes written $\omega \sin\theta = m\lambda$, $m = \pm 1, \pm 2, \pm 3, \ldots$, where the plus and minus numbers are used to indicate dark bands on opposite sides of the central maximum.

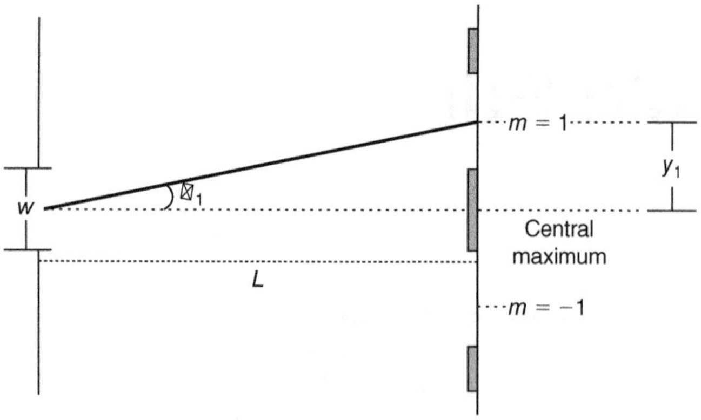

CI Figure 35.2 Geometry of the single-slit diffraction pattern. The first-order minimum, $m = 1$, is at a distance y_1 from the central maximum. In experimental conditions, L is much larger than ω and y_1 $(L \gg y_1)$.

where y_1 is the distance between the center of the central maximum and the center of the first-order minimum, and so on (see CI Fig. 35.2).

B. Double-Slit Interference

When light passes through two slits, the diffraction pattern is again bright-and-dark regions, but regions smaller than those seen with the single slit. These small dots are usually called fringes. ● CI Fig. 35.3 shows a diagram of the fringes of this interference pattern.

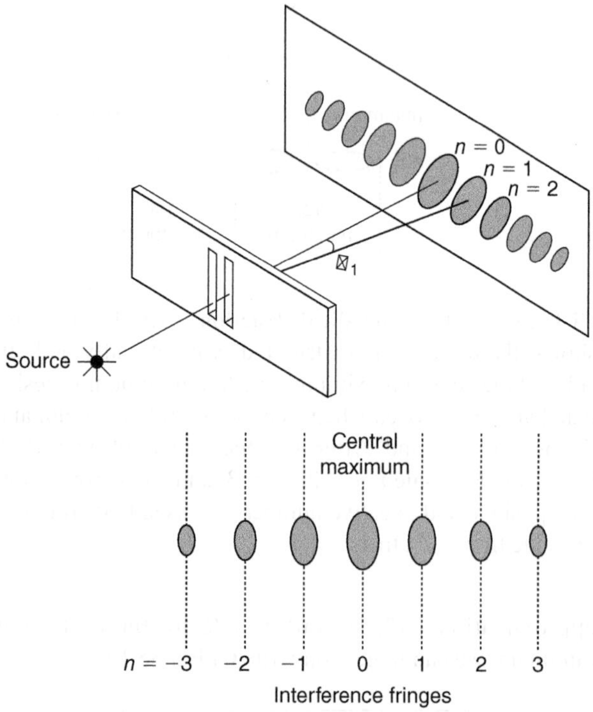

CI Figure 35.3 Double-slit interference pattern. The interference pattern from two slits produces a smaller and sharper set of bright and dark fringes than the diffraction pattern from a single slit.

From the geometry, it can be shown that the positions of the bright fringe maxima are given by

$$d \sin\theta = n\lambda \quad n = 1, 2, 3, \ldots \tag{CI 35.3}$$

(condition for bright fringes)

where d is the distance between the double slits, θ is the angular distance between the central maximum and another bright fringe of order n, and λ is the wavelength of the light.

Using a small-angle approximation as before, we find that CI Eq. 35.3 becomes

$$y_n = \frac{nL\lambda}{d} \quad n = 0, 1, 2, 3, \ldots \tag{CI 35.4}$$

(lateral distances to bright fringes, small angles only)

SETTING UP DATA STUDIO

1. Open Data Studio, and choose "Create Experiment."
2. The Experiment Setup window will open, and you will see a picture of the Science Workshop interface. There are seven channels to choose from. (Digital channels 1, 2, 3, and 4 are the small buttons on the left; analog channels A, B, and C are the larger buttons on the right.)
3. Click on the channel A button in the picture. A window with a list of sensors will open.
4. Choose the Light Sensor from the list, and press OK.
5. Connect the sensor to channel A of the interface, as shown on the computer screen.
6. In the same window, under Measurement, select Light Intensity, and deselect all others. Set the Sample Rate to 20 Hz.
7. Now click on the Channel 1 button in the picture to access the list of sensors again.
8. Choose the Rotary Motion Sensor (RMS) from the list, and press OK.
9. Connect the RMS to channels 1 and 2 of the interface, as shown on the computer screen.
10. On the same window, adjust the properties of the RMS as follows:

    First Measurements tab: select Position, Ch 1&2 and deselect all others.

    Rotary Motion Sensor tab: set the Resolution to high (1440 divisions/rotations); and set the Linear Scale to Rack & Pinion.

    Set the Sample Rate to 20 Hz.

    The Data list on the left of the screen should now have three icons: one for voltage, one for light intensity, and one for the position data.

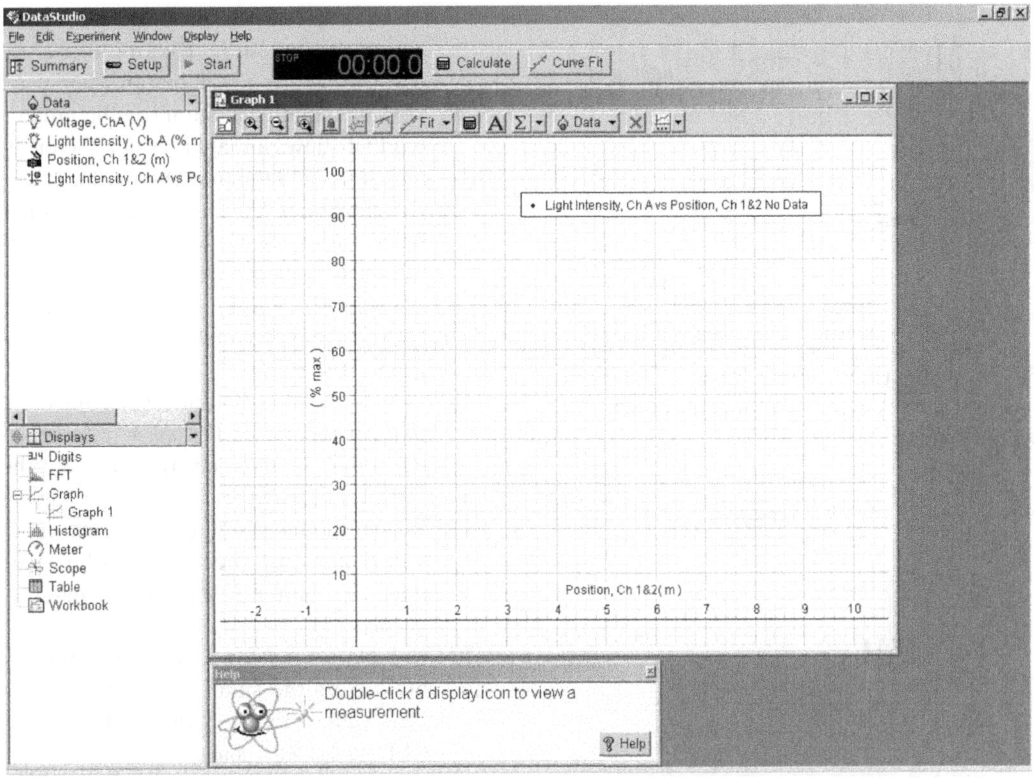

CI Figure 35.4 Data Studio setup. A light sensor, together with a rotary motion sensor, will be used to produce a plot of light intensity versus position for diffraction and interference patterns. (Reprinted courtesy of PASCO Scientific.)

11. Create a graph by dragging the Light Intensity icon from the Data list and dropping it on top of the Graph icon of the Displays list. A graph of intensity versus time will open in a window called Graph 1.
12. Now drag the Position icon from the data list, and drop it on top of the horizontal axis of the graph. The horizontal axis will change to measure position instead of time. ● CI Fig. 35.4 shows what the screen should look like after the setup is complete.

CI EQUIPMENT SETUP

1. Mount the single-slit accessory to the optics bench. The slit disks are mounted on a ring that snaps into an empty lens holder. Rotate the ring in the lens holder so that the slits at the center of the ring are vertical in the holder. Then tighten the screw on the holder. (See ● CI Fig. 35.5.)
2. Align the laser beam with the slit.
   a. Mount the diode laser at one end of the bench. Put the slit holder a few centimeters away from the laser, with the disk side closer to the laser. Plug in the laser, and turn it on.
   b. Adjust the position of the beam from left-to-right and up-to-down until the beam is centered on the slit. The knobs to do this are on the back of the diode laser. (See ● CI Fig. 35.6.)

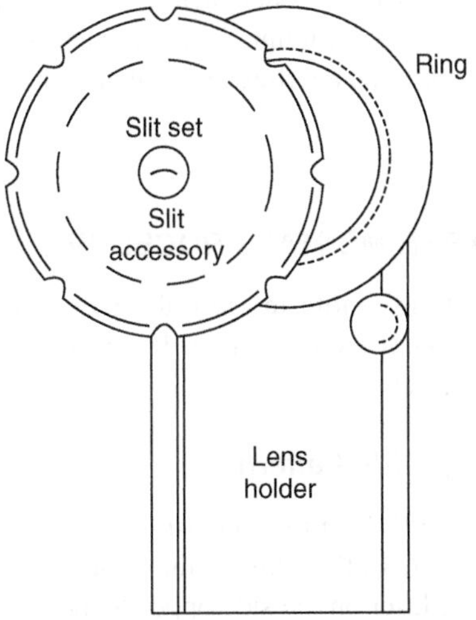

CI Figure 35.5 Slit accessory on lens holder. The slit accessory is mounted on a ring that snaps into the lens holder.

3. Prepare the rotary motion sensor and the light sensor.
   a. Mount the RMS in the rack of the linear translator. Then mount the linear translator to the end of the optics bench.
   b. The light sensor with the aperture bracket (set to slit 6) is mounted on the RMS rod clamp. (See ● CI Fig. 35.7.)
4. Plug the RMS and the light sensor into the interface, as shown in the Setup window on the computer screen.

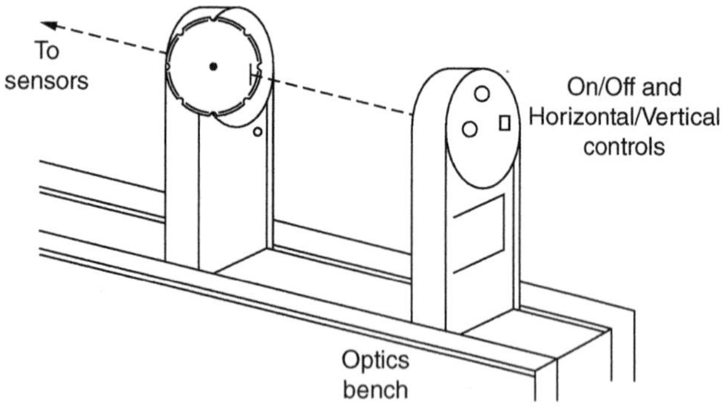

CI Figure 35.6 Diode laser and slits on track. The diode laser is placed in the track a few centimeters behind the lens holder with the slits.

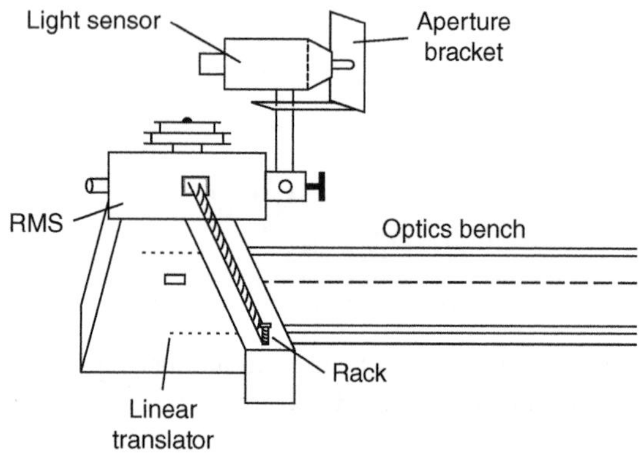

CI Figure 35.7 The RMS mounted in the rack with the light sensor. The RMS is mounted on the linear translator. The light sensor with the apertures is mounted on the RMS.

CI EXPERIMENTAL PROCEDURE

Start by making a note in the Laboratory Report of the wavelength (λ) of the laser light. It is printed on the back of the diode laser.

A. The Single-Slit Pattern

1. Select the 0.04-mm wide single slit from the disk. Make a note of the value in the Laboratory Report.
2. Place the laser on the side opposite the light sensor on the track. The slit disk should be a few centimeters in front of the laser. Record in the Laboratory Report the distance (L) between the slit and the sensor.
3. Set the light sensor aperture bracket to slit 6.
4. Turn the laser on, and set the gain switch to $\times 10$. If the light intensity goes offscale when you are measuring, turn it down to $\times 1$.
5. The diffraction pattern should be visible on the aperture bracket of the light sensor. Move the light sensor to one side of the pattern.
6. Turn the classroom lights off.
7. Press the START button, and *slowly* move the sensor across the pattern by rotating the large pulley of the RMS. Click the STOP button when you are finished.
8. Use the magnifier button (on the graph toolbar, a button that shows a magnifier lens) to enlarge the central maximum and the first maximum on each side.
9. Use the Smart-Tool (on the graph toolbar, a button labeled with xy-axes) to measure the distance between the centers of the first minima on the two sides of the central maximum. That is, measure the distance between $m = -1$ and $m = 1$. Record the value in CI Data Table 1.
10. Determine the distance y_m from the center of the pattern to one of the $m = 1$ minima by dividing the previous distance by 2. Record the result in CI Data Table 1.
11. Repeat the measurements for the second-, the third-, and if possible the fourth-order minima. Use the magnifier button to enlarge the parts of the graph as needed.
12. Calculate $\sin\theta$ for each case, using the derived formula from the small-angle approximation. (See CI Data Table 1 for the formula.) Enter the values in both CI Data Tables 1 and 2.
13. To check how well the observed pattern matches the theory, use the known wavelength of the light and the known width of the slit to calculate the theoretical value of $\sin\theta$ for each case. Compare the theory to the experiment by taking percent differences. Record all results in CI Data Table 1.
14. To demonstrate that the experimental data can also be used to find the wavelength of the light, use the data in CI Data Table 2 with CI Eq. 35.1 to calculate the wavelength of the light for each case; then find an average. Compare the average to the expected value by taking a percent error.
15. Cancel all zooms, and fix up the graph window so that all data collected can be seen. Print the graph and label each minimum, on both sides of the center, with the appropriate m value. Title this graph "Graph 1. Single-Slit Pattern, $\omega = 0.04$ mm." If no printer is available, make a careful sketch of the graph, paying attention to the location of the minima along the horizontal axis. Attach the graph to the Laboratory Report.

B. The Double-Slit Pattern

1. Change the slit accessory to a multiple-slit disk and realign the laser, if needed. Choose the double slit with slit separation 0.25 mm and slit width 0.04 mm.
2. Set the light sensor aperture bracket to slit 4.
3. The interference pattern should be visible on the aperture bracket of the light sensor. Move the light sensor to one side of the pattern.
4. Turn the classroom lights off.

5. Press the START button, and *slowly* move the sensor across the pattern by rotating the large pulley of the RMS. Click the STOP button when you are finished.
6. Use the magnifier button to enlarge the central maximum and the first maximum on each side.
7. Use the Smart-Tool to measure the distance between the first maxima on the two sides of the central maximum. That is, measure the distance between $n = 1$ and $n = -1$. Record the value in CI Data Table 3.
8. Determine the distance y_n to one of the $n = 1$ fringes by dividing the previous distance by 2. Record the result in CI Data Table 3.
9. Repeat the measurements for the second-, third-, and if possible all the way to the sixth-order maxima. Use the magnifier button to enlarge the parts of the graph as needed.
10. Calculate $\sin\theta$ for each case, using the derived formula from the small-angle approximation. Enter the values in both CI Data Tables 3 and 4.
11. To check how well the observed pattern matches the theory, use the known wavelength of the light and the known separation of the slits to calculate the theoretical value of $\sin\theta$ for each case. Compare the theory to the experiment by taking percent differences. Record all results in CI Data Table 3.
12. To demonstrate that this experimental data can also be used to find the wavelength of the light, use the data in CI Data Table 4 to calculate the wavelength of the light for each case; then find an average. Compare the average to the expected value by taking a percent error.
13. Cancel all zooms, and fix up the graph window so that all data collected can be seen. Print the graph and label each maximum, on both sides of the center, with the appropriate n value. Title this graph "Graph 2. Double-Slit Pattern, $d = 0.25$ mm, $\omega = 0.04$ mm." If no printer is available, make a careful sketch of the graph, paying attention to the location of the maxima along the horizontal axis. Attach the graph to the Laboratory Report.

C. Comparing Single-Slit Pattern to Double-Slit Pattern

1. Change the double-slit set to a set with slit separation 0.25 mm and slit width 0.08 mm.
2. Collect data as before, and print the graph. Label on the graph the maxima with their appropriate n values. Title this graph "Graph 3. Double-Slit Pattern, $d = 0.25$ mm, $\omega = 0.08$ mm."
3. Repeat with the double-slit set of slit separation 0.50 mm and slit width 0.04 mm. This time, title the graph "Graph 4. Double-Slit Pattern, $d = 0.50$ mm, $\omega = 0.04$ mm."

Name ______ Section ______ Date ______

Lab Partner(s) ______

CI EXPERIMENT 35

Single-Slit and Double-Slit Diffraction

CI Laboratory Report

Wavelength of light λ ______

A. The Single-Slit Pattern

Slit width ω ______

Distance between slit and pattern L ______

CI DATA TABLE 1

Purpose: To compare the experimental single-slit pattern with the pattern predicted by theory.

Order of distance	Distance from $-m$ to m	Distance y_m (from center to m)	Calculated $\sin\theta \approx \frac{y_m}{L}$	Predicted $\sin\theta = \frac{m\lambda}{\omega}$	Percent difference
$m = 1$					
$m = 2$					
$m = 3$					
$m = 4$					

CI DATA TABLE 2

Purpose: To determine the wavelength of light using a single-slit diffraction pattern.

Order of minimum	Calculated $\sin\theta \approx \frac{y_m}{L}$	Calculated λ
$m = 1$		
$m = 2$		
$m = 3$		
$m = 4$		
	Average =	
	Percent error =	

Be sure to attach a copy of the graph to the report.

Don't forget units

(continued)

B. The Double-Slit Pattern

Slit width ω ____________

Separation of slits d ____________

Distance between slit and pattern L ____________

CI DATA TABLE 3

Purpose: To compare the experimental double-slit pattern with the pattern predicted by theory.

Order of minimum	Distance from $-n$ *to* n	Distance y_n (from center to n)	Calculated $\sin\theta \approx \frac{y_n}{L}$	Predicted $\sin\theta = \frac{n\lambda}{d}$	Percent difference
$n = 1$					
$n = 2$					
$n = 3$					
$n = 4$					
$n = 5$					
$n = 6$					

CI DATA TABLE 4

Purpose: To determine the wavelength of light using a double-slit interference pattern.

Order of minimum	Calculated $\sin\theta \approx \frac{y_m}{L}$	Calculated λ
$n = 1$		
$n = 2$		
$n = 3$		
$n = 4$		
	Average =	
	Percent error =	

Be sure to attach a copy of the graph to the report.

C. Comparing Single-Slit Pattern to Double-Slit Pattern

Attach the graphs to the report. Don't forget to label them appropriately so that it is easy to distinguish between them.

Name ______________________ Section ______________ Date ______________

Lab Partner(s) ____

EXPERIMENT 35 **Single-Slit and Double-Slit Diffraction** *Laboratory Report*

CI QUESTIONS

1. Comparison between Graphs 1 and 2:
   (a) What parameters of the experiment were kept constant in producing Graphs 1 and 2? What parameters were changed?

   (b) Compare the locations of the first minima of diffraction ($m = 1$ and $m = -1$ on Graph 1) to the same positions along the x-axis on Graph 2. Are the positions also minima in Graph 2?

   (c) In Graph 2, how many interference fringes (bright) are in between the locations of $m = 1$ and $m = -1$ of the single-slit pattern?

2. Comparison between Graphs 2 and 3:
   (a) What parameters of the experiment were kept constant in producing Graphs 2 and 3? What parameters were changed?

   (b) Describe all things that look different between Graphs 2 and 3. What is the effect of changing the slit width?

(continued)

3. Comparison between Graphs 2 and 4:
   (a) What parameters of the experiment were kept constant in producing Graphs 2 and 4? What parameters were changed?

   (b) Describe all things that look different between Graphs 2 and 4. What is the effect of changing the separation between the slits?

Name ______ Section ______ Date ______

Lab Partner(s) ______

EXPERIMENT 36

Detection of Nuclear Radiation: The Geiger Counter

TI Advance Study Assignment

Read the experiment and answer the following questions.

1. What is the principle of operation of the Geiger tube?

2. Define each of the following: (a) threshold voltage, (b) cumulative ionization, (c) plateau, and (d) dead time.

3. Are any radiations counted when the tube voltage is below the threshold voltage? Explain.

(continued)

4. Approximately how many volts above the threshold voltage is the normal operating voltage of the Geiger tube, and why is the operating voltage selected this way?

5. What is background radiation?

6. How does the count rate vary with distance from a point source? If the counter is moved twice the distance from the source, how is the count rate affected?

NUCLEAR SAFETY RULES

Radioactive sources will be used in the next few experiments. Some sources are solids and are encapsulated to prevent contact. However, liquid sources may also be used and transferred during an experiment. Some general safety precautions for the use of radioactive materials follow:

1. Radioactive materials should be used only by or under the supervision of a person properly informed about the nature of the material.
2. Care should be taken to avoid unnecessary handling or contact with the skin.
3. Mouth pipetting is strictly prohibited.
4. Eating, drinking, and smoking should not be permitted in any area where radioactive materials are being used.
5. Protective gloves or forceps should be used when the material is handled or transferred.
6. All persons working with radioactive material should thoroughly wash their hands immediately afterward.
7. When not in use, radioactive materials should be stored in an appropriately labeled container and in a place of limited access.
8. Should an accident occur (particularly if it involves radioactive materials), it should be reported immediately to the laboratory instructor.
9. If you are pregnant, make your instructor aware of this, and do not go to the laboratory.

PRE-LAB DEMONSTRATION | **Instructor's Choice**

EXPERIMENT 36

Detection of Nuclear Radiation: The Geiger Counter

INTRODUCTION AND OBJECTIVES

Nuclear radiation (alpha particles, beta particles, and gamma rays) cannot be detected directly by our senses. Hence, some observable detection method employing the interaction of nuclear decay particles with matter must be used. There are several methods, but the most common is the **Geiger tube**.* In a Geiger tube, the particles from radioactive decay ionize gas molecules, giving rise to electrical pulses that can be amplified and counted. The total instrument is referred to as a Geiger counter.

In this experiment, the characteristics of a Geiger tube and the inverse-square relationship for nuclear radiation will be investigated.

After performing this experiment and analyzing the data, you should be able to:

1. Explain the principle of operation of the Geiger counter and its major disadvantage.
2. Describe how the count rate of a Geiger counter varies with its distance from a radioactive source.

*Sometimes referred to as a *Geiger-Müller tube* (or G-M tube). A prototype was developed in 1913 by the German physicist Hans Geiger (1882–1945), who worked in England on experiments that led to our present nuclear model of the atom. The tube was improved in 1928 in collaboration with the German physicist S. Müller.

EQUIPMENT NEEDED

- Geiger counter (rate meter or scaler type)
- Radioactive source [for example, Cs-137 (beta-gamma)]
- Laboratory timer or stopwatch
- Calibrated mounting board or meterstick
- 2 sheets of Cartesian graph paper [or 1 sheet of Cartesian and (*optional*) 1 sheet of log (log-log) graph paper (3-cycle)]

THEORY

The three types of nuclear radiation—alpha, beta, and gamma—are all capable of ionizing a gas. The degree of ionization depends on the energy of the particles and the amount of radiation absorbed by the gas. The ionization of gas molecules by nuclear radiation is the principle of the Geiger tube.

A **Geiger tube** consists of a fine wire running axially through a metal cylinder filled with a gas, usually argon, at a pressure of about 0.1 atm (● Fig. 36.1). A potential difference or voltage is maintained between the central wire and the cylinder, the central wire being at a positive potential (+) with respect to the cylinder (−).

Energetic nuclear particles (ionizing radiation) passing through the cylinder and entering the tube ionize the gas molecules. The freed electrons are attracted toward the wire and the positive ions toward the cylinder. If the voltage between the wire and cylinder is great enough, the accelerated electrons acquire enough energy to ionize other gas molecules on their way to the positive wire. The electrons from the secondary ionizations produce additional ionizations. This process is called **cumulative ionization**.

As a result, an "avalanche" discharge sets in, and a current is produced in the resistor. This reduces the potential difference between wire and cylinder to the point where cumulative ionization does not occur. After the momentary current pulse, which lasts on the order of microseconds, the potential difference between the wire and the cylinder resumes its original value.

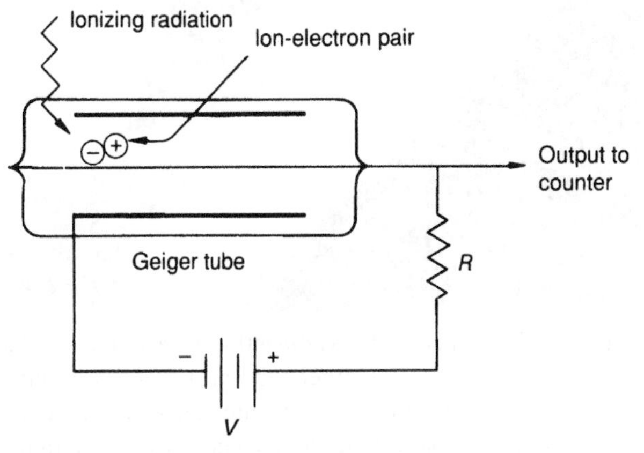

Figure 36.1 Geiger tube. A schematic diagram of the Geiger tube and circuit. See text for description.

A finite time is required for the discharge to be cleared from the tube. During this time, the voltage of the tube is less than that required to detect other radiation that might arrive. This recovery time is referred to as the **dead time** of the tube. If a large amount of radiation arrives at the tube, the counting rate (counts per minute, or cpm) as indicated on the counting equipment will be less than the true value.

There are two common types of Geiger tubes—a "normal" or side-window tube and an "end-window" tube. The side-window tube has a relatively thick wall that may not be penetrated by less penetrating radiation such as alpha particles (● Fig. 36.2). The end-window tube has a thin end window, usually of mica, and may be thin enough to be penetrated by very energetic alpha particles.

The brief change in the potential that occurs when a discharge takes place in the tube produces a voltage pulse that can be detected and counted by appropriate instrumentation. Common instruments used for counting are scalers and count-rate meters.

A *scaler* displays the cumulative number of counts on a lighted panel. By using a separate timer, the number of counts per minute (cpm) can be obtained. Some scalers have internal timers that stop the counting after a preset time interval.

A *rate meter* displays the average counting rate directly via a dial needle (Fig. 36.2). The needle reading fluctuates back and forth. This is due to the electronic averaging of the number of counts received during a short period of time. A scaler timer is usually preferred over a rate meter because of this effect.

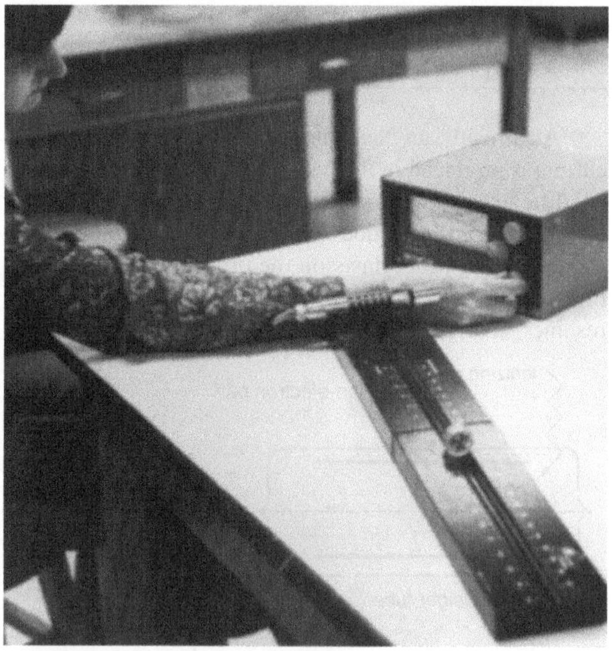

Figure 36.2 Apparatus for radioactive experiments. The standard side-window Geiger tube probe on the mounting board is connected to a count-rate meter. A radioactive source is on the board in the foreground. Notice the radioactivity warning sign on the source. (© Cengage Learning.)

A. Tube Voltage and Count Rate

When a Geiger tube is in the vicinity of a radiation source with particles of varying energy and there is no voltage on the tube, no counts are observed on the counter. (Counters usually have a loudspeaker circuit so that the counts may also be heard as audible "clicks.") If the tube voltage is slowly increased from zero, then at some applied voltage, counts will be observed. The lowest applied voltage that will produce a count in the instrument is called the *starting voltage* or **threshold voltage** (● Fig. 36.3).

As the tube voltage is increased above the threshold voltage, the number of counts per minute increases rapidly. In this region (about 50 V wide, beginning at about 600 V to 700 V, depending on the tube), the count rate is almost linearly proportional to the voltage. This is because as the voltage increases, more of the less energetic particles are counted. Hence, in this region, the tube discriminates between particles of different energy. At a given voltage, only particles above a certain energy are detected. The tube then acts as a proportional counter—the voltage being proportional to the energies of the incident particles.

Eventually, as the voltage is increased, the number of counts per minute becomes almost independent of the applied voltage (the level region in Fig. 36.3). This region (about 200 V wide) is called the **plateau** of the tube. A change in voltage has little effect on the number of counts detected. Normally, the Geiger tube is operated at a voltage in about the middle of the plateau. Fluctuations in the applied voltage from the power supply will then have little effect on the counting rate.

The tube voltage should never be raised to a value far above that of the end of the plateau. At such high voltages, a continuous discharge sets in, and if allowed to persist, this may destroy the tube.

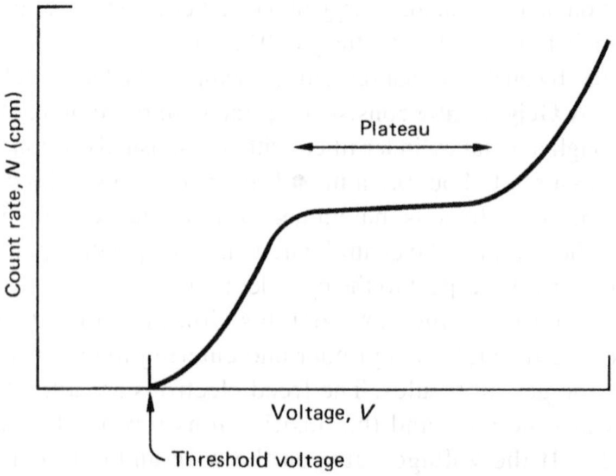

Figure 36.3 Count rate versus tube voltage. A typical graph showing how the count rate varies with Geiger tube voltage. Normal operation is the plateau region. See text for description.

B. Inverse-Square Relationship

In normal operation, the count rate depends on the number of particles per unit time entering the Geiger tube. Hence, the count rate depends on the distance of the tube from the source. For a point source emitting a total of N_o particles/min, the particles are emitted in all directions. The number of particles/min, N', passing through a unit area of a sphere of radius r is

$$N' = \frac{N_o}{A} = \frac{N_o}{4\pi r^2} \quad \text{(counts/min/area)} \tag{36.1}$$

where $A = 4\pi r^2$ is the circular cross-sectional area of the sphere.

A Geiger tube with a window area A' at a distance r from a point source then intercepts or receives N counts/min, given by

$$N = N'A' = \frac{N_o A'}{4\pi r^2} \tag{36.2}$$

Although the effective area A' of the Geiger tube is usually not known, the equation shows that the count rate is inversely proportional to r^2 (inverse-square form):

$$N \propto \frac{1}{r^2} \tag{36.3}$$

Hence, for a point source, the count rate "falls off" as $1/r^2$ with the distance from the source.

EXPERIMENTAL PROCEDURE

***Caution:** Review the radiation safety procedures before performing this experiment.*

1. Connect the Geiger tube probe to the counter by means of the coaxial cable. Before plugging the counter into an ac outlet, familiarize yourself with the controls, particularly the high-voltage control.

Scaler: Set the high-voltage control to the minimum setting.

Rate meter: Set the high-voltage control to the minimum setting. The off-on switch is commonly on the high-voltage control. A selector switch is labeled with volts and counts per minute multiplier positions ($\times 1$, $\times 10$, etc.). When the Geiger tube voltage is adjusted by means of the high-voltage control, the selector switch should *always* be set on "volts."

The selector switch is then turned to the appropriate count multiplier range for counting. The meter display scale usually has dual calibrations in volts and counts per minute.

2. Plug in and turn on the counter. Place the radioactive source near the Geiger tube, with the source facing the probe opening as in Fig. 36.2. (A tube mount may be available for an end-window tube. The source is placed at the bottom of the tube mount in this case. *Note:* The end window is very fragile and can be punctured easily.)

Slowly increase the tube voltage by means of the high-voltage control until the first indication of counting is observed. Then increase the voltage to about 75 V to 100 V above this value.

3. Set the counter to the counting mode, and adjust the distance of the source from the tube (or add aluminum sheets to the end-window tube mount) so that the count rate is several thousand (3000 to 5000) counts per minute. The Geiger tube is then operating normally, and the dead time will not cause serious error.

A. Tube Voltage and Count Rate

4. Lower the high-voltage control to the minimum setting. Then, slowly raise the voltage until the first indication of counting is observed (rate meter selection on "volts"). Record this threshold voltage in Data Table 1.

5. Increase the voltage to 25 V above the threshold voltage, and record the tube voltage. Measure and record the number of counts per minute at this voltage setting. (A rate meter is switched to a counting position. Because the meter needle fluctuates, it is best to watch the meter for 30 s and note the highest and lowest meter readings. The count rate is then taken as the mean or average of these readings.)

6. Continue to repeat Procedure 5, increasing the voltage by 25 V each time. Record the voltage and the corresponding count rate for each voltage setting. You will notice that the count rate first increases rapidly with voltage. It then levels off, increasing only slightly with increases in voltage. This is the plateau region of the Geiger tube.

Eventually, with a particular voltage step, a sharp increase in the count rate will be observed. *Do not increase the voltage above this value.* Quickly lower the tube voltage to the minimum setting after this reading to avoid damaging the tube.

7. Plot the count rate N (counts/min) versus voltage V on Cartesian graph paper. Include the threshold voltage. Draw a smooth curve that best fits the data.

B. Background Radiation

8. Remove the source several meters (across the room) from the Geiger tube, and apply the midplateau voltage to the tube as determined from the graph. (If using an end-window tube with a tube mount, remove the tube from the mount and lay the tube on the table.)

You will observe an occasional count on the counter. This is due to background radiation arising from cosmic rays and radioactive elements in the environment (for example, in building materials).

Let the counter run for a measured time, for example, 4 min to 5 min, and determine the background count rate in counts per minute and record this value in Part B of the Laboratory Report. If the background count rate is small compared to the source count, it may be considered negligible.

C. Inverse-Square Relationship

9. Bring the source toward the Geiger tube, and locate the source at a distance from the tube where the counting rate begins to increase significantly over background. Record the distance r and the count rate N in Data Table 2. Record this r as the farthest distance.

10. Then bring the source relatively close to the tube, and determine the distance from the source that gives a full-scale count rate. Record the count rate and distance (closest).

11. Divide the length between the two measured distances into eight intervals or steps. Measure and record the count rate and source distance from the tube for each step as the source is moved away from the tube.

12. The inverse-square relationship $N = A/r^2$ (where A is a constant) can be put into linear form by taking the logarithm of both sides:

$$\log N = \log(Ar^{-2}) = \log r^{-2} + \log A$$

or

$$\boxed{\log N = -2 \log r + \log A} \qquad \textbf{(36.4)}$$

where log is the common logarithm (base 10). Note that Eq. 36.4 has the form of a straight line: $y = mx + b$.

Take the logs of the r and N values in Data Table 2. On Cartesian graph paper, plot $\log N$ versus $\log r$, and draw a straight line that best fits the data. Determine the slope of the line, and compare it to the theoretical value by finding the percent error.

(Optional) Your instructor may wish to introduce you to log-log graph paper. This special graph paper automatically takes the log values. See Appendix D for a discussion of graphing on log-log and semi-log graph papers. (There is optional use of the latter in Experiments 37 and 38.)

Name ______________________ Section ____________ Date ____________
Lab Partner(s) ______________________

EXPERIMENT 36

Detection of Nuclear Radiation: The Geiger Counter

TI *Laboratory Report*

A. Tube Voltage and Count Rate

DATA TABLE 1

Purpose: To determine dependence of the count rate on tube voltage.

	Tube voltage	Count rate (cpm)
Threshold voltage		X

B. Background Radiation

Number of counts ____________

Counting time (min) ____________

Counts/min ____________

Don't forget units

(continued)

C. Inverse-Square Relationship

DATA TABLE 2

Purpose: To determine the count rate versus distance from source.

	Source-to-counter distance r ( )	Count rate N (cpm)
Closest distance		
Farthest distance		

log r	log N

Calculations (show work)

Slope of graph _______________

Theoretical value _______________

Percent error _______________

Name ______________________ Section ______________ Date ______________

Lab Partner(s) ____

TI QUESTIONS

1. What is the average percent increase in the count rate over the voltage range of the Geiger tube plateau? (Obtain from a graph of the data.)

2. If (a) dead time and (b) background radiation corrections were taken into account, how would each correction affect the graph of N versus V?

3. Give possible reasons why the experimental result of N versus r is not exactly an inverse-square relationship.

4. A count rate of 8000 cpm is recorded at a distance of 5.0 cm from a point source. What would be the observed count rate at a distance of 20 cm?

Name ________________ Section ________________ Date ________________

Lab Partner(s) ________________

EXPERIMENT 37

Radioactive Half-Life

TI Advance Study Assignment

Read the experiment and answer the following questions.

1. What is the significance of the half-life of a radioactive isotope in terms of (a) the amount of sample or number of nuclei, and (b) the activity of the sample?

2. What is the *decay constant?* Is it the same for each decay process? What are the units of the decay constant?

3. How is the half-life related to the decay constant of a radioactive process?

4. What is meant by *milking a cow?* Give the technical terms for *milking* and *cow.*

(continued)

5. Ba-137m is a nuclear isomer of Ba-137. Explain what this means.

6. If a particular radioactive sample undergoes four half-lives, what fraction of the original material remains?

EXPERIMENT 37

Radioactive Half-Life

INTRODUCTION AND OBJECTIVES

The decrease in the activity of a radioactive isotope is characterized by its half-life. This is the time required for one-half of the nuclei of a sample to decay. Of course, the nuclei of a sample cannot be counted directly, but when one-half of the sample has decayed, the activity, or the rate of emission of nuclear radiation, has also decreased by one-half. Thus, as we monitor the sample with a Geiger counter, when the count rate (counts per minute, cpm) has decreased by one-half, one half-life has elapsed.

In this experiment, the half-life of a radioactive isotope will be determined.

After performing this experiment and analyzing the data, you should be able to:

1. Explain what is meant by the half-life of a radioactive isotope.
2. Distinguish between radioactive half-life and time constant.
3. Describe how the half-life of a short-lived radioactive isotope can be measured.

EQUIPMENT NEEDED

- Geiger counter (rate meter with clip mount or scaler type with tube mount)
- Cesium-137/Barium-137m Minigenerator with solution
- Laboratory timer or stopwatch
- Disposable planchet (small, metal, cuplike container to hold radioactive sample)
- 2 sheets of Cartesian graph paper [or (*optional*) 1 sheet of Cartesian and 1 sheet of semi-log graph paper (3-cycle)]

THEORY

The **activity** of a radioactive isotope is proportional to the quantity of isotope present, and the radioactive decay process is described by an exponential function:

$$\boxed{N = N_o e^{-\lambda t} = N_o e^{-t/\tau}} \qquad \textbf{(37.1)}$$

where N is the number of nuclei present at time t, N_o is the original number of nuclei present (at $t = 0$), λ is the decay constant of the process, and the time constant $\tau = 1/\lambda$. The variable N can also represent the activity (cpm) of an isotope sample.

The **half-life $t_{1/2}$** is the time it takes for the number of nuclei present, or activity, to decrease by one-half ($N = N_o/2$). Hence,

$$\frac{N}{N_o} = \frac{1}{2} = e^{-t_{1/2}/\tau}$$

Taking the ln of both sides,

$$\ln(1/2) = \ln\left(e^{-t_{1/2}/\tau}\right) = \frac{-t_{1/2}}{\tau}$$

and

$$\ln(1/2) = -0.693$$

So,

$$t_{1/2} = 0.693\tau$$

and

$$\boxed{t_{1/2} = 0.693\tau = \frac{0.693}{\lambda}} \qquad \textbf{(37.2)}$$

Thus the half-life can be computed if the time constant or the decay constant is known.

Example 37.1 A radioactive sample has an activity of 4000 cpm. What is the observed activity after three half-lives?

Solution After one half-life, the activity decreases by $\frac{1}{2}$, and after another half-life by another $\frac{1}{2}$, and so on. Hence, after three half-lives, the initial activity decreases by a factor of $\frac{1}{2} \times \frac{1}{2} \times \frac{1}{2} = \frac{1}{8}$. With $N_o = 4000$ cpm,

$$N = \frac{1}{8} N_o = \frac{1}{8}(4000) = 500 \text{cpm}$$

Notice that, in general,

$$N = \frac{N_o}{2^n}$$

where n is the number of half-lives.

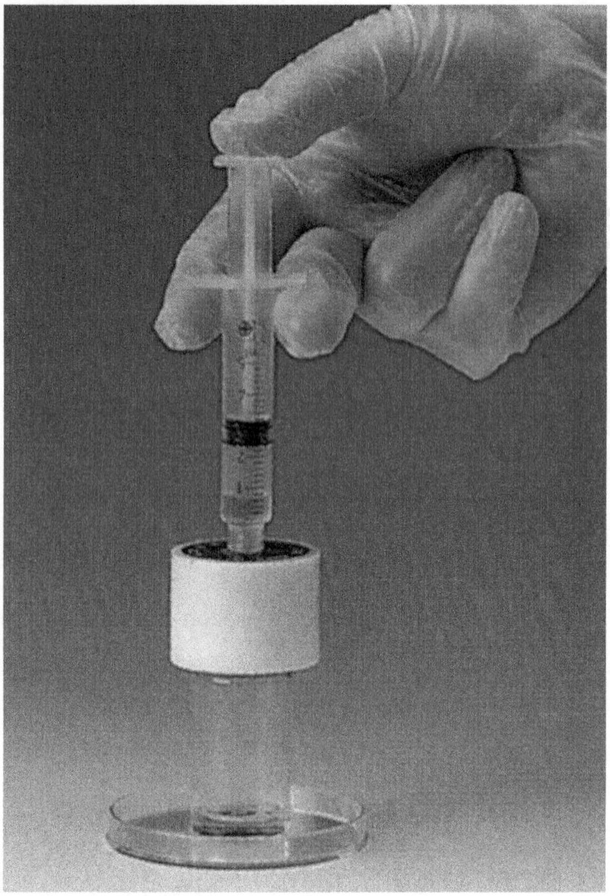

Figure 37.1 The Cesium-137/Barium-137m Minigenerator system. See text for description. (Fisher Scientific Company, LLC.)

Theory of Minigenerator

The Cesium-137/Barium-137m Minigenerator* (● Fig. 37.1) is an eluting system, in which a short-lived daughter radioactive isotope is eluted (separated by washing) from a long-lived parent isotope. A small "generator" contains radioactive Cs-137, which has a half-life of 30 years. Cs-137 beta-decays into Ba-137m, which is an isomeric (excited) state of the stable nucleus Ba-137. The excited isomer Ba-137m has a relatively short half-life and gamma-decays into Ba-137.

The nuclear equation for the decay is

$$\underset{\text{(cesium)}}{{}^{137}_{55}\text{Cs}} \longrightarrow \underset{\substack{\text{(barium)}\\\text{(excited)}}}{{}^{137}_{56}\text{Ba}^*} + \underset{\text{(electron)}}{{}^{0}_{-1}\text{e}}$$

$$\hookrightarrow \underset{\text{(barium)}}{{}^{137}_{56}\text{Ba}} + \underset{\text{(gamma ray)}}{\gamma}$$

where the asterisk (*) indicates an excited state. The original nucleus, ^{137}Cs, is commonly referred to as the *parent* nucleus, and the resulting nucleus (^{137}Ba) is called the *daughter* nucleus.*

The Ba-137m is washed or eluted from the generator by passing a hydrochloric acid–saline solution through the generator. Because of this process, the generator is commonly referred to as a "cow," and the Ba-137m is said to be "milked" from the cow. The generator "cow" may be milked many times, but as with an actual cow, a time interval must elapse between milkings.

Eluting removes the Ba-137m from the generator, and time is required for the "regeneration" of Ba-137m from the decay of Cs-137. Normally, the parent and daughter isotopes exist in equilibrium, with equal activities. After eluting (or milking the cow), it takes about 12 min for the Ba-137m to build up and again reach equilibrium with the Cs-137.

EXPERIMENTAL PROCEDURE

***Caution:** Review the radiation safety procedures at the beginning of Experiment 36.*

1. Before the Minigenerator is eluted, turn on the Geiger counter. Apply the appropriate tube voltage for normal operation (see Experiment 36). Over a period of 4 min or 5 min, measure and record the count of the background radiation, as was done in Experiment 36.

2. Mount the Geiger probe so that a planchet with the radioactive Ba-137m sample can be quickly and carefully placed below and near the probe opening at a fixed distance.

3. The counting procedure is as follows. When the sample is in place, the laboratory timer is started ($t = 0$) and allowed to run continuously. Simultaneously with the starting of the timer, the activity is measured on the Geiger counter for 15 s, and the count rate (cpm), together with the time elapsed on the timer, is recorded in the data table.

   Note: If using a rate meter, take the average of the high and low meter readings over the 15-s interval as the count rate. If a scaler is used (with or without an internal timer), the count rate in cpm must be computed. For example, suppose that 500 counts are observed for the 15 s ($\frac{1}{4}$-min) interval. The count rate is then 500 counts/$\frac{1}{4}$ min $= 500 \times 4 = 2000$ cpm.

   Repeat the 15-s count of activity at the beginning of each minute of elapsed time for 10 min to 12 min. A dry run of the counting procedure is helpful.

4. The instructor will "milk the cow" or supervise you in doing so. Only a few (2 to 3) drops of the eluate (milk) are needed.

*Registered trademark, Union Carbide.

*There doesn't seem to be a *son* nucleus in nuclear physics.

Caution: *Care should be taken in handling the sample. The milking should be done over a sheet of paper that can be discarded in case of a spill, and* ***if you should come in contact with the sample, immediately wash your hands****.*

The instructor may wish to give you a sample for a trial run of the counting procedure.

5. When given the actual data sample, carry out the counting procedure as described above. Record in the Data Table.

6. Correct for background radiation, if necessary. Plot the sample activity (N) in cpm versus the elapsed time (t) in minutes on Cartesian graph paper, and note the shape of the curve.

From the graph, make two determinations of the half-life by finding the time required for the sample activity to decay from its initial value to $\frac{1}{2}$ of the initial value, and from $\frac{1}{2}$ to $\frac{1}{4}$ of the initial value. Average and compare with the half-life for Ba-137m in Appendix Table A9 by computing the percent error. Also compute the decay constant (λ) from the average value of the half-life.

7. The decay constant may be found graphically by putting the exponential function, $N = N_o e^{-\lambda t}$, into linear form by taking the natural logarithm (base e) of both sides:

$$\ln N = \ln(N_o e^{-\lambda t}) = \ln(e^{-\lambda t}) + \ln N_o$$

or

$$\boxed{\ln N = -\lambda t + \ln N_o} \quad \textbf{(37.3)}$$

Note that Eq. 37.3 has the form of a straight line: $y = mx + b$.

Find ln N for each value of N in the Data Table. (Make a column for these to the right of the table).

Plot ln N versus t on Cartesian graph paper, and draw a straight line that best fits the data. Determine the slope of the line, and compare it to the value of the decay constant computed in the preceding procedure by finding the percent difference.

(*Optional*) Your instructor may wish to introduce you to semi-log graph paper. This special graph paper automatically takes the log values of the variable plotted on the Y-axis. See Appendix D for a discussion of graphing on semi-log (and log-log) graph paper.

Name ______________________ Section ______________ Date ______________

Lab Partner(s) ____

EXPERIMENT 37

Radioactive Half-Life

TI *Laboratory Report*

Background count (cpm) ______________

DATA TABLE

Purpose: To determine the half-life of Ba-137m.

t Elapsed time (min)	*N* Observed activity (cpm)	Corrected for background radiation (cpm)

Calculations (show work)

Half-life Measurements for Ba-137m

Time from full to $\frac{1}{2}$ activity ______________

Time from $\frac{1}{2}$ to $\frac{1}{4}$ activity ______________

Average experimental half-life ______________

Accepted half-life ______________

Percent error ______________

Decay constant from average half-life ______________

Decay constant from graph ______________

Percent difference ______________

Don't forget units

(continued)

TI QUESTIONS

1. In the experiment, if the Ba-137m sample were placed closer to the Geiger tube, the measured activity would be greater (inverse-square relationship). Would this affect the result of the half-life? Explain how this would affect the N-versus-t graph on Cartesian graph paper (or on semi-log paper).

2. A cobalt-60 source has a measured activity of 12,000 cpm. After how long would the observed activity be 750 cpm? (The half-life of Co-60 is 5.27 years.)

3. An instructor buys a 10-μCi Cs-137 source for laboratory experiments. After 5 years, what is (a) the strength of the source in μCi; (b) the activity of the source in disintegrations per second? (1 Ci $= 3.70 \times 10^{10}$ disintegrations/s.) (c) What is the strength of the source in becquerels (Bq)? (The becquerel is the official SI unit: 1 Ci $= 3.7 \times 10^{10}$ Bq.)

4. Cesium-136 is also radioactive and decays into barium-136. Write the nuclear equation for this reaction.

Name ______________________ Section ______________ Date ______________

Lab Partner(s) ______________________

EXPERIMENT 38

The Absorption of Nuclear Radiation

TI Advance Study Assignment

Read the experiment and answer the following questions.

1. On what parameters does the absorption of nuclear radiation depend?

2. Do the three basic types of nuclear radiation have definite ranges of penetration in materials? Explain. What is meant by *half-thickness*?

3. What is the mass absorption coefficient, and what are its units? Are there any advantages to using the mass absorption coefficient rather than the linear absorption coefficient? Explain.

(continued)

4. Explain how a source that has only one radioactive isotope can emit both beta and gamma radiation.

5. Why is a beta-gamma source that is shielded with a relatively thin sheet of aluminum effectively a gamma source?

EXPERIMENT 38

The Absorption of Nuclear Radiation

INTRODUCTION AND OBJECTIVES

The observed activity of a radioactive source of a given strength depends on several factors—for example, the distance of the counter from the source. For a point source, the observed activity varies inversely with the distance from the source (inverse-square relationship). This decrease is due to the geometrical spreading of the emitted nuclear radiation outward from the source.

If a Geiger probe is a fixed distance from a long-lived source, the observed activity is relatively constant. However, if a sheet of material is placed between the source and the counter, a decrease in the activity may be observed. That is, the nuclear radiation is absorbed by the material. The amount of absorption depends on the type and energy of the radiation and on the kind and density of the absorbing material.

The absorption or degree of penetration of nuclear radiation is an important consideration in applications such as medical radioisotope treatment and nuclear shielding (for example, around a nuclear reactor). Also, in industrial manufacturing processes, the absorption of nuclear radiation is used to monitor and control automatically the thickness of metal and plastic sheets and films.

In this experiment, the absorption properties of various materials for different kinds of nuclear radiation will be investigated.

After performing this experiment and analyzing the data, you should be able to:

1. Describe the parameters on which the penetration of nuclear radiation in a material depends.
2. Explain the linear absorption coefficient, "half-thickness," and stopping range.
3. Explain the mass absorption coefficient.

EQUIPMENT NEEDED

- Geiger counter (rate meter or scaler type)
- Calibrated mounting board (or meterstick)
- Beta-gamma source (Cs-137 suggested)
- Set of cardboard, aluminum, and lead sheets (about 1 mm thick, 10 sheets of each)
- Laboratory timer or stopwatch
- Micrometer caliper
- 3 sheets of Cartesian graph paper [or (*optional*) 2 sheets of Cartesian graph paper and 1 sheet of semi-log graph paper (3-cycle)]

THEORY

The three types of nuclear radiation (alpha, beta, and gamma) are absorbed quite differently by different materials. The electrically charged alpha and beta particles interact with the material and produce ionizations along their paths. The greater the charge and the slower the particle, the greater the linear energy transfer (LET) and ionization along the path, and this determines the degree of penetration of the radiation. The absorption or degree of penetration of the radiation also depends on the density of the material.

Alpha particles are easily absorbed. A few centimeters of air and even a sheet of paper will almost completely absorb them. Hence, alpha particles do not generally penetrate the walls or window of an ordinary Geiger tube and so are not counted by this method.

Beta particles can travel a few meters in air or a few millimeters in aluminum before being completely absorbed. Beta radiation, then, does penetrate a Geiger tube. Both alpha and beta particles of a given energy therefore have a definite *range* of penetration in a particular material.

● Figure 38.1 illustrates the radiation intensity (in counts per minute, cpm) versus absorber thickness for a relatively low-density absorber for radiation from a beta-gamma source. The "bend" in the curve indicates the range of the beta radiation. The penetration for thickness beyond this is due to gamma radiation.

Gamma rays, which consist of electromagnetic radiation of very short wavelength, are not readily absorbed. A significant number of high-energy gamma rays can penetrate 1 cm or more of a dense material such as lead. In a given material, a beam of gamma rays is absorbed exponentially.

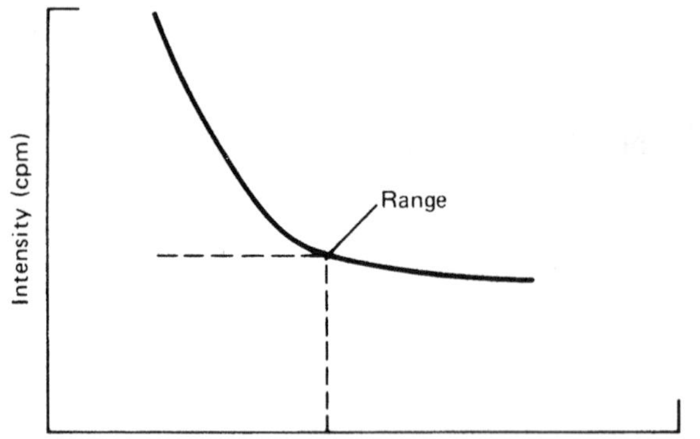

Figure 38.1 Radiation intensity versus absorber thickness. A typical graph of radiation intensity versus absorber thickness for beta-gamma radiation by a low-density absorber. The range is that of the beta radiation.

The intensity I (in cpm) of the beam after passing through a certain thickness x of a material is given by

$$\boxed{I = I_o e^{-\mu x}} \quad \textbf{(38.1)}$$

where I_o is the original intensity (at $x = 0$) and the decay constant μ is called the **linear absorption coefficient.** The absorption coefficient is characteristic of the absorbing material (and the wavelength or energy of the gamma radiation). Notice that the unit of μ is inverse length (such as 1/cm or cm^{-1}).

The absorption of gamma radiation of a given wavelength or energy is related to the atomic number of a substance and, macroscopically, to the density ρ of the material. Thus it is convenient to define a **mass absorption coefficient** μ_m:

$$\boxed{\mu_m = \frac{\mu}{\rho}} \quad \textbf{(38.2)}$$

The mass absorption coefficient provides a "standardized" coefficient. Samples of a particular absorbing material may have different densities. Each sample would have a different linear absorption coefficient μ, but the mass absorption coefficient μ_m would have the same value for all the samples.

Notice from Eq. 38.2 that the units of μ_m are cm^2/g:

$$\mu_m = \frac{\mu\ (1/\text{cm})}{\rho\ (\text{g/cm}^3)} = \frac{\mu}{\rho}\ (\text{cm}^2/\text{g})$$

If μ_m is used in Eq. 38.1 in place of μ, then

$$\boxed{I = I_o e^{-\mu x} = I_o e^{-(\mu/\rho)(x\rho)} = e^{-\mu_m x'}} \quad \textbf{(38.3)}$$

and the absorber thickness $x' = x\rho$ is in g/cm^2. Absorber thicknesses are frequently expressed in these units.

A beta-gamma source will be used to study the absorption of nuclear radiations. The decay scheme of the suggested Cs-137 source is illustrated in ● Fig. 38.2.

The chief decay mode (94%) is beta decay to the excited (isomeric) state of Ba-137. This decays by gamma

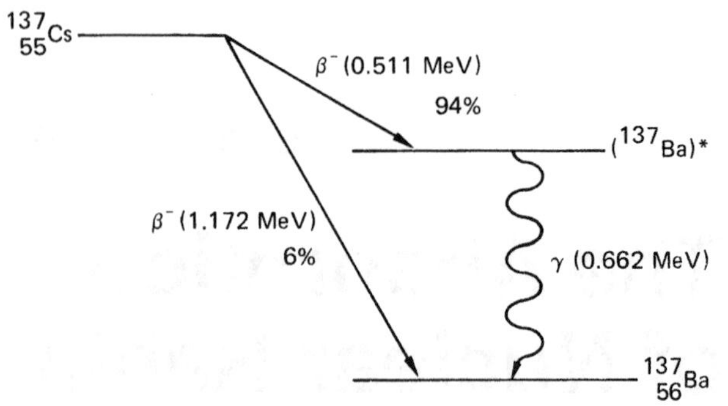

Figure 38.2 Decay scheme of Cs-137. Most of the cesium-137 (Cs-137) nuclei (94%) decay to an excited state of barium-137 ($^{137}Ba^*$), which then gamma-decays to a stable state.

emission to the stable ground state of Ba-137. Only 6% of the Cs-137 beta decays directly to ground-state Ba-137. Hence, for the most part, Cs-137 is a beta-gamma source of 0.511-MeV beta particles and 0.662-MeV gamma rays. (The emitted beta particles actually have a spectrum of energies from 0 MeV to 0.511 MeV.)

Since the gamma intensity decays exponentially, there is no definite penetrating or stopping range as there is in the case of beta radiation. Hence, it is convenient to speak in terms of a **half-thickness $x_{1/2}$**, the material thickness required to reduce the intensity by one-half (that is, $I_{1/2} = I_o/2$ or $I_{1/2}/I_o = \frac{1}{2}$). Then, by Eq. 38.1,

$$\frac{I_{1/2}}{I_o} = e^{-\mu x_{1/2}} = \frac{1}{2}$$

Taking the logarithm (base e) of both sides of the equation,

$$\ln(e^{-\mu x_{1/2}}) = \ln\frac{1}{2}$$

or

$$-\mu x_{1/2} = -\ln 2$$

and

$$\boxed{x_{1/2} = \frac{\ln 2}{\mu} = \frac{0.693}{\mu}} \quad \textbf{(38.4)}$$

Hence, knowing the absorption coefficient of a material, the half-thickness can be calculated.

EXPERIMENTAL PROCEDURE

Caution: *Review the radiation safety procedures at the beginning of Experiment 36.*

1. A radioactivity setup is shown in ● Fig. 38.3. First, measure the individual thickness of three different sheets of (a) cardboard, (b) aluminum, and (c) lead with the micrometer, and determine the average sheet thickness of each. Record in Data Table 1.

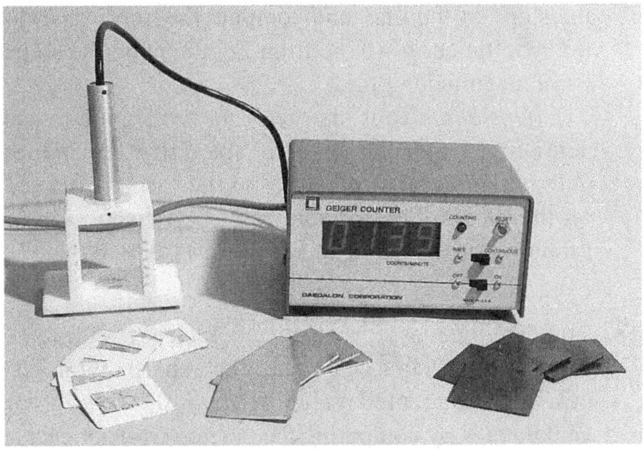

Figure 38.3 Geiger counter setup. (Fisher Scientific Company, LLC.)

2. Set up the Geiger counter with the probe on the mounting board (see Fig. 38.3). If an end-window tube is used, lay the tube in the mounting board groove and tape it down to immobilize it (or tape it to a meterstick). Turn on the counter. Place the radioactive source near the probe, and adjust the tube voltage to the plateau operating voltage.

A. Absorption of Beta Radiation

3. Adjust the distance of the source from the probe so that the observed count rate is about 8000 cpm. (For a rate meter, the count rate is taken as the average of the high and low meter readings for 30-s time intervals.) Record the count rate I_o in the cardboard column in Data Table 2.

   Place a sheet of cardboard between the source and the probe, and measure and record the count rate. (Allow a rate meter to come to equilibrium before taking a 30-s reading.)

4. Add cardboard sheets between the source and the probe one at a time, measuring and recording the count rate after the addition of each sheet. Continue until the count rate is relatively constant with the addition of four successive sheets.

5. Remove the cardboard sheets, and repeat the procedure with aluminum sheets.

6. Without recording data, repeat the procedure with lead sheets, and mentally note the degree of beta absorption or penetration in lead.

7. Plot the intensity I (in cpm) versus the number n of absorber sheets for both cardboard and aluminum on the same Cartesian graph. Dual label the ordinate (Y) axis so that the curve for each absorber occupies most of the graph paper.

   Determine the range of beta absorption for each absorber in sheet units from the graph, and record. Multiply each range (in sheet units) by the respective average sheet thickness to determine the range in length units.

B. Absorption of Gamma Radiation

8. Using the result of the range of beta absorption in aluminum from Procedure 7, place in front of the probe the minimum number of sheets of aluminum that will completely absorb the beta radiation. Then move the source toward the probe until the intensity observed on the Geiger counter is 700 cpm to 800 cpm.

   Record this intensity, I_o, in Data Table 3. The observed intensity is then almost solely due to gamma radiation. Why?

9. Leaving the aluminum sheet(s) in place, insert lead sheets one at a time between the aluminum sheets and the source. Measure and record the count rate after each sheet is inserted. Be careful not to move the source. Insert a total of 10 sheets of lead. After the sixth sheet, two sheets may be inserted at a time.

10. Remove all the sheets. Remove the source several meters (across the room) from the probe, and measure the background radiation intensity, I_b, over a 4-min to 5-min interval. (See Experiment 36 for a description of the procedure, if necessary.)

11. Subtract the background count rate from each reading for the lead sheets to obtain the corrected intensities. To find the half-thickness, plot the corrected intensity (I_c) versus the number (n) of lead sheets on Cartesian graph paper, and note the shape of the curve.

    From the graph, make a determination of the number of sheets ($n_{1/2}$) needed to reduce the intensity from its initial value to half of the initial value. (Try to express this number to the nearest 0.05 of a sheet.) The half-thickness is $x_{1/2} = x_i n_{1/2}$, where x_i is the thickness of an individual sheet. From the half-thickness, compute the linear absorption coefficient μ from Eq. 38.4, and record this value in Data Table 3.

12. The absorption coefficient may be found graphically by putting the exponential Eq. (38.1) into linear form by taking the natural (base e) logarithm of both sides. But first note that in terms of the number of sheets (n), Eq. 38.1 has the form

$$I = I_o e^{-\mu x} = I_o e^{-\mu(n x_i)} = I_o e^{-(\mu x_i)n} \quad \textbf{(38.5)}$$

where x_i is the individual sheet thickness and the absorber thickness is $x = n x_i$.

Then, taking the natural log of both sides of Eq. 38.5 yields

$$\ln I = \ln I_o e^{-(\mu x_i)n} = \ln e^{-(\mu x_i)n} + \ln I_o$$

or

$$\boxed{\ln I = -(\mu x_i)n + \ln I_o} \qquad \textbf{(38.6)}$$

Note that Eq. 38.6 has the form of a straight line: $y = mx + b$.

Find $\ln I_c$ for each value of I_c in Data Table 3. (Make a column for these to the right of the table.)

Plot $\ln I_c$ versus n on Cartesian graph paper, and draw a straight line that best fits the data. Determine the slope of the line, and compute the linear absorption coefficient μ. (Note from Eq. 38.6 that the slope has a magnitude of μx_i.)

(*Optional*) Your instructor may wish you to use semi-log graph paper. This special graph paper automatically takes in values of the variable plotted on the *y*-axis. See Appendix D for a discussion of graphing on semi-log (and log-log) graph paper.

13. Compute the mass absorption coefficient μ_m for lead, $\rho_{pb} = (11.3 \text{ g/cm}^3)$. Compare the experimental values to the accepted value of $\mu_m = 0.10 \text{ cm}^2/\text{g}$ for gamma rays by computing the percent error for each.

Name ______________________ Section ______________ Date ______________
Lab Partner(s) ______________________

EXPERIMENT 38

The Absorption of Nuclear Radiation

TI Laboratory Report

DATA TABLE 1

Purpose: To determine sheet thicknesses.

	Cardboard ( )	Aluminum ( )	Lead ( )
Average sheet thickness			

Calculations (show work)

Don't forget units

(continued)

A. Absorption of Beta Radiation

DATA TABLE 2

Purpose: To determine the relation between intensity and thickness.

Cardboard		Aluminum	
Number of sheets n	Intensity I (cpm)	Number of sheets n	Intensity I (cpm)
0	(I_o)	0	(I_o)
Range of beta radiation (cm)		Range of beta radiation (cm)	

Calculations
(show work)

Name ______________________ Section ______________ Date ______________

Lab Partner(s) ____

EXPERIMENT 38 **The Absorption of Nuclear Radiation** *Laboratory Report*

B. Absorption of Gamma Radiation

DATA TABLE 3

Purpose: To determine the relationship of intensity and thickness.

Number of lead sheets n	Intensity I (cpm)	Corrected intensity $I_c = I - I_o$
0	(I_o)	

Calculations
(show work)

Background Radiation

Number of counts ______________

Time interval ______________

Intensity I_b (cpm) ______________

(continued)

Absorption Coefficient Measurements for Gamma Rays

Number of sheets to reduce initial intensity to one-half $n_{1/2}$ ______________

Half-thickness $x_{1/2}$ ______________

Linear absorption coefficient μ ______________

Mass absorption coefficient μ_m ______________

Percent error ______________

Slope of graph x_i ______________

Linear absorption coefficient μ ______________

Mass absorption coefficient μ_m ______________

Percent error ______________

TI QUESTIONS

1. Was there a large difference in the percent errors of the experimental mass absorption coefficients? If so, why do you think this was the case?

2. Compute what percent of an incident beam of 0.662-MeV gamma rays is absorbed while passing through 2.5 mm of lead.

Name ______________________ Section ______________ Date ______________

Lab Partner(s) ____

3. Would the Cartesian graph of ln I_c versus n (or x) be a straight line if the gamma radiation contained gamma rays of two different energies? Explain.

4. The mass absorption coefficient of iron is 0.058 for 1.24-MeV gamma rays. What percentage (if any) of the beam of such gamma rays is transmitted through an iron plate 3 cm thick? ($\rho_{Fe} = 7.86\ g/cm^3$.)

Material Properties

Table A1 Densities of Materials

Substance	(g/cm^3)	(kg/m^3)
Solids		
Aluminum	2.7	2.7×10^3
Brass	8.4	8.4×10^3
Copper	8.9	8.9×10^3
Glass		
crown	2.5–2.7	$2.5–2.7 \times 10^3$
flint	3.0–3.6	$3.0–3.6 \times 10^3$
Gold	19.3	19.3×10^3
Iron and steel (general)	7.88	7.88×10^3
Lead	11.3	11.3×10^3
Nickel	8.8	8.8×10^3
Silver	10.5	10.5×10^3
Wood		
oak	0.60–0.90	$0.60–0.90 \times 10^3$
pine	0.35–0.50	$0.35–0.50 \times 10^3$
Zinc	7.1	7.1×10^3
Liquids		
Alcohol		
ethyl	0.79	0.79×10^3
methyl	0.81	0.81×10^3
Carbon tetra-chloride	1.60	1.60×10^3
Gasoline	0.68–0.75	$0.68–0.75 \times 10^3$
Glycerine	1.26	1.26×10^3
Mercury	13.6	13.6×10^3
Turpentine	0.87	0.87×10^3
Water	1.00	1.00×10^3
Gases (at STP):		
Air	0.001293	0.001293×10^3
Carbon dioxide	0.001975	0.001975×10^3
Helium	0.000179	0.000179×10^3
Hydrogen	0.000089	0.000089×10^3
Nitrogen	0.000125	0.000125×10^3
Oxygen	0.00143	0.00143×10^3

Table A2 Young's Modulus for Some Metals

Metals	(N/m^2)
Aluminum	6.5×10^{10}
Brass	9.0×10^{10}
Copper	12.0×10^{10}
Iron	
cast	9.0×10^{10}
wrought	19.0×10^{10}
Steel	19.2×10^{10}

Table A3 Coefficients of Linear Thermal Expansion

Substance	(1/°C)
Aluminum	24.0×10^{-6}
Brass	18.8×10^{-6}
Copper	16.8×10^{-6}
Glass	
window	8.5×10^{-6}
Pyrex	3.3×10^{-6}
Iron	11.4×10^{-6}
Lead	29.4×10^{-6}
Nickel	12.8×10^{-6}
Silver	18.8×10^{-6}
Steel	13.4×10^{-6}
Tin	26.9×10^{-6}
Zinc	26.4×10^{-6}

Table A4 Specific Heats

Substance	kcal/(kg-°C) or cal/(g-°C)	J/(kg-°C)
Aluminum	0.22	921
Brass	0.092	385
Copper	0.093	389
Glass	0.16	670
Iron	0.11	460
Lead	0.031	130
Mercury	0.033	138
Nickel	0.11	460
Silver	0.056	234
Steel	0.11	460
Tin	0.054	226
Water	1.00	4186
Zinc	0.093	389

Table A5 Color Code for Resistors (Composition Type)

Bands A and B		(ohms, Ω) Band C		Band D	
Color	Significant figure	Color	Multiplier	Color	Resistance tolerance (percent)
Black	0	Black	1	Silver	±10
Brown	1	Brown	10	Gold	±5
Red	2	Red	100	Red	±2
Orange	3	Orange	1000		
Yellow	4	Yellow	10,000		
Green	5	Green	100,000		
Blue	6	Blue	1,000,000		
Purple (violet)	7				
Gray	8	Silver	0.01		
White	9	Gold	0.1		

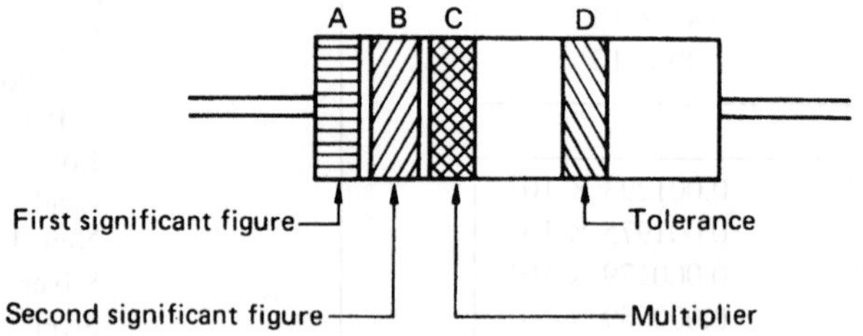

For example, if the bands on a resistor are red (A), black (B), orange (C), the resistance is 20 × 1000 = 20,000 Ω, or 20 kΩ.

TABLE A6 Resistivities and Temperature Coefficients

Substance	Resistivity ρ (Ω-cm)	Temperature coefficient (1/°C)
Aluminum	2.8×10^{-6}	0.0039
Brass	7×10^{-6}	0.002
Constantan	49×10^{-6}	0.00001
Copper	1.72×10^{-6}	0.00393
German silver (18% Ni)	33×10^{-6}	0.0004
Iron	10×10^{-6}	0.005
Manganin	44×10^{-6}	0.00001
Mercury	95.8×10^{-6}	0.00089
Nichrome	100×10^{-6}	0.0004
Nickel	7.8×10^{-6}	0.006
Silver	1.6×10^{-6}	0.0038
Tin	11.5×10^{-6}	0.0042

TABLE A7 Wire Sizes [American Wire Gauge (AWG)]

Gauge No.	Diameter	
	in.	cm
0000	0.4600	1.168
000	0.4096	1.040
00	0.3648	0.9266
0	0.3249	0.8252
1	0.2893	0.7348
2	0.2576	0.6543
3	0.2294	0.5827
4	0.2043	0.5189
5	0.1819	0.4620
6	0.1620	0.4115
7	0.1443	0.3665
8	0.1285	0.3264
9	0.1144	0.2906
10	0.1019	0.2588
11	0.09074	0.2305
12	0.08081	0.2053
13	0.07196	0.1828
14	0.06408	0.1628
15	0.05707	0.1450
16	0.05082	0.1291
17	0.04526	0.1150
18	0.04030	0.1024
19	0.03589	0.09116
20	0.03196	0.08118
21	0.02846	0.07229
22	0.02535	0.06439
23	0.02257	0.05733
24	0.02010	0.05105
25	0.01790	0.04547
26	0.01594	0.04049
27	0.01419	0.03604
28	0.01264	0.03211
29	0.01126	0.02860
30	0.01003	0.02548
31	0.008928	0.02268
32	0.007950	0.02019
33	0.007080	0.01798
34	0.006304	0.01601
35	0.005614	0.01426
36	0.005000	0.01270
37	0.004453	0.01131
38	0.003965	0.01007
39	0.003531	0.008969
40	0.003145	0.007988

TABLE A8 Major Visible Spectral Lines of Some Elements

Element	Wavelength (nm)	Color	Relative intensity
Helium	388.9	Violet	1000
	396.5 (near)	Violet	50
	402.6 (near)	Violet	70
	438.8	Blue-violet	30
	447.1	Dark blue	100
	471.3	Blue	40
	492.2	Blue-green	50
	501.5	Green	100
	587.6	Yellow	1000
	667.8	Red	100
	706.5	Red	70
Mercury	404.7	Violet	300
	407.8	Violet	150
	435.8	Blue	500
	491.6	Blue-green	50
	546.1	Green	2000
	577.0	Yellow	200
	579.0	Yellow	1000
	690.7	Red	125
Sodium	449.4	Blue	60
	449.8	Blue	70
	466.5	Blue	80
	466.9	Blue	200
	498.3	Green	200
	514.9	Green	400
	515.3	Green	600
	567.0	Green	100
	567.5	Green	150
	568.3	Green	80
	568.8	Green	300
	589.0	Yellow-orange	9000
	589.6	Yellow-orange	5000
	615.4	Orange	500
	616.1	Orange	500

Wavelengths of various colors

Color	Representative (nm)	General ranges (nm)
Red	650.0	647.0–700.0
Orange	600.0	584.0–647.0
Yellow	580.0	575.0–585.0
Green	520.0	491.2–575.0
Blue	470.0	424.0–491.2
Violet	410.0	400.0–420.0

Visible spectrum ≈ 400.0–700.0 nm

TABLE A9 Radioisotopes

Isotope	Half-life	Principal Radiations (MeV)		
		Alpha	Beta	Gamma
Barium-133	10.4 years			0.356
Bismuth-210	5.01 days	4.654, 4.691	1.161	
Carbon-14	5730 years		0.156	
Cesium-137	30.1 years		0.512, 1.173	
Barium-137m	2.6 min			0.662
Cobalt-60	5.26 years		0.315	
Iodine-131	8.07 days		0.606	
Lead-210	22.3 years		0.017, 0.061	0.0465
Manganese-54	312.5 days			0.835
Phosphorus-32	14.3 days		1.710	
Polonium-210	138.4 days	5.305		
Potassium-42	12.4 hours		3.52 1.97	
Radium-226	1600 years	4.781 4.598		0.186
Sodium-22	2.60 years	0.545 1.82		1.275
Strontium-90	28.1 years		0.546	
Thallium-204	3.78 years		0.763	
Uranium-238	4.5×10^6 years	4.195		0.48
Yttrium-90	64.0 hours		2.27	
Zinc-65	243.6 days		0.329	1.116

TABLE A10 Elements: Atomic Numbers and Atomic Weights

The atomic weights are based on ^{12}C = 12.0000. If the element does not occur naturally, the mass number of the most stable isotope is given in parentheses.

	Symbol	Atomic number	Atomic weight		Symbol	Atomic number	Atomic weight
Actinium	Ac	89	(227)	Mercury	Hg	80	200.59
Aluminum	Al	13	26.9815	Molybdenum	Mo	42	95.94
Americium	Am	95	(243)	Neodymium	Nd	60	144.24
Antimony	Sb	51	121.75	Neon	Ne	10	20.179
Argon	Ar	18	39.948	Neptunium	Np	93	(237)
Arsenic	As	33	74.9216	Nickel	Ni	28	58.71
Astatine	At	85	(210)	Niobium	Nb	41	92.9064
Barium	Ba	56	137.34	Nitrogen	N	7	14.0067
Berkelium	Bk	97	(247)	Nobelium	No	102	(253)
Beryllium	Be	4	9.01218	Osmium	Os	76	190.2
Bismuth	Bi	83	208.9806	Oxygen	O	8	15.9994
Boron	B	5	10.81	Palladium	Pd	46	106.4
Bromine	Br	35	79.90	Phosphorus	P	15	30.9738
Cadmium	Cd	48	112.40	Platinum	Pt	78	195.09
Calcium	Ca	20	40.08	Plutonium	Pu	94	(224)
Californium	Cf	98	(251)	Polonium	Po	84	(209)
Carbon	C	6	12.011	Potassium	K	19	39.102
Cerium	Ce	58	140.12	Praseodymium	Pr	59	140.9077
Cesium	Cs	55	132.9055	Promethium	Pm	61	(145)
Chlorine	Cl	17	35.453	Protactinium	Pa	91	(231)
Chromium	Cr	24	51.996	Radium	Ra	88	(226)
Cobalt	Co	27	58.9332	Radon	Rn	86	(222)
Copper	Cu	29	63.545	Rhenium	Re	75	186.2
Curium	Cm	96	(247)	Rhodium	Rh	45	102.9055
Dysprosium	Dy	66	162.50	Rubidium	Rb	37	85.4678
Einsteinium	Es	99	(254)	Ruthenium	Ru	44	101.07
Erbium	Er	68	167.26	Rutherfordium	Rf	104	(257)
Europium	Eu	63	151.96	Samarium	Sm	62	150.4
Fermium	Fm	100	(253)	Scandium	Sc	21	44.9559
Fluorine	F	9	18.9984	Selenium	Se	34	78.96
Francium	Fr	87	(223)	Silicon	Si	14	28.086
Gadolinium	Gd	64	157.25	Silver	Ag	47	107.868
Gallium	Ga	31	69.72	Sodium	Na	11	22.9898
Germanium	Ge	32	72.59	Strontium	Sr	38	87.62
Gold	Au	79	196.967	Sulfur	S	16	32.06
Hafnium	Hf	72	178.49	Tantalum	Ta	73	180.9479
Hahnium	Ha	105	(260)	Technetium	Tc	43	(99)
Helium	He	2	4.00260	Tellerium	Te	52	127.60
Holmium	Ho	67	164.9303	Terbium	Tb	65	158.9254
Hydrogen	H	1	1.0080	Thallium	Tl	81	204.37
Indium	In	49	114.82	Thorium	Th	90	232.0381
Iodine	I	53	126.9045	Thulium	Tm	69	168.9342
Iridium	Ir	77	192.22	Tin	Sn	50	118.69
Iron	Fe	26	55.847	Titanium	Ti	22	47.90
Krypton	Kr	36	83.80	Tungsten	W	74	183.85
Lanthanium	La	57	138.9055	Uranium	U	92	238.029
Lawrencium	Lr	103	(257)	Vanadium	Vy	23	50.9414
Lead	Pb	82	207.12	Xenon	Xe	54	131.30
Lithium	Li	3	6.941	Ytterbium	Yb	70	173.04
Lutetium	Lu	71	174.97	Yttrium	Y	39	88.9059
Magnesium	Mg	12	24.305	Zinc	Zn	30	65.37
Manganese	Mn	25	54.9380	Zirconium	Zr	40	91.22
Mendelevium	Md	101	(256)				

APPENDIX B

Mathematical and Physical Constants

TABLE B1 Metric Prefixes

Multiple		Name	Abbreviation
1,000,000,000,000,000,000	10^{18}	exa	E
1,000,000,000,000,000	10^{15}	peta	P
1,000,000,000,000	10^{12}	tera	T
1,000,000,000	10^{9}	giga	G
1,000,000	10^{6}	mega	M
1,000	10^{3}	kilo	k
100	10^{2}	hecto	h
10	10^{1}	deka	da
1	1	—	—
0.1	10^{-1}	deci	d
0.01	10^{-2}	centi	c
0.001	10^{-3}	milli	m
0.000001	10^{-6}	micro	μ
0.000000001	10^{-9}	nano	n
0.000000000001	10^{-12}	pico	p
0.000000000000001	10^{-15}	femto	f
0.000000000000000001	10^{-18}	atto	a

TABLE B2 Physical Constants

Acceleration due to gravity	g	$9.8\ \text{m/s}^2 = 980\ \text{cm/s}^2 = 32.2\ \text{ft/s}^2$
Universal gravitational constant	G	$6.67 \times 100^{-11} \dfrac{\text{N-m}^2}{\text{kg}^2}$
Electron charge	e	$1.60 \times 10^{-19}\ \text{C}$
Speed of light	c	$3.0 \times 10^{8}\ \text{m/s} = 3.0 \times 10^{10}\text{cm/s} = 1.86 \times 10^{5}\ \text{mi/s}$
Boltzmann's constant	k	$1.38 \times 10^{-23}\ \text{J/K}$
Planck's constant	h	$6.63 \times 10^{-34}\ \text{J-s} = 4.14 \times 10^{-15}\ \text{eV-s}$
	$\hbar$	$h/2\pi = 1.05 \times 10^{-34}\ \text{J-s} = 6.58 \times 10^{-16}\ \text{eV-s}$
Electron rest mass	m_e	$9.11 \times 10^{-31}\ \text{kg} = 5.49 \times 10^{-4}\ \text{u} \leftrightarrow 0.511\ \text{MeV}$
Proton rest mass	m_p	$1.672 \times 10^{-27}\ \text{kg} = 1.00783\ \text{u} \leftrightarrow 938.3\ \text{MeV}$
Neutron rest mass	m_n	$1.674 \times 10^{-27}\ \text{kg} = 1.00867\ \text{u} \leftrightarrow 939.1\ \text{MeV}$
Coulomb's law constant	k	$1/4\pi\varepsilon_o = 9.0 \times 10^{9}\ \text{N-m}^2/\text{C}^2$
Permittivity of free space	ε_o	$8.85 \times 10^{-12}\ \text{C}^2/\text{N-m}^2$
Permeability of free space	μ_o	$4\pi \times 10^{-7} = 1.26 \times 10^{-6}$ Wb/A-m (T-M/A)
Astronomical and Earth data		
Radius of the Earth		
equatorial		$3963\ \text{mi} = 6.378 \times 10^{6}\ \text{m}$
polar		$3950\ \text{mi} = 6.357 \times 10^{6}\ \text{m}$
average		6.4×10^{3} km (for general calculations)
Mass of the Earth		$6.0 \times 10^{24}\ \text{kg}$
the Moon		$7.4 \times 10^{22}\ \text{kg} = \frac{1}{81}$ mass of the Earth
the Sun		$2.0 \times 10^{30}\ \text{kg}$
Average distance of the Earth from the Sun		$93 \times 10^{6}\ \text{mi} = 1.5 \times 10^{8}\ \text{km}$
Average distance of the Moon from the Earth		$2.4 \times 10^{5}\ \text{mi} = 3.8 \times 10^{5}\ \text{km}$
Diameter of the Moon		$2160\ \text{mi} \approx 3500\ \text{km}$
Diameter of the Sun		$864{,}000\ \text{mi} \approx 1.4\ 10^{6}\ \text{km}$

TABLE B3 Conversion Factors

Mass	$1\ \text{g} = 10^{-3}\ \text{kg}$ $1\ \text{kg} = 10^{3}\ \text{g}$ $\quad$ $1\ \text{metric ton} = 1000\ \text{kg}$ $1\ \text{u} = 1.66 \times 10^{-24}\ \text{g} = 1.66 \times 10^{-27}\ \text{kg}$
Length	$1\ \text{cm} = 10^{-2}\ \text{m} = 0.394\ \text{in.}$ $1\ \text{m} = 10^{-3}\ \text{km} = 3.28\ \text{ft} = 39.4\ \text{in.}$ $1\ \text{km} = 10^{3}\ \text{m} = 0.621\ \text{mi}$ $1\ \text{in.} = 2.54\ \text{cm} = 2.54 \times 10^{-2}\ \text{m}$ $1\ \text{ft} = 12\ \text{in.} = 30.5\ \text{cm} = 0.305\ \text{m}$ $1\ \text{mi} = 5280\ \text{ft} = 609\ \text{m} = 1.609\ \text{km}$
Area	$1\ \text{cm}^2 = 10^{-4}\ \text{m}^2 = 0.1550\ \text{in}^2 = 1.08 = 10^{-3}\ \text{ft}^2$ $1\ \text{m}^2 = 10^{4}\ \text{cm}^2 = 10.76\ \text{ft}^2 = 1550\ \text{in}^2$ $1\ \text{in}^2 = 6.94 \times 10^{-3}\ \text{ft}^2 = 6.45\ \text{cm}^2 = 6.45 \times 10^{-4}\ \text{m}^2$ $1\ \text{ft}^2 = 144\ \text{in}^2 = 9.29 \times 10^{-2}\ \text{m}^2 = 929\ \text{cm}^2$
Volume	$1\ \text{cm}^3 = 10^{-6}\ \text{m}^3 = 3.53 \times 10^{-5}\ \text{ft}^3 = 6.10 \times 10^{-2}\ \text{in}^3$ $1\ \text{m}^3 = 10^{6}\ \text{cm}^3 = 10^{3}\ \text{liters} = 35.3\ \text{ft}^3 = 6.10 \times 10^{4}\ \text{in}^3 = 264\ \text{gal}$ $1\ \text{liter} = 10^{3}\ \text{cm}^3 = 10^{-3}\ \text{m}^3 = 1.056\ \text{qt} = 0.264\ \text{gal}$ $1\ \text{in}^3 = 5.79 \times 10^{-4}\ \text{ft}^3 = 16.4\ \text{cm}^3 = 1.64 \times 10^{-5}\ \text{m}^3$ $1\ \text{ft}^3 = 1728\ \text{in}^3 = 7.48\ \text{gal} = 0.0283\ \text{m}^3 = 28.3\ \text{liters}$ $1\ \text{qt} = 2\ \text{pt} = 946.5\ \text{cm}^3 = 0.946\ \text{liter}$ $1\ \text{gal} = 4\ \text{qt} = 231\ \text{in}^3 = 3.785\ \text{liters}$
Time	$1\ \text{h} = 60\ \text{min} = 3600\ \text{s}$ $1\ \text{day} = 24\text{h} = 1440\ \text{min} = 8.64 \times 10^{4}\ \text{s}$ $1\ \text{year} = 365\ \text{days} = 8.76 \times 10^{3}\ \text{h} = 5.26 \times 10^{5}\ \text{min} = 3.16 \times 10^{7}\ \text{s}$
Angle	$360° = 2\pi\ \text{rad}$ $180° = \pi\ \text{rad}$ $\quad$ $1\ \text{rad} = 57.3°$ $90° = \pi/2\ \text{rad}$ $60° = \pi/3\ \text{rad}$ $\quad$ $1° = 0.0175\ \text{rad}$ $45° = \pi/4\ \text{rad}$ $30° = \pi/6\ \text{rad}$
Speed	$1\ \text{m/s} = 3.6\ \text{km/h} = 3.28\ \text{ft/s} = 2.24\ \text{mi/h}$ $1\ \text{km/h} = 0.278\ \text{m/s} = 0.621\ \text{mi/h} = 0.911\ \text{ft/s}$ $1\ \text{ft/s} = 0.682\ \text{mi/h} = 0.305\ \text{m/s} = 1.10\ \text{km/h}$ $1\ \text{mi/h} = 1.467\ \text{ft/s} = 1.609\ \text{km/h} = 0.447\ \text{m/s}$ $60\ \text{mi/h} = 88\ \text{ft/s}$
Force	$1\ \text{newton (N)} = 10^{5}\ \text{dynes} = 0.225\ \text{lb}$ $1\ \text{lb} = 4.45\ \text{N}$ Equivalent weight of 1-kg mass on the Earth's surface $= 2.2\ \text{lb} = 9.8\ \text{N}$
Pressure	$1\ \text{Pa}\ (\text{N/m}^2) = 1.45 \times 10^{-4}\ \text{lb/in}^2 = 7.4 \times 10^{-3}\ \text{torr (mm Hg)}$ $1\ \text{tor (mm Hg)} = 133\ \text{Pa}\ (\text{N/m}^2) = 0.02\ \text{lb/in}^2$ $1\ \text{atm} = 14.7\ \text{lb/in}^2 = 1.013 \times 10^{5}\ \text{N/m}^2\ \text{(Pa)}$ $= 30\ \text{in. Hg} = 76\ \text{cm Hg}$ $1\ \text{bar} = 10^{5}\ \text{N/m}^2\ \text{(Pa)}$ $1\ \text{millibar} = 10^{2}\ \text{N/m}^2\ \text{(Pa)}$
Energy	$1\ \text{J} = 10^{7}\ \text{ergs} = 0.738\ \text{ft-lb} = 0.239\ \text{cal} = 9.48 \times 10^{-4}\ \text{Btu} = 6.24 \times 10^{18}\ \text{eV}$ $1\ \text{kcal} = 4186\ \text{J} = 3.968\ \text{Btu}$ $1\ \text{Btu} = 1055\ \text{J} = 778\ \text{ft-lb} = 0.252\ \text{kcal}$ $1\ \text{cal} = 4.186\ \text{J} = 3.97 = 10^{-3}\ \text{Btu} = 3.09\ \text{ft-lb}$ $1\ \text{ft-lb} = 1.356\ \text{J} = 1.29 \times 10^{-3}\ \text{Btu}$ $1\ \text{eV} = 1.60 \times 10^{-19}\ \text{J}$
Power	$1\ \text{W} = 0.738\ \text{ft-lb/s} = 1.34 \times 10^{-3}\ \text{hp} = 3.41\ \text{Btu/h}$ $1\ \text{ft-lb/s} = 1.36\ \text{W} = 1.82 \times 10^{-3}\ \text{hp}$ $1\ \text{hp} = 550\ \text{ft-lb/s} = 745.7\ \text{W} = 2545\ \text{Btu/h}$
Rest mass-energy equivalents	$1\ \text{u} = 1.66 \times 10^{-27}\ \text{kg} \leftrightarrow 931\ \text{MeV}$ $1\ \text{electron mass} = 9.11 \times 10^{-31}\ \text{kg} = 5.49 \times 10^{-4}\ \text{u} \leftrightarrow 0.511\ \text{MeV}$ $1\ \text{proton mass} = 1.672 \times 10^{-27}\ \text{kg} = 1.00728\ \text{u} \leftrightarrow 938.3\ \text{MeV}$ $1\ \text{neutron mass} = 1.674 \times 10^{-27}\ \text{kg} = 1.00867\ \text{u} \leftrightarrow 939.6\ \text{MeV}$

TABLE B4 Trigonometric Relationships

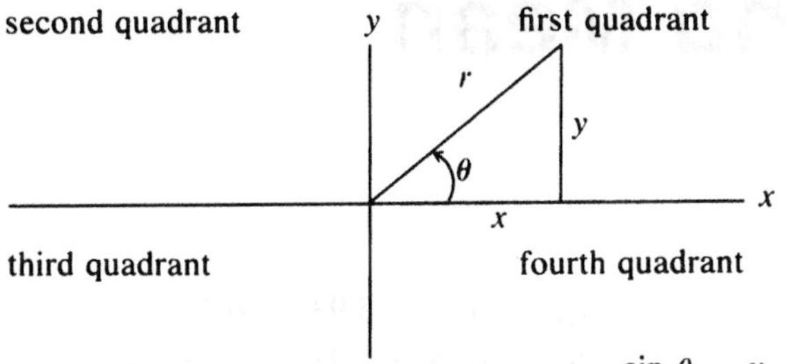

$\sin\theta = \frac{y}{r}$ $\quad\cos\theta = \frac{x}{r}$ $\quad\tan\theta = \frac{\sin\theta}{\cos\theta} = \frac{y}{x}$

θ (rad)	$\sin\theta$	$\cos\theta$	$\tan\theta$
0° (0)	0	1	0
30° ($\pi/6$)	0.500	0.866	0.577
45° ($\pi/4$)	0.707	0.707	1.00
60° ($\pi/3$)	0.866	0.500	1.73
90° ($\pi/2$)	1	0	$\rightarrow\infty$

The sign of trigonometric functions depends on the quadrant, or sign of x and y, for example, in the second quadrant $(-x, y)$, $-x/r = \cos\theta$ and $x/r = \sin\theta$, or by:

Reduction Formulas

	(θ in second quadrant)	(θ in third quadrant)	(θ in fourth quadrant)
$\sin\theta$ =	$\cos\theta\ (\theta - 90°)$	$= -\sin(\theta - 180°)$	$= -\cos(\theta - 270°)$
$\cos\theta$ =	$-\sin(\theta - 90°)$	$= -\cos(\theta - 180°)$	$= \sin(\theta - 270°)$

Fundamental Identities

$\sin^2\theta + \cos^2\theta = 1$

$\sin 2\theta = 2\sin\theta\cos\theta$

$\cos{}^2\theta = \cos^2\theta - \sin^2\theta = 2\cos^2\theta - 1 = 1 - 2\sin^2\theta$

$\sin^2\theta = \frac{1}{2}(1 - \cos 2\theta)$

$\cos^2\theta = \frac{1}{2}(1 + \cos 2\theta)$

For half-angle ($\theta/2$) identities, replace $\theta/2$, for example,

$\sin^2\theta/2 = \frac{1}{2}(1 - \cos\theta)$ $\qquad \cos^2\theta/2 = \frac{1}{2}(1 + \cos\theta)$

$\sin(\alpha \pm \beta) = \sin\alpha\cos\beta \pm \cos\alpha\sin\beta$

$\cos(\alpha \pm \beta) = \cos\alpha\cos\beta \mp \sin\alpha\sin\beta$

For very small angles:

$\cos\theta \approx 1$

$\sin\theta \approx \theta$ (radians) $\qquad \tan\theta = \frac{\sin\theta}{\cos\theta} \approx \theta$

Law of sines:

$$\frac{a}{\sin\alpha} = \frac{b}{\sin\beta} = \frac{c}{\sin\gamma}$$

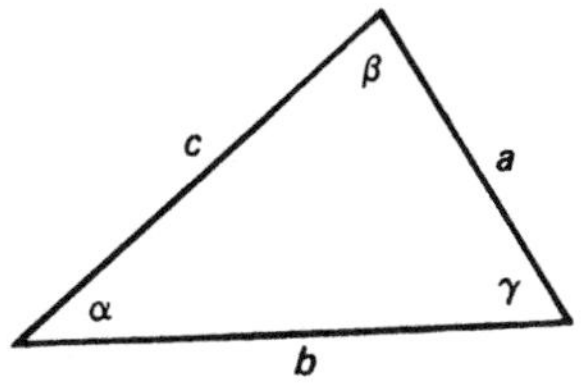

Law of cosines:

$a^2 = b^2 + c^2 - 2bc\cos\alpha$

$b^2 = a^2 + c^2 - 2ac\cos\beta$

$c^2 = a^2 + b^2 - 2ab\cos\gamma$

Absolute Deviation and Mean Absolute Deviation

ABSOLUTE DEVIATION

Having obtained a set of measurements and determined the mean value, it is helpful to report how widely the individual measurements are scattered from the mean. A quantitative description of this scatter, or dispersion, of measurements will give an idea of the precision of the experiment.

The **absolute deviation** $|d_i|$ is the absolute difference between a measured value (x_i) and the mean $(\bar{x})$ of a set of measurements.

$$|d_i| = |x_i - \bar{x}| \tag{C.1}$$

Mean Absolute Deviation

To obtain what is called the **mean** (or **average**) **absolute deviation** of a set of N measurements, the absolute deviations $|d_i|$ are determined (Eq. C.1).

The *mean absolute deviation* $\bar{d}$ is then

$$\bar{d} = \frac{|d_1| + |d_2| + |d_3| + \cdots + |d_N|}{N} = \frac{1}{N}\sum_{i=1}^{N} |d_i| \tag{C.2}$$

(The mean absolute deviation is sometimes referred to as simply the mean deviation.)

Example C.1 What is the mean deviation of the set of numbers given in Example 2.6 in Experiment 2?

Solution First find the absolute deviation of each of the numbers in Example 2.6, using the determined mean of 5.93.

$$|d_1| = |5.42 - 5.93| = 0.51$$
$$|d_2| = |6.18 - 5.93| = 0.25$$
$$|d_3| = |5.70 - 5.93| = 0.23$$
$$|d_4| = |6.01 - 5.93| = 0.08$$
$$|d_5| = |6.32 - 5.93| = 0.39$$

Then

$$\bar{d} = \frac{1}{N}\sum_{i=1}^{N} |d_i|$$
$$= \frac{0.51 + 0.25 + 0.23 + 0.08 + 0.39}{5} = 0.29$$

The mean absolute deviation is a measure of the dispersion of experimental measurements about the mean (that is, a measure of precision). It is common practice to report the experimental value E of a quantity in the form

$$E = \bar{x} \pm \bar{d}$$

In Example C.1, this would be $E = 5.93 \pm 0.29$. The $\pm$ term gives a measure of the *precision* of the experimental value. The *accuracy* of the mean value of a set of experimental measurements (5.93 in the example above) may be expressed in terms of percent error or percent difference.

The dispersion of an experimental measurements may be expressed by other means (such as the standard deviation; see Appendix D), so the method should be specified when reporting.

Standard Deviation and Method of Least Squares

STANDARD DEVIATION

To avoid the problem of negative deviations and absolute values, it is statistically convenient to use the square of the deviation.

The **variance σ^2** of a set of measurements is the average of the squares of the deviations:

$$\sigma^2 = \frac{(x_1 - \bar{x})^2 + (x_2 - \bar{x})^2 + (x_3 - \bar{x})^2 + \cdots + (x_N - \bar{x})^2}{N}$$
$$= d_1^2 + d_2^2 + d_3^2 + \cdots d_N^2$$
$$= \frac{1}{N}\sum_{i=1}^{N}(x_i - \bar{x})^2 = \frac{1}{N}\sum_{i=1}^{N} d_i^2 \qquad \textbf{(D.1)}$$

The square root of the variance σ is called the **standard deviation***:

$$\sigma = \sqrt{\frac{1}{N}\sum_{i=1}^{N}(x_i - \bar{x})^2} = \sqrt{\frac{1}{N}\sum_{i=1}^{N} d_i^2} \qquad \textbf{(D.2)}$$

Because we take the average of the squares of the deviations and then the square root, the standard deviation is sometimes called the **root-mean-square deviation**, or simply the **root mean square**. Notice that σ always has the same units as x_i and that it is always positive.

Example D.1 What is the standard deviation of the set of numbers given in Example 1.6 in Experiment 1?

Solution First find the square of the deviation of each of the numbers.

$$d_1^2 = (5.42 - 5.93)^2 = 0.26$$
$$d_2^2 = (6.18 - 5.93)^2 = 0.06$$
$$d_3^2 = (5.70 - 5.93)^2 = 0.05$$
$$d_4^2 = (6.01 - 5.93)^2 = 0.01$$
$$d_5^2 = (6.32 - 5.93)^2 = 0.15$$

Then

$$\sigma = \sqrt{\frac{1}{N}\sum_{i=1}^{5} d_i^2}$$
$$= \left(\frac{0.26 + 0.06 + 0.05 + 0.01 + 0.15}{5}\right)^{1/2}$$
$$= 0.33$$

The experimental value E is then commonly reported as

$$E = \bar{x} \pm \sigma = 5.93 \pm 0.33$$

The standard deviation is used to describe the precision of the mean of a set of measurements. For a normal distribution of random errors, it can be statistically shown that the probability that an individual measurement will fall within 1 standard deviation of the mean, which is assumed to be the true value, is 68% (● Fig. D.1). The

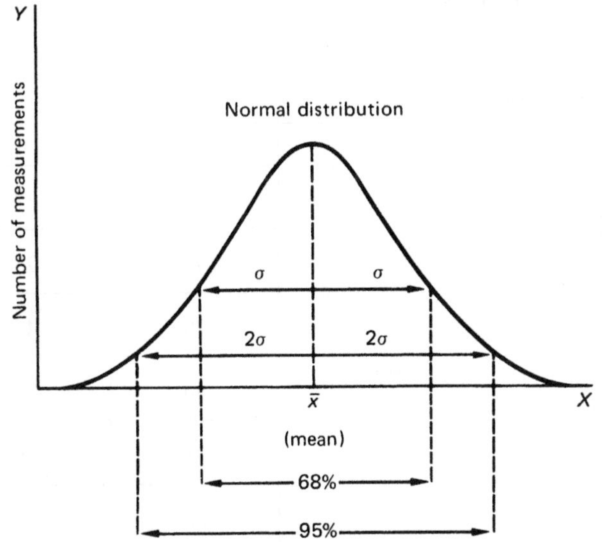

Figure D.1 Normal Distribution The *normal*, or *Gaussian*, distribution is represented by a "bell-shaped" curve. That is, the scatter, or dispersion, of the measurements is assumed to be symmetric about the true value of a quantity. The probability that an individual measurement will fall within 1 standard deviation (σ) of the mean, which is assumed to be the true value, is 68%, and within 2 standard deviations (2σ) is 95%.

*For a small number of measurements, it can be statistically shown that a better value of the standard deviation is given by $\sigma = \sqrt{[1/(N-1)]\Sigma d_i^2}$, where N is replaced by $N - 1$. Your instructor may want you to use this form of the standard deviation.

probability of a measurement falling within 2 standard deviations is 95%.

METHOD OF LEAST SQUARES

Let $y' = m'x + b'$ be the predicted equation of the best-fitting straight line for a set of data. The vertical deviation of the ith data point from this line is then $(y_i - y_i')$.

The principle of least squares may be stated as follows: The "best-fitting" straight line is the one that minimizes the sum of the squares of the deviations of the measured y values from those of the predicted equation $y' = m'x + b'$.

The numerical values of the slope m' and intercept b' that minimize the sum of the squares of the deviations, $\sum_{i=1}^{N}(y_i - y_i')^2$, may be found using differential calculus. The results are as follows:

$$m' = \frac{M_{xy}}{M_{xx}}$$

and

$$b' = \bar{y} - m'\bar{x}$$

where $\bar{x}$ and $\bar{y}$ are the mean values, $\bar{x} = \sum_{i=1}^{N} x_i$ and $\bar{y} = \sum_{i=1}^{N} y_i$,

and

$$M_{xy} = \sum_{i=1}^{N}(x_i - \bar{x})(y_i - \bar{y})$$

$$= \sum_{i=1}^{N} x_i y_i - \frac{\left(\sum_{i=1}^{N} x_i\right)\left(\sum_{i=1}^{N} y_i\right)}{N}$$

$$M_{xx} = \sum_{i=1}^{N}(x_i - \bar{x})^2 = \sum_{i=1}^{N} x_i^2 - \frac{\left(\sum_{i=1}^{N} x_i\right)^2}{N}$$

where the sums of the deviations, for example, $\sum_{i=1}^{N}(x_i - \bar{x})$, are zero.

Exercises

1. Plot the data given in Data Table 1 on a sheet of graph paper, and draw the straight line you judge to fit the data best.
2. Using the method of least squares, find the slope and intercept of the "best-fitting" straight line, and compare them with the slope and intercept of the line you drew in Exercise 1. Plot this "best-fitting" line on the graph. (Recall that the slope of a line is the change in y for a 1-unit increase in x.)

DATA TABLE 1

	y_i	x_i	x_i^2	$x_i\, y_i$
	25	12		
	44	28		
	78	47		
	80	70		
	43	16		
	58	53		
	95	72		
	67	38		
Sums (Σ)				

APPENDIX E

Graphing Exponential Functions

In some cases, exponential functions of the form

$$N = N_o e^{\lambda t} \quad (\text{or } y = Ae^{ax}) \tag{E.1}$$

are plotted on Cartesian coordinates in linear form by first taking the natural, or Naperian, logarithm (base e) of both sides of the equation. For example, $N = N_o e^{\lambda t}$,

$$\ln N = \ln (N_o e^{\lambda t}) = \ln N_o + \ln e^{\lambda t} = \ln N_o + \lambda t$$

or

$$\ln N = \lambda t + \ln N_o \tag{E.2}$$

Similarly, for $y = Ae^{ax}$,

$$\ln y = \ln A + \ln e^{ax} = \ln A + ax$$

or

$$\ln y = ax + \ln A \tag{E.3}$$

These equations have the general form of a straight line when plotted on a Cartesian graph ($y = mx + b$). For example, when we plot $\ln N$ versus t as Cartesian coordinates, the slope of the line is λ and the intercept is $\ln N_o$. The value of N_o is obtained by taking the *antilog* of the intercept value $\ln N_o$. (For a decaying exponential, $N = N_o e^{-\lambda t}$, the slope would be negative. Note that before plotting $\ln N$ versus t on Cartesian graph paper, we must find $\ln N$ for each value of N.)

Because logarithmic functions occur quite often in physics, special graph paper, called *semi-log graph paper*, is printed with graduations along the y or ordinate axis that are spaced logarithmically rather than linearly. The x or abscissa axis is graduated linearly. (Look at a sheet of semi-log graph paper.)

If a quantity is plotted on the ordinate axis of semi-log paper, the logarithmic graduated scale automatically takes the logarithm, so it is not necessary to look up the logarithm for each y value. However, commercial logarithmic graph paper is set up for common (base 10) logarithms rather than natural (base e) logarithms. Exponential functions may be treated as follows. Taking the (common) log of each side of $y = Ae^{ax}$ yields

$$\begin{aligned} \log y &= \log A + \log e^{ax} \\ &= \log A + ax \log e \\ &= \log A + (0.4343)ax \end{aligned} \tag{E.4}$$

where $\log e = 0.4343$.

Hence the slope of the resulting straight line is (0.4343)*a* rather than simply *a*.

The logarithmic ordinate scale is called "one-cycle," "two-cycle," and so on, depending on the number of powers of 10 covered on the axis. The beginnings of the cycles are consecutively labeled in multiples of 10 (for example, 0.1, 1.0, 10, or 1.0, 10, 100, etc.), depending on the range (cycles) of the function. (Common logarithms can also be plotted on semi-log paper.)

Care must be taken in determining the slope of the line on a semi-log plot. On an ordinary Cartesian graph, the slope of a line is given by $\Delta y/\Delta x = (y_2 - y_1)/(x_2 - x_1)$. However, on a semi-log graph, the slope of a line is given by

$$\text{slope} = \frac{\Delta \log y}{\Delta x} \left(\text{or } \frac{\Delta \log N}{\Delta t} \right) \tag{E.5}$$

On a semi-log plot, the listed ordinate values are y, not $\ln y$. Hence, one must explicitly take the logs of the ordinate values of the endpoints of the slope interval, y_2 and y_1, or the log of their ratio:

$$\begin{aligned} \text{slope} &= \frac{\Delta \log y}{\Delta x} = \frac{\log y_2 - \log y_1}{x_2 - x_1} \\ &= \frac{\log y_2/y_1}{x_2 - x_1} \end{aligned} \tag{E.6}$$

The value of N_o can be read directly from the y-intercept of the graph.

Another common equation form in physics is

$$y = ax^n \tag{E.7}$$

For example, the electric field, $E = kq/r^2 = kqr^{-2}$, is of this form, with $a = kq$ and $n = -2$. By plotting y versus x^n on Cartesian graph paper, we obtain a straight line with a slope of a. However, in an experiment, the measured values are usually y and x, so computation of the x^n's is required.

But in some instances the exponent n may not be known. This constant, along with the constant a, may be found by plotting y versus x on log graph paper. (This is commonly called *log-log graph paper* because of the logarithmic graduations on both axes. Look at a sheet of log-log graph paper.)

At logarithmic graduations on axes, we again automatically take the logarithms of x and y. Working with common logarithms (base 10) in this instance, we find that the log-log plot of y versus x yields a straight line, as can be seen by taking the (common) log of both sides of Eq. E.7.

$$\begin{aligned} \log y &= \log (ax^n) = \log a + \log x^n \\ &= \log a + n \log x \end{aligned}$$

or

$$\log y = n \log x + \log a \qquad \textbf{(E.8)}$$

which has the general form of a straight line with a slope of n and an intercept of $\log a$. For the electric field example, this would be

$$E = \frac{kq}{r^2} = kqr^{-2}$$

$$\log E = -2 \log r + \log kq$$

Again, care must be taken in determining the slope of a straight line on a log-log graph. In this case,

$$\begin{aligned} \text{slope} &= \frac{\Delta \log y}{\Delta \log x} \\ &= \frac{\log y_2 - \log y_1}{\log x_2 - \log x_1} = \frac{\log y_2/y_1}{\log x_2/x_1} \end{aligned} \qquad \textbf{(E.9)}$$

and the logs of the endpoints of the slope interval or their ratio must be found explicitly. (The ordinate and abscissa values on the log-log plot are y and x, *not* $\log y$ and $\log x$.)

As in the case of the semi-log plot, the value of a in $y = ax^n$ can be read directly from the y-intercept of the graph. However, in this case, the intercept is not at $x = 0$ *but* at $x = 1$, since the intercept $\log y = \log a$ requires that $\log x = 0$ and $\log 1 = 0$.